FLUORINE IN
BIOORGANIC CHEMISTRY

FLUORINE IN BIOORGANIC CHEMISTRY

John T. Welch

State University of New York at Albany
Albany, New York

Seetha Eswarakrishnan

PPG Industries, Inc.
Monroeville, Pennsylvania

A WILEY-INTERSCIENCE PUBLICATION

JOHN WILEY & SONS

New York / Chichester / Brisbane / Toronto / Singapore

In recognition of the importance of preserving what has
been written, it is a policy of John Wiley & Sons, Inc. to
have books of enduring value published in the United
States printed on acid-free paper, and we exert our best
efforts to that end.

Library of Congress Cataloging in Publication Data:

Welch, John T.
  Flourine in bioorganic chemistry / John T. Welch, Seetha
Eswarakrishnan.

      p.   cm.
  Includes bibliographical references and index.
  ISBN 0-471-50649-4
  1. Organofluorine compounds—Synthesis. 2. Fluorination.
3. Bioorganic chemistry—Technique. I. Eswarakrishnan, Seetha.
II. Title.

QP801.P5W45   1991   90-39152
574.19′24—dc20   CIP

Printed in the United States of America

10 9 8 7 6 5 4 3 2 1

Dedicated To
Shekar
Jan and Juliann

PREFACE

This book was completed while one of us (JTW) was a guest of Professor Dieter Seebach at the Eidgenössiche Technische Hochschule Zürich. It is a pleasure to acknowledge Dr. Seebach's hospitality, the stimulating atmosphere of his group, and the support and assistance of his co-workers, especially Mr. Albert Beck and Mr. Christoph von dem Bussche.

This book is devoted to a description of methods for the preparation of selectively fluorinated, biologically active molecules. It is organized by compound class; each class is then arranged by the reagent or building block employed in the preparation of the subject materials. For setting the precedent in organizing the subject in this manner as well as his encouragement to proceed with this project, we acknowledge Professor Nobuo Ishikawa.

We would like to recognize the generosity of PPG Industries in allowing us the use of their excellent library facilities. Finally, we apologize in advance to any author whose work we have inadvertently overlooked.

Special recognition must be afforded our families, especially our respective spouses, Jinne and Eswar, for their indispensible patience and understanding throughout the course of this project.

JOHN T. WELCH
SEETHA ESWARAKRISHNAN

Albany, New York
Monroeville, Pennsylvania
August 1990

CONTENTS

1 Introduction   1

  1.1   Historical Perspective / 1
  1.2   Fluorine in Biological Chemistry / 2
  1.3   Effects of Fluorine on Chemical Reactivity / 2
  1.4   Effects of Fluorination on Biological Reactivity / 3
  1.5   Methods for the Introduction of Fluorine / 4
      References / 4

2 Fluorinated Amino Acids   7

  2.1   Introduction / 7
  2.2   Synthesis / 8
      2.2.1   Sulfur Tetrafluoride / 8
          Synthesis of 5,5-Difluorolysine / 10
          Synthesis of 4,4-Difluoroproline / 11
      2.2.2   Diethylaminosulfur Trifluoride / 12
          Asymmetric Synthesis of β-Fluorovaline and
             phenylalanine / 12
          Asymmetric Synthesis of L-4-Fluorothreonine / 13
      2.2.3   Yarovenko Reagent / 14
          Synthesis of 4-Fluoroglutamic acid / 14
      2.2.4   Hydrogen Fluoride / 14

Synthesis of β-Fluoroaspartic Acid / 15
Synthesis of Difluoroborane Analogs of Fluorinated
    Amino Acids / 15

2.2.5   Hydrogen Fluoride–Pyridine / 16

Aziridine Ring-Opening Reactions / 16
Azirine Carboxylates with Hydrogen Fluoride–
    Pyridine / 18
Epoxide Ring-Opening Reactions / 18

2.2.6   Trifluoromethyl Hypofluorite / 19

Photofluorination / 19
Fluorodesulfurization / 20

2.2.7   Molecular Fluorine / 21

Fluorination with Replacement of Hydrogen / 21
Fluorodesulfurization Using Fluorine in Helium / 22

2.2.8   Perchlorylfluoride / 23

Synthesis of 4-Fluoroglutamic Acid / 23
Synthesis of β,β-Difluoroaspartic Acid and β,β-
    Difluoroasparagine / 24
Synthesis of β-Difluoroaspartic Acid / 24

2.2.9   Fluoroboric Acid / 25

Synthesis of 2-Fluorotyrosine / 26

2.2.10   Metal Fluorides as Fluorinating Agents (KF, AgF, HgF,
    SbF_3, KHF_2) / 27

Synthesis of 3-Fluoroalanine with Potassium
    Fluoride / 27
Synthesis of (E)-β-Fluoromethylene Glutamic Acid with
    Potassium Fluoride / 28
Synthesis of 4,4′-Difluorovaline with Potassium
    Fluoride / 28
Synthesis of cis- and trans-4-Fluoro-L-proline with
    Potassium Fluoride / 29
Synthesis of 4-Amino-5-fluoropentanoic Acid with Silver
    Fluoride / 29
Synthesis of 4-Fluoroglutamic Acid with Mercuric
    Fluoride / 30
Synthesis of N-Acetyl-4-fluoroglutamic Acid with
    Antimony Trifluoride / 30
Synthesis of (E)-β-Fluoromethylene-m-tyrosine with
    Potassium Bifluoride / 30

2.2.11   Xenon Difluoride / 31

Synthesis of Fluoromethionine / 31

2.2.12   Freons as Building Blocks (CH_2FCl, $CHFCl_2$, CHF_2Cl,
    CF_3Br) / 31

> *Amino Malonates* / *31*
> *Schiff Bases* / *31*
> *Alkylation of Sodium Enolate* / *33*
> *Alkylation of Malonate Esters* / *33*

2.2.13   Perfluoroalkyl Iodides / 34
2.2.14   Fluoroacetonitrile / 35

> *Synthesis of 3-Amino-4-fluorobutanoic Acid* / *35*
> *Synthesis of α-Fluoromethyldehydroornithine* / *36*

2.2.15   *N*-Acyl Trifluoroacetaldimines / 36

> *Synthesis of 3,3,3-Trifluoroalanine* / *36*

2.2.16   Ethyl Fluoroacetate / 37

> *Synthesis of α-Fluoro-β-alanine* / *37*
> *Synthesis of 4-Amino-2-fluorocrotonic Acid and*
>   *5-Amino-2-fluoropentenoic Acid* / *37*
> *Synthesis of 4-Fluoroglutamic Acid* / *38*
> *Synthesis of 3-Fluoro-*D,L-*alanine, 2d* / *38*

2.2.17   *t*-Butyl Fluoroacetate / 38

> *Synthesis of erythro- and threo-β-Fluoroaspartic Acids*
>   *and erythro-β-Fluoroasparagine* / *38*

2.2.18   Ethyl Bromodifluoroacetate / 39
2.2.19   Ethyl Trifluoroacetate / 40
2.2.20   Diethyl Fluoromalonate / 40

> *Synthesis of 4-Amino-2-fluorobutanoic Acid* / *40*
> *Synthesis of 4-Amino-2-fluorohexanoic Acid* / *40*

2.2.21   1,1,1-Trifluoro-, 1,1'-Difluoro-, and
         Hexafluoroacetone / 41

> *Synthesis of 3,3,3-Trifluoroalanine Using*
>   *Hexafluoroacetone* / *41*
> *Synthesis of Hexafluorovaline Using*
>   *Hexafluoroacetone* / *41*
> *Synthesis of 5,5,5-Trifluoroleucine Using*
>   *1,1,1-Trifluoroacetone* / *41*
> *Synthesis of 5-Methyl-6,6,6-Trifluoroleucine Using*
>   *1,1,1-Trifluoroacetone* / *42*
> *Synthesis of α-Amino-β,β-difluoroisobutyric Acid Using*
>   *1,1'-Difluoroacetone* / *42*

2.2.22   3,3,3-Trifluoropropene / 42
2.2.23   Hexafluoropropene / 45
2.2.24   Ethyl α-Fluoroacrylate / 45

> *Synthesis of 4-Fluoroglutamic Acid* / *45*

2.2.25   3-Fluoro-2-butanone / 46

> *Synthesis of γ-Fluoroisoleucine* / *46*

2.2.26   Ethyl α-Fluorocrotonate / 46

2.2.27 4,4,4-Trifluorobutanoic Acid / 47

Synthesis of 5,5,5-Trifluoronorvaline / 47
Synthesis of 6,6,6-Trifluoronorleucine / 48

2.2.28 Trifluoroacetic Anhydride / 48

Synthesis of 2-Trifluoromethyl-L-histidine / 48
Synthesis of 3,3,3-Trifluoroalanine / 49

2.2.29 Ethyl Trifluoroacetoacetate / 50

Synthesis of 2-Amino-3-hydroxy-
4,4,4-trifluoro-3-butanoic Acids / 50

2.2.30 3-Trifluoromethyl-γ-butyrolactone / 51

Synthesis of 5,5,5-trifluoroleucine / 51

2.2.31 Pentafluorostyrene / 52

Synthesis of 5,5,5-trifluoroleucine / 51

2.2.32 Fluoroindole / 52

2.3 Biological Activity / 54

References / 61

3 Fluorinated Amines   **66**

3.1 Introduction / 66
3.2 Synthesis / 66

3.2.1 Sulfur Tetrafluoride / 66
3.2.2 Diethylaminosulfur Trifluoride / 67

Synthesis of Monofluoroputrescine / 67

3.2.3 Hydrogen Fluoride–Pyridine / 67

Ring-Opening Reactions of Aziridines / 67
Addition of Hydrogen Fluoride–Pyridine to
1-Azirines / 69
Epoxide Ring Opening with Hydrogen Fluoride–
Pyridine / 71
Fluorodehydroxylation Using Hydrogen Fluoride–
Pyridine / 73
Fluorodediazoniation with Hydrogen Fluoride–
Pyridine / 73

3.2.4 Trifluoromethylhypofluorite / 73
3.2.5 Molecular Fluorine / 74
3.2.6 Fluoroboric Acid / 74

Synthesis of 4-Fluorohistamine / 74
Synthesis of 2-Fluorohistamine / 75
Synthesis of 3-Fluorotyramine / 75
Synthesis of 3,5-Difluorotyramine / 76
Synthesis of 6-Fluorodopamine and
2-Fluorodopamine / 76

3.2.7   Chlorodifluoromethane / 77

Synthesis of α-Difluoromethyl Dopamine / 77
Synthesis of α-Difluoromethyl Putrescine / 78
Synthesis of α-Difluoromethyl Dehydroputrescine / 78
Synthesis of (E)-2-(3,4-Dihydroxyphenyl)-
3-fluoroallylamine / 78
Synthesis of (Z)-2-Phenyl-3-fluoroallylamine / 79

3.2.8   Dibromodifluoromethane / 80

Synthesis of 2-Phenyl-3,3-difluoroallylamine / 80

3.2.9   Trifluoromethyl Iodide / 81

Synthesis of 2- and 4-Trifluoromethylhistamine / 81
Synthesis of Trifluoromethylurocanic Acids / 81

3.2.10 Fluoroacetonitrile / 82

Synthesis of α-Fluoromethyl Dehydroputrescine / 82

3.2.11 Trifluoroacetic Anhydride / 82
3.2.12 Perfluorosuccinamide / 83

3.3   Enzymatic Synthesis / 83

3.3.1   Synthesis of 4-Fluoro-L-histamine / 83
3.3.2   Synthesis of (E)-β-Fluoromethylene-*meta*-tyramine / 83
3.3.3   Synthesis of (E)-β-Fluoromethylene GABA / 83
3.3.4   Synthesis of 4-Fluorourocanic Acid / 84
3.3.5   Synthesis of 2-Fluorourocanic Acid / 84

3.4   Biological Activity / 84
References / 85

4 Fluorinated Acids and Esters   **87**

4.1   Synthesis of Fluorination Reaction / 87
4.1.1   Diethylaminosulfur Trifluoride / 87
4.1.2   Hydrogen Fluoride / 87
4.1.3   Hydrogen Fluoride—Pyridine / 88

Ring Opening of Glycidic Esters / 88
Deaminative Fluorination / 88

4.1.4   Potassium Fluoride / 91

4.2   Synthesis with Fluorinated C_1 Units / 91
4.2.1   Dichlorofluoromethane / 91
4.2.2   Chlorodifluoromethane / 92

4.3   Synthesis with Fluorinated C_2 Units / 92
4.3.1   Ethyl Fluoroacetate / 92

Synthesis of Fluoropyruvic Acid / 92
Synthesis of 2-Fluorocitrate Esters / 93
Synthesis of 2-Fluoro-2-alkenoic Acids / 93

Synthesis of 3-Fluoro-Coumarin / 94
Synthesis of 4-Fluoro-2,6-dimethyl
resorcinol / 94

4.3.2 Ethyl Bromodifluoroacetate / 94
4.3.3 Methyl Iododifluoroacetate / 95

4.4 Synthesis with Fluorinated C₃ Units / 95

4.4.1 Fluoropyruvic Acid / 95
4.4.2 Trifluoropropene / 96
Hydrocarboxylation and Hydroesterification / 96
Bromination and Carboxylation / 96

4.5 Synthesis with Fluorinated C₄ Units / 97

4.5.1 1-Fluoro-3-bromo- and 1-Fluoro-4-bromo-butane / 97
4.5.2 Perfluoro-2-methylpropene / 97
4.5.3 4,4,4-Trifluorocrotonate / 98
4.5.4 4,4,4-Trifluoro-3-oxobutanoate / 99

4.6 Synthesis with Larger Fluorinated Synthons / 99

4.6.1 3-Methyl-4,4,4-trifluorobutyrate / 99
4.6.2 6-Benzyloxy-3-trifluoromethyl-
hex-4-enoic acid / 100
4.6.3 Pentafluorostyrene / 101
4.6.4 ω-Fluoro-Alcohols, Acids & Nitriles / 101

4.7 Enzymatic Methods / 102
4.8 Biological Activity / 102
References / 107

5 Fluorinated Alcohols   **109**

5.1 Chemical Synthesis / 109

5.1.1 1-Phenylsulfonyl-3,3,3-trifluoropropene / 109
5.1.2 3-Fluoro-1,2-propanediol / 109

5.2 Enzymatic Reductions of Fluorinated Ketones and
Ketoesters / 110
5.3 Catalytic Reduction of Fluorinated Olefins / 111
5.4 Biological Activity / 111
References / 111

6 Fluorinated Phosphonic Acids and Derivatives   **113**

6.1 Synthesis of Difluorophosphonoacetic Acid and
Derivatives / 113

6.1.1 Dibromodifluoromethane / 113
6.1.2 Fluoromethane Phosphonates / 113
6.1.3 Acylation / 114

6.2   Synthesis of Monofluorophosphonate / 115
    6.2.1   Perchloryl Fluoride / 115
6.3   Synthesis of Fluoromethylene bisphosphonates / 115
    6.3.1   Phosphorylation / 115
    6.3.2   Carboxylation / 115
    6.3.3   Electrophilic Displacement / 116
    6.3.4   Perchlorylfluoride / 117
    6.3.5   Bromodifluoromethylphosphonate / 117
6.4   Biological Activity / 117
    References / 118

7   Fluorinated Ketones   **119**

7.1   Hydrogen Fluoride–Pyridine / 119
    7.1.1   Diazoketones / 119
    7.1.2   α-Haloketones / 119
    7.1.3   Azirine Ring Opening / 120
7.2   Molecular Fluorine / 120
7.3   Xenon Difluoride / 121
7.4   Acetyl Hypofluorite / 121
7.5   2,2,2-Trifluorethanol / 122
7.6   3,3,3-Trifluoropropionic Acid / 122
7.7   Methyl 2,2-Difluorobutyrate / 122
7.8   4-Methyl Iodobenzenedifluoride / 122
7.9   Perfluorocarboxylic Acid Esters / 124
7.10 Biological Activity / 126
    References / 130

8   Fluorinated Sugars   **131**

8.1   Introduction / 131
8.2   Displacement Reactions / 132
    8.2.1   Sulfonate Esters / 132
         Trifluoromethanesulfonates / 132
         Imidazolesulfonates / 139
         Cyclic Sulfite or Sulfate Esters / 139
         4-Bromobenzenesulfonates / 140
         Methanesulfonates / 141
8.3   Ring-Opening Reactions / 142
    8.3.1   2-Fluoro-L-Daunosamine / 144
8.4   Additions to Olefins / 145
    8.4.1   Molecular Fluorine / 146
    8.4.2   Acetyl Hypofluorite / 146

8.4.3   Xenon Difluoride / 146
8.4.4   Trifluoromethylhypofluorite / 147
8.4.5   Fluorinative Dehydroxylations / 147

8.5   Total Synthesis / 152
8.5.1   Claisen Condensations / 152
8.5.2   Aldol Condensations / 152
8.5.3   3,3-Sigmatropic Rearrangements / 154

8.6   Enzymatic Methods / 155
8.6.1   Enzymatic Transformations of Fluorinated
         Carbohydrates / 155
8.6.2   Aldolase- and Isomerase-Catalyzed Reactions / 157

8.7   Glycosyl Fluorides / 157
       References / 159

9   Fluorinated Prostaglandins   **163**

9.1   Introduction / 163
9.2   Fluorinative Dehydroxylation / 165
9.2.1   Fluoroalkylamine Reagents / 165
9.2.2   Diethylaminosulfur Trifluoride / 166
9.2.3   Sulfonate Displacement Reactions / 168

9.3   Epoxide Ring-Opening Reactions / 170
9.3.1   Potassium Bifluoride / 170
9.3.2   Hydrogen Fluoride–Pyridine / 171

9.4   Difluoromethylene Groups / 172
9.4.1   Molybdenum Hexafluoride / 172
9.4.2   Sulfur Tetrafluoride / 173

9.5   Electrophilic Fluorination of Enols and Enolates / 174
9.5.1   Perchlorylfluoride / 174

9.6   Olefination Reactions / 177
9.6.1   Horner-Emmons–Type Reagents / 177
9.6.2   Chlorodifluoromethane / 179
9.6.3   Fluorosulfoximines / 180
9.6.4   Chlorodifluoroacetic Acid / 181

9.7   Reformatsky-Type Reactions / 182
9.7.1   Reformatsky Reagents / 182
9.7.2   Fluorinated Lithium Enolates / 183
9.7.3   Fluorinated Alane Reagents / 184
9.7.4 α-Fluorinated Carbonyls as Electrophilic Building
         Blocks / 185
         References / 186

10 Fluorinated Steroids   **187**

10.1   Introduction / 187
10.2   Direct Fluorination / 189
10.3   Fluorinative Dehydroxylation / 190

    10.3.1   Diethylaminosulfur Trifluoride / 190
    10.3.2   Chlorotrifluoroethyl(diethyl)amine / 193
    10.3.3   Phenyltetrafluorophosphine / 193

10.4   Displacement Reactions / 194

    10.4.1   Toluenesulfonates / 194
    10.4.2   Trifluoromethanesulfonates / 195

10.5   Balz-Schiemann Reactions / 197
10.6   Epoxide Opening Reactions / 198

    10.6.1   Hydrogen Fluoride / 198
    10.6.2   Potassium Fluoride–Potassium Hydrogen
            Difluoride / 199

10.7   Geminal Difluorination Reactions / 201

    10.7.1   Diethylaminosulfur Trifluoride / 201
    10.7.2   Iodine Monofluoride / 203

10.8   Additions to Olefins / 203

    10.8.1   Molecular Fluorine / 204
    10.8.2   Bromofluorination / 204

10.9   Addition to Enol Ethers / 205

    10.9.1   Perchlorylfluoride / 207
    10.9.2   Xenon Difluoride and p-Iodotoluene Difluoride / 208
    10.9.3   Diethylaminosulfur Trifluoride / 209
    10.9.4.   Difluorocarbene / 210

10.10   Fluorinated Building Blocks / 211

    10.10.1   3-Trifluoromethylphenol / 211
    10.10.2   4-Methoxy-α,α,α-trifluoroacetophenone / 212
    10.10.3   Ethyl Trifluoroacetate / 213
    10.10.4   Hexafluoroacetone / 214
    10.10.5   (R)-3-Fluoro-5-iodo-2-methyl-2-pentanol / 215
            References / 217

11 Fluorinated Purines and Pyrimidines   **220**

11.1   Introduction / 220
11.2   5-Fluoropyrimidines / 220

    11.2.1   5-Fluorouracil / 220
    11.2.2   6-O-Dicyclo-5,5-difluoro-5,6-dihydrouracils / 223
    11.2.3   5-Fluoro-2',3'-dideoxy-3'-fluorouridine / 223

11.2.4   5'-Deoxy-4',5-difluorouridine / 223

11.2.5   6-Aza-5-fluorouracil / 224

11.2.6   4-O-(Difluoromethyl)-5-fluorouracil / 224

11.3   5-Fluoroalkyl-Substituted Pyrimidines / 225

11.3.1   5-Trifluoromethyluracil / 225

11.3.2   5-Trifluoromethyluridine / 226

11.3.3   5-Monofluoromethyl- and 5-Difluoromethyluracil / 226

11.3.4   Total Synthesis / 227

11.3.5   5-Fluoroalkyl and 5-Fluoroalkenyl Pyrimidines / 228

11.3.6   (E)-5-(3,3,3-Trifluoro-1-propenyl)-2'-deoxyuridine / 229

11.4   Purines / 230

11.4.1   2-Fluoroadenine / 230

11.4.2   2-Fluoro-8-aza-adenosine / 230

11.4.3   2-Fluoroformycin / 230

11.4.4   8-Trifluoromethyl Adenosine and Inosine / 231

References / 232

12 Fluorinated Aromatics   **234**

12.1   Introduction / 234

12.2   General Methods for the Fluorination of Aromatics / 234

12.2.1   Background / 234

12.2.2   Balz-Schiemann Reactions / 235

12.2.3   Direct Fluorination / 240

Molecular Fluorine / 240

Trifluoroacetyl Hypofluorite / 241

Acetyl Hypofluorite / 241

N-Fluorosulfonimides / 241

12.3   Fluorinated Aromatic Building Blocks / 242

12.3.1   Fluorinated Phenols / 242

12.3.2   Fluorinated Aldehydes / 243

12.3.3.   Fluorinated Acids / 243

12.3.4   Fluorinated Anilines / 246

12.3.5   Fluorinated Chalcones / 246

12.4   Fluoroalkyl Aromatics / 246

12.4.1   Trifluoromethylated Aromatic Building Blocks / 246

12.4.2   Introduction of the Trifluoromethyl Group / 247

References / 248

Index   **251**

FLUORINE IN
BIOORGANIC CHEMISTRY

1

INTRODUCTION

The current level of interest in the preparation of selectively fluorinated compounds is indicated by the increasing number of publications in this area, although it has been known for some time that fluorine can have profound and unexpected results on biological activity. Selective fluorination has been an extremely effective synthetic tool for modifying reactivity. Replacement of hydrogen or hydroxyl by fluorine in a biologically important molecule may yield an analog of that substance with improved selectivity or a modified spectrum of activity.[1] Valuable monographs and general reference works are available for guiding synthetic chemists in the organic chemistry of fluorine.[2]

1.1 HISTORICAL PERSPECTIVE

Although hydrogen fluoride was discovered by Scheele in 1771,[2b] molecular fluorine itself was not prepared until 1886 by Moissan.[3] The availability of fluorine was indeed the key to the development of organofluorine chemistry. Organofluorine chemistry began to unfold with the work of Swarts[4] on the preparation of fluorinated materials by metal fluoride–promoted halogen–fluoride exchange reactions. The commercial utility of organofluorine compounds as refrigerants, developed by Midgley and Henne,[5] further accelerated the growth of the field by virtue of the economic incentives involved. Wartime requirements stimulated research on thermally stable and chemically resistant materials, which led to a renewed interest in perfluorinated substances. Many of the exciting developments of this period have been reviewed elsewhere.[6] A synopsis of the growth of organofluorine chemistry could not be complete without recognition of the enormous impact of electrochemical fluorination techniques. The electrolysis of organic compounds in anhydrous hydrogen fluoride was discovered and developed by Simons.[7]

1.2   FLUORINE IN BIOLOGICAL CHEMISTRY

It was the pioneering work of Fried on the preparation of 9α-fluoro-hydrocortisone acetate[8] that led to the first significant report on the successful application of selective fluorination to modify biological activity. This public-ation marked the beginning of a new era when medicinal chemists and biochemists routinely introduced fluorine as a substituent to modify reactivity.

9α-Fluoro-hydrocortisone acetate

The attractiveness and utility of fluorine as a substituent in a biologically active molecule results from the pronounced electronic effects that may result on fluorination as well as on the fact that fluorine is not a sterically demanding substituent. With its small van der Waals radius (1.35 Å), fluorine closely resembles hydrogen (van der Waals radius 1.20 Å). The carbon–fluorine bond length, 1.39 Å, is comparable to that of the carbon–oxygen bond, 1.43 Å. The electronegativity of fluorine (4.0 vs. 3.5 for oxygen) can have pronounced effects on the electron distribution in the molecule, affecting the acidity or basicity of neighboring groups, dipole moments within the molecule, and overall reactivity and stability. Fluorine can function as a hydrogen bond acceptor because of its available electron density.[9]

1.3   EFFECTS OF FLUORINE ON CHEMICAL REACTIVITY

Important differences in chemical reactivity may commonly be expected based on the difference in electronegativity between fluorine and hydrogen as well as the higher carbon–fluorine bond strength versus the strength of the carbon–hydrogen bond, and, somewhat less commonly, on the ability of fluorine to participate in hydrogen bonding as an electron pair donor. The effects of fluorination have been thoroughly summarized by Chambers,[2c] Smart,[2a] and others,[10] but it may be worthwhile to restate them.

Fluorine has a pronounced electron-withdrawing effect by relay of an induced dipole along the chain of bonded atoms, a sigma-withdrawing effect, I_σ, or, by through-space electrostatic interaction, a field effect.[11] However, the electronic effect of fluorine directly attached to a pi system can be complex, as electrons from fluorine may be donated back to the pi system in an I_π repulsive interaction.

These repulsive interactions are most important in the reactions of α-fluorinated anions and radicals.

In summary, fluorine may be understood to stabilize α cations by the interaction of the vacant p orbital of the carbocation with the filled orbitals of fluorine. Fluorine bonded directly to a carbanionic center is generally destabilizing by I_π repulsive interactions but when not bonded directly, as in a trifluoromethyl-substituted carbanion, can be stabilizing. Whereas fluorine bonded to a radical center may have a profound effect on the geometry of that center, the effect of fluorination on the stability of the radical is difficult to assess.

1.4 EFFECTS OF FLUORINATION ON BIOLOGICAL REACTIVITY

Systematic substitution of fluorine can help establish the effect of hydroxylation on the activity of a molecule. This principle has been successfully applied in the synthesis of fluorinated vitamin D_3 analogs. Once introduced, the high carbon–fluorine bond strength (108 kcal/mole) renders the fluorine substituent resistant to many such metabolic transformations. Fluorine may also be employed as a leaving group in addition–elimination processes where its superior leaving group ability relative to hydrogen is important. Such applications have led to the development of very effective mechanism-based enzyme inhibitors. As previously mentioned, fluorine can also be a useful probe of hydrogen bonding interactions, since it may act as a electron pair donor but is otherwise unable to participate in such bonding interactions. Finally, fluorine also has been identified as improving the lipophilicity of a molecule and hence its distribution within an organism.

Fluorine has a useful short-lived isotope, fluorine-18 (^{18}F), which decays by positron emission. Positron emission tomography (PET), an especially useful noninvasive technique for the survey of living tissue, complements traditional methods such as X-ray studies by allowing real time analysis of metabolic processes.[12] Introduction of ^{18}F-substituted materials into living tissue is an essential part of PET. While other isotopes such as ^{11}C, ^{13}N, and ^{15}O have half-lives of 20, 10, and 2 minutes, ^{18}F conveniently has a half-life of 110 minutes, sufficient for synthesis and for administration of the radiolabeled materials.[13] One application of ^{18}F positron emission tomography is in the brain imaging of Parkinson's disease patients. By using ^{18}F-labeled fluorodopa, new insights into the chemistry and metabolism of brain have been revealed. In yet another

6α-[^{18}F]-Fluoro-L-dopa   6α-[^{18}F]-Fluoro-L-dopamine example, ^{18}F-labeled estrogen may be useful in diagnosing breast tumors by positron emission tomography.

16α-[^{18}F]-Fluoroestradiol-17β

1.5 METHODS FOR THE INTRODUCTION OF FLUORINE

For an introduction to the techniques of organic fluorine chemistry and preparative methods, the best source of information is the excellent treatise by Hudlicky.[2b] Other valuable works include those of Houben-Weyl,[14] Sheppard and Sharts[2d] and Chambers.[2c] More recently, methods for the fluorination of organic molecules have been reviewed by several authors.[15] One review discusses the synthesis of α-fluorinated carbonyl compounds,[16] a functional relationship which appears quite often in biologically active molecules. An especially rapidly growing area of interest is fluorination with molecular fluorine or with reactive species prepared by in situ reaction of molecular fluorine where the fluorine is relayed to the target molecule.[17] Methods for the construction of more exhaustively fluorinated molecules have also been collated recently.[18] The utility of selectively fluorinated molecules as enzyme substrates has been reviewed,[19] as has the preparation of fluorinated analogs as insect juvenile hormones and pheromones.[20] A progress report on the preparation and biological activity of fluorinated analogs of vitamin D_3 has been published,[21] as has a study on the preparation and biochemistry of fluorinated carbohydrates.[22]

This book is organized by compound class; each class is then arranged by the reagent or building block employed in the preparation of the subject materials. This organization should facilitate the use of the book as a tool for designing new syntheses of fluorinated materials.

REFERENCES

1. (a) P. Goldman, *Science* **164**, 1123, 1969; (b) F. A. Smith, *Chem. Tech.* **1973**, 422; (c) R. Filler, *Chem. Tech.*, **1974**, 752; (d) *Ciba Foundation Symposium, Carbon–Fluorine Compounds, Chemistry, Biochemistry and Biological Activities*, Elsevier, New York, 1972; (e) R. Filler, in *Organofluorine Chemicals and Their Industrial Applications* (ed. R. E. Banks), Ellis Horwood, Chichester, 1979; (f) R. Filler and Y. Kobayashi, eds., *Biomedical Aspects of Fluorine Chemistry*, Kodansha, Tokyo, 1982; (g) R. Filler, ed.,

Biochemistry Involving Carbon–Fluorine Bonds, American Chemical Society, Washington, D.C., 1976; (h) M. R. C. Gerstenberger and A. Haas, *Angew. Chem. Int. Ed.* **20**, 647, 1981; (i) J. T. Welch, *Tetrahedron* **43**, 3123–3197, 1987; (j) N. Ishikawa, ed., *Synthesis and Reactivity of Fluorocompounds*, Vol. 3, CMC, Tokyo, 1987; (k) R. E. Banks, ed., *Preparation, Properties and Industrial Applications of Organofluorine Compounds*, Ellis Horwood, Chichester, 1982.

2. (a) B. Smart, in *Chemistry of Functional Groups*, Supplement D, *The Chemistry of Halides, Pseudohalides and Azides* (ed. S. Patai and Z. Rappoport), Wiley, New York, 1983, pp. 603–655; (b) M. Hudlicky, *Chemistry of Organic Fluorine Compounds*, 2nd ed., Ellis Horwood, Chichester, 1976; (c) R. D. Chambers, *Fluorine in Organic Chemistry*, Wiley, New York, 1973; (d) W. A. Sheppard and C. M. Sharts, *Organic Fluorine Chemistry*, Benjamin, New York, 1968.

3. (a) H. Moissan, *Compte Rendu* **102**, 1543, 1886; (b) H. Moissan, *Compte Rendu* **103**, 203, 1886.

4. F. Swarts, *Bull. Acad. Roy. Belg.* **24**, 474, 1892.

5. T. Midgley and A. L. Henne, *Ind. Eng. Chem.* **22**, 542, 1930.

6. (a) J. H. Simons, ed., *Fluorine Chemistry*, Vol 1., Academic, New York 1950; (b) C. Slesser and S. R. Schram, eds., *Preparation, Properties and Technology of Fluorine and Organic Fluorine Compounds*, McGraw-Hill, New York, 1951.

7. (a) J. H. Simons, *J. Electrochem. Soc.* **95**, 47, 1949; J. H. Simons, H. T. Francis, and J. A. Hogg, *J. Electrochem. Soc.* **95**, 53, 1949; (c) J. H. Simons and W. J. Harland, *J. Electrochem. Soc.* **95**, 55, 1949; (d) J. H. Simons, J. H. Pearlson, W. H. Brice, W. A. Watson, and R. D. Dresdner, *J. Electrochem. Soc.* **95**, 59, 1949.

8. J. Fried and E. F. Sabo, *J. Am. Chem. Soc.* **76**, 1455, 1954.

9. (a) A. W. Baker and A. T. Shulgin, *Nature (London)* **206**, 712, 1965; (b) S. Doddrell, E. Wenkert, and P. V. Demarco, *J. Mol. Spectrosc.* **32**, 162, 1969.

10. (a) J. F. Liebman, A. Greenberg, and W. R. Dolbier, Jr., eds., *Fluorine-Containing Molecules*, VCH, Deerfield Beach, Fla., 1988; (b) *L'Actualité-Chimique* 135–188, **1987**, (May).

11. (a) R. D. Topson, *Prog. Phys. Org. Chem.* **12**, 1, 1976; (b) L. S. Levitt and H. F. Widing, *Prog. Phys. Org. Chem.* **12**, 119, 1976.

12. *Chem. Eng. News*, Aug. 15, 1988, pp. 26–29.

13. M. Reivich and A. Alavi, *Positron Emission Tomography*, Alan R. Liss, New York, 1985.

14. E. Forche, W. Hahn, and R. Stroh, in *Methoden der Organischen Chemie (Houben-Weyl)*, Vol. V/3, (ed. E. Müller) Georg Thieme, Stuttgart, 1962.

15. (a) J. Mann, *J. Chem. Soc. Rev.* **16**, 381, 1987; (b) S. Rozen, *Acc. Chem. Res.* **21**, 307, 1988; (c) Chia-Lin J. Wang, *Org. React.* **34**, 319–400, 1985; (d) M. Hudlicky, *Org. React.* **35**, 513–637 1987; (e) L. German and S. Zemskov, eds., *New Fluorinating Agents in Organic Synthesis*, Springer, Berlin, 1989.

16. S. Rozen and R. Filler, *Tetrahedron* **41**, 1111, 1985.

17. (a) H. Vyplel, *Chimia* **39**, 305, 1985; (b) A. Haas and M. Lieb, *Chimia* **39**, 134, **1985**; (c) S. Purrington, B. S. Kagen, and T. B. Patrick, *Chem. Rev.* **86**, 997, 1986.

18. I. L. Knunyants and G. G. Yacobson, eds., *Synthesis of Fluoroorganic Compounds*, Springer, Berlin, 1985.

19. C. Walsh, *Tetrahedron* **38**, 871, 1982.
20. G.D. Prestwich, *Pestic. Sci.* **37**, 430, 1986.
21. N. Ishikawa, *Kagaku To Seibatsu* **22**, 93, 1984.
22. N. F. Taylor, ed., *Fluorinated Carbohydrates, Chemical and Biochemical Aspects*, American Chemical Society, Washington, D.C., 1988.

2

FLUORINATED AMINO ACIDS

2.1 INTRODUCTION

Fluorinated analogs of naturally occurring biologically active compounds exhibit unique physiological activities. Introduction of fluorine into a pharmacologically active substance often leads to the development of more potent agonists or antagonists.[1] Fluorine-containing amino acids have been synthesized and studied as potential enzyme inhibitors and therapeutic agents.[2] β-Fluorine-substituted amino acids are generally regarded as potential irreversible inhibitors of pyridoxal phosphate-dependent enzymes.[2,3] Since the fluorine atom is comparable to hydrogen on the size scale of the enzymes, the enzyme accepts the fluoroamino acid as a substrate but cannot metabolize it properly. The β-fluoro group lowers the pK_a of the carboxyl group of the amino acid by about 2 pK_a units and consequently affects the biological function.

α-Monofluoromethyl and difluoromethyl amino acids have been recognized as potent enzyme-activated irreversible inhibitors of parent α-amino acid decarboxylases.[2] The importance of enzymatic decarboxylations of amino acids in biosynthetic pathways suggests the utility of specific inhibitors of these decarboxylation enzymes in studying their pathways. Typical of the physiologically important amines formed by decarboxylation are dopamine, 5-hydroxytryptamine (serotonin), histamine, tyramine, and gamma-aminobutyric acid (GABA).[3c] The catecholamines are important in peripheral and central control of blood pressure.[4] Elevated histamine levels are observed in allergies, hypersensitivity, gastric ulcers, inflammation, and similar conditions.[5] High putrescine levels are associated with rapid cell development, including tumor growth.[6] The irreversible enzyme inhibitors are extremely selective and have great practical value in elucidating physiological roles of specific enzymes.

Amino acids containing trifluoromethyl group have been observed to be potential antimetabolites.[7] The apparent similarity in the size of the methyl and trifluoromethyl groups allows the duplication of the relative size and configuration of the natural amino acid. However, due to the high electron density of the trifluoromethyl group, the compound might participate in forming strong hydrogen bonds with enzymes, thereby blocking a metabolite from forming an enzyme–substrate complex. Other attractive features of this group are its relative nontoxicity and stability compared to monofluoromethyl and difluoromethyl analogs.

Several fluorinated amino acids exhibiting antibacterial, cancerostatic,[8] cytotoxic, or other inhibitory activities have been reported. Some of these fluorocompounds may be useful in the treatment of disorders of the central nervous system.

2.2 SYNTHESIS

The syntheses of various fluorinated amino acids are organized based on the synthon size: the fluorinating agents (C_0 synthons), followed by C_1 synthons, C_2, C_3, and so on.

2.2.1 Sulfur Tetrafluoride

Fluorodehydroxylation is a general method for the selective transformation of hydroxyamino acids into fluoroamino acids.[9,10] The method involves the use of sulfur tetrafluoride in liquid hydrogen fluoride at $-78°C$ and atmospheric pressure. Reactivity of sulfur tetrafluoride toward a variety of alcoholic hydroxyl groups is dramatically and selectively in increased when liquid hydrogen fluoride is used as solvent. Surprisingly, the reactivity of sulfur tetrafluoride with carbonyl compounds and carboxylic acids is not concomitantly increased. Moreover, liquid hydrogen fluoride protects the amino group by protonation against electrophilic reaction with sulfur tetrafluoride (see Fig. 2.1).

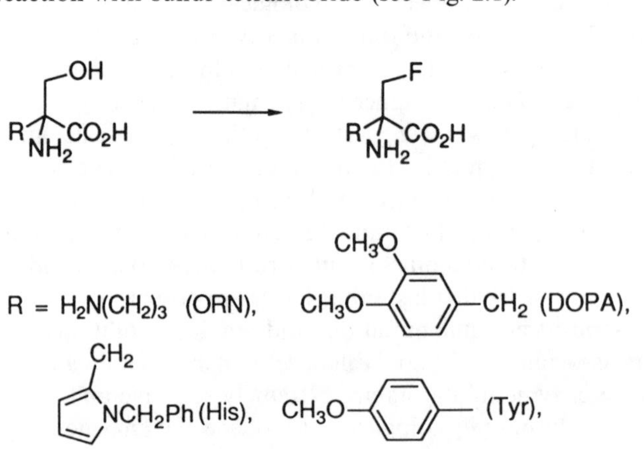

FIGURE 2.1

The mechanism of R—OH to R—F transformation with a sulfur tetrafluoride–hydrogen fluoride system is represented by the following equations:[9,10]

$$SF_4 + HF \rightleftharpoons SF_3^+ + HF_2^- \qquad (2.1)$$

$$ROH + SF_3^+ \longrightarrow ROSF_3 + H^+ \qquad (2.2)$$

$$ROSF_3 + HF \longrightarrow ROSF_2^+ + HF_2^- \qquad (2.3)$$

$$ROSF_2^+ + F^- \longrightarrow RF + SOF_2 \qquad (2.4)$$

The mechanism involves the initial generation of a powerful leaving group at the alcoholic carbon. Hydrogen fluoride plays an important role in the dissociation of sulfur tetrafluoride to the much more electrophilic SF_3^+ and also activates the ionization of alkoxy–sulfur trifluoride intermediate.

The mechanism of fluorine substitution (Eq. 2.4) depends on the alcohol. S_N1 processes lead to stable carbocations and S_N2-type processes take place in primary and secondary alkyl systems, which usually require direct displacements.

Fluorodehydroxylation of both isomers of ephedrine yielded the same mixture of fluoroisomers, implying that bond breaking precedes bond making in these good carbonium ion precursors. The reaction occurred with inversion of configuration with threonine,[8,10] allothreonine,[10] hydroxy aspartates,[11] and hydroxy glutamates.[12a] A higher percentage of inversion was observed with the threo isomer: 92% for threonine versus 78% for allothreonine and 90% for *threo*-3-hydroxy aspartate versus 90% for the erythro isomer.[12a]

Fluorodehydroxylation of *erythro*- and *threo*-3-hydroxyglutamic acids (**1**) by sulfur tetrafluoride in hydrogen fluoride[12a] led only to the formation of fluorinated lactams (**2**). However, fluorodehydroxylation of N-acetyl-3-hydroxyglutamic acids (**3**) under the same conditions suppressed the formation of the lactams, and N-acetyl-3-fluoroglutamic acids (**4**) were obtained. The erythro and threo acids were separated and deacetylated with acylase to give **5**.

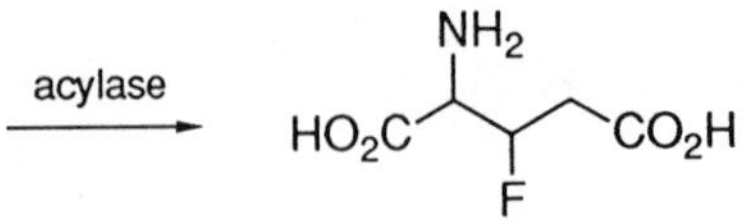

5

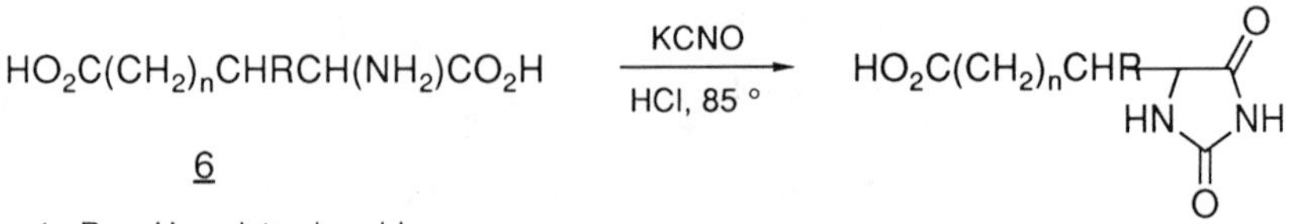

6

n = 1, R = H glutamic acid
n = 0, R =CH₃ β - methyl aspartic acid
n = 1, R = H 5,5,5-trifluoronorvaline
n = 0, R =CH₃ 4,4,4-trifluorovaline

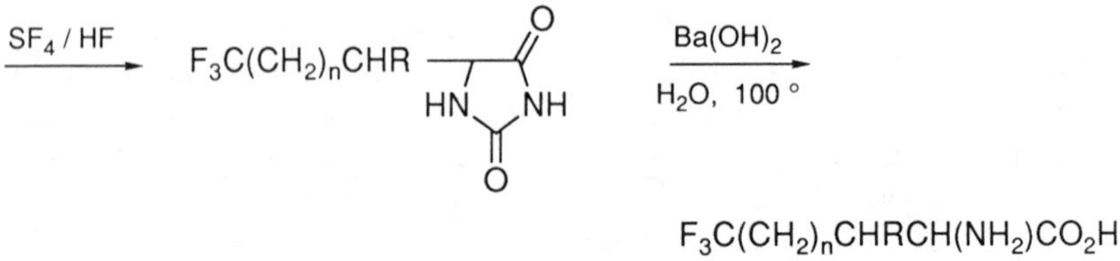

$$F_3C(CH_2)_nCHRCH(NH_2)CO_2H$$

7

Sulfur tetrafluoride–hydrogen fluoride can also be used for the conversion of a carboxyl as in **6** into a trifluoromethyl group as in **7**.[12b] A probable mechanism is

$$RCO_2H + SF_4 \longrightarrow RCOF + HF + SOF_2$$
$$RCOF + SF_4 \longrightarrow RCF_3 + SOF_2$$

Fluorination of ketone precursors using sulfur tetrafluoride leads to geminal difluorocompounds.[13,14]

Synthesis of 5,5-Difluorolysine (9)

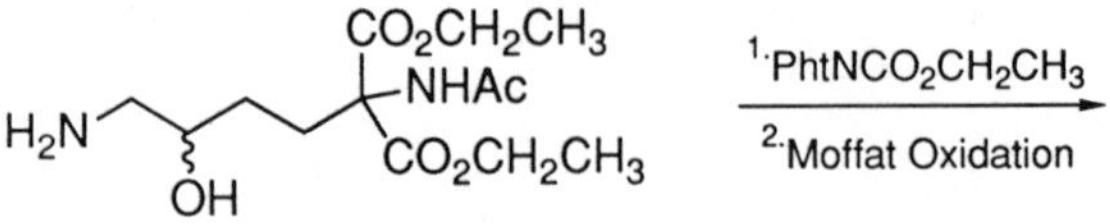

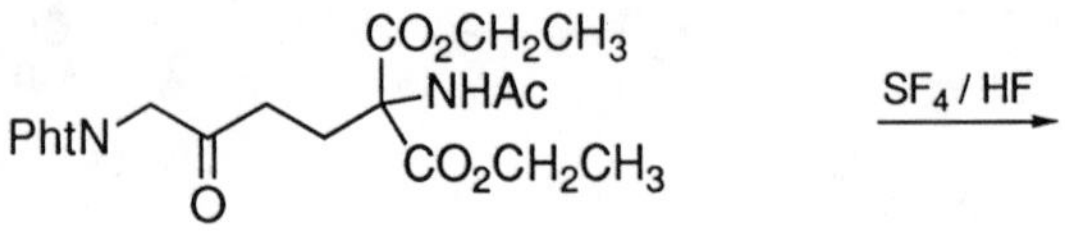

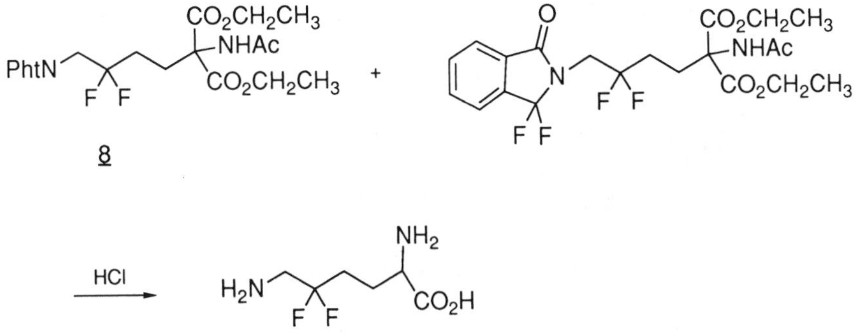

$$\text{8}$$

$$\text{9}$$

Synthesis of 4,4-Difluoroproline (12)[14]

The amino and carboxyl groups of trans-4-hydroxy-L-proline (**10**) are simultaneously protected as the diketopiperazine (**11**). The lowered reactivity of sulfur tetrafluoride–hydrogen fluoride toward the amide carbonyl relative to the ketone carbonyl at temperature below 60°C allows the use of sulfur tetrafluoride as the fluorinating agent.

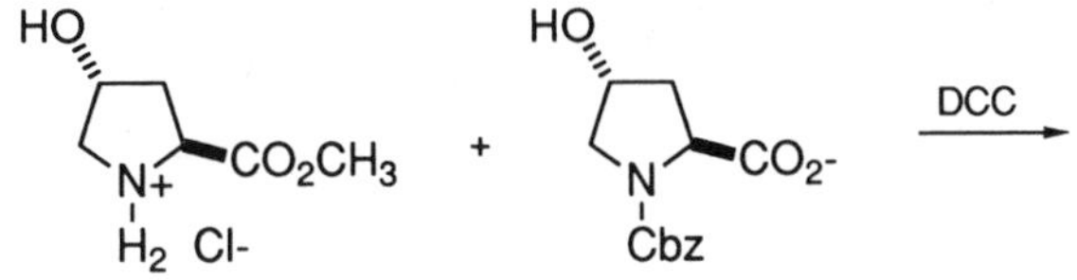

$$\text{10}$$

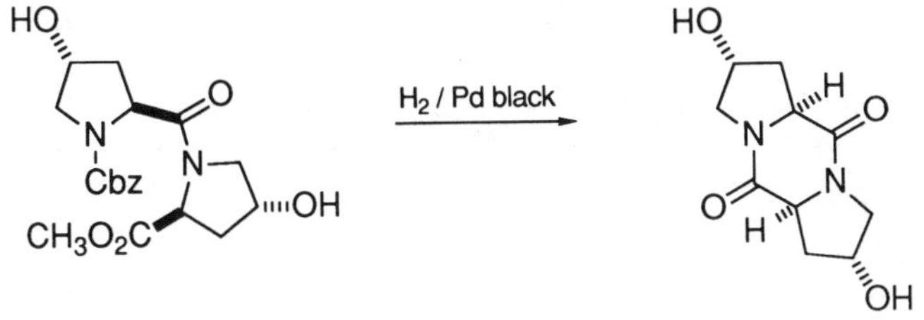

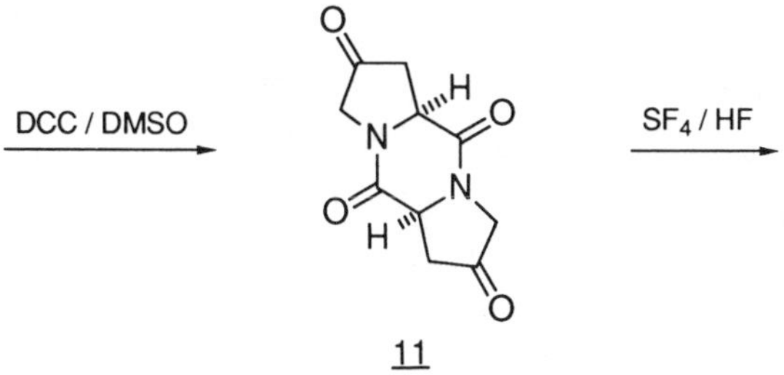

$$\text{11}$$

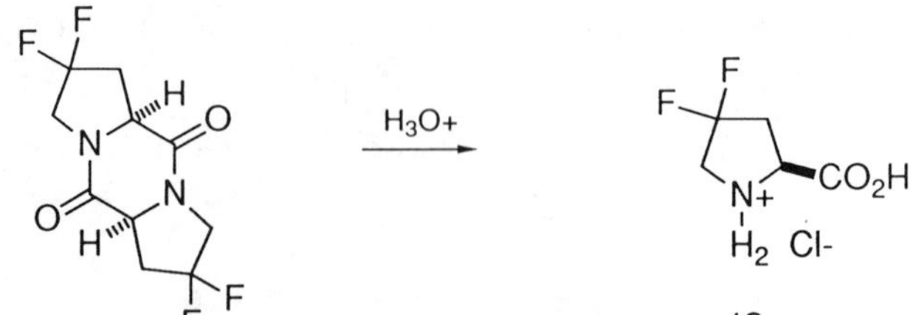

12

Although sulfur tetrafluoride–hydrogen fluoride is a very useful system for fluorodehydroxylation, the toxicity of sulfur tetrafluoride (comparable to that of phosgene) necessitates extreme caution in handling.

2.2.2 Diethylaminosulfur Trifluoride

Diethylaminosulfur trifluoride (DAST) is a useful reagent for replacing hydroxyl and the oxygen of a carbonyl group by fluorine under very mild conditions.[15] The ease of handling DAST also makes this a better reagent for fluorodehydroxylation than sulfur tetrafluoride. It is particularly useful for fluorinating sensitive alcohols which might be rearranged in the presence of hydrogen fluoride.

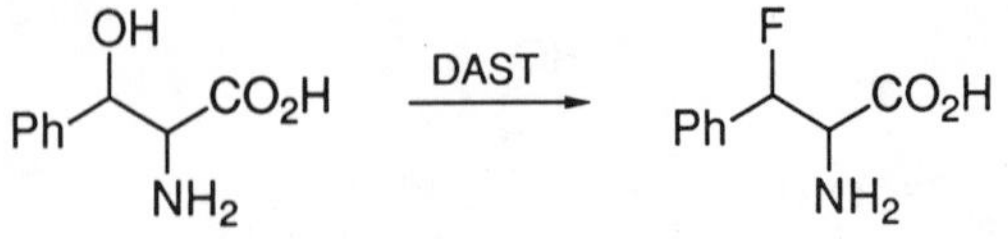

13   14

Fluorodehydroxylation of *threo*-β-phenyl serine with DAST gave two isomers in equal amounts. The reaction was nonselective.[16,17] These fluorination reactions are observed to have considerable carbonium ion character.

Asymmetric Synthesis of β-Fluorovaline and phenylalanine

In asymmetric synthesis of β-fluorovaline (**15**) and phenylalanine (**16**) esters,[18] the

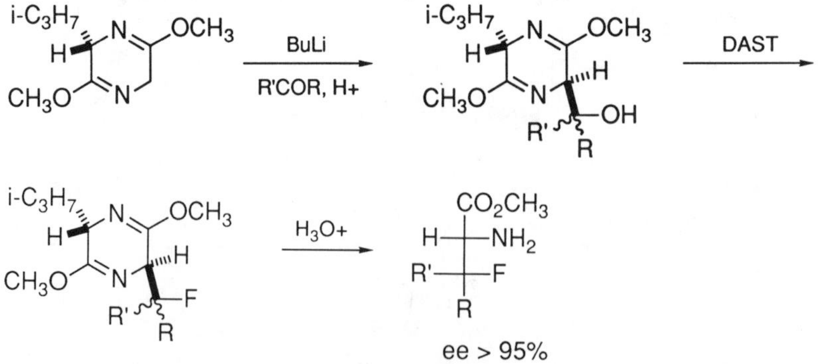

ee > 95%

15 R' = R = CH₃ (val)
16 R = H, R' = Ph (phe)

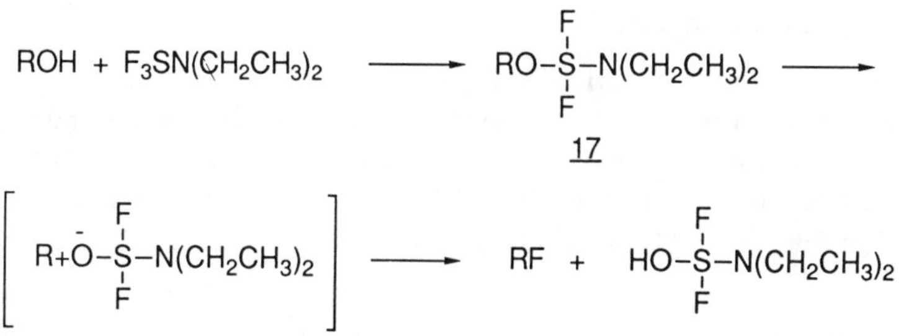

$$ROH + F_3SN(CH_2CH_3)_2 \longrightarrow RO-\underset{F}{\overset{F}{S}}-N(CH_2CH_3)_2 \longrightarrow$$

17

$$\left[R \overset{+}{} \overset{-}{O}-\underset{F}{\overset{F}{S}}-N(CH_2CH_3)_2 \right] \longrightarrow RF + HO-\underset{F}{\overset{F}{S}}-N(CH_2CH_3)_2$$

reaction goes through an unisolated intermediate (**17**) in which one of the fluorides on sulfur has been replaced by an alkoxide group. This could then dissociate and give an ion pair consisting of a carbonium ion and a sulfur oxide anion.

Asymmetric Synthesis of L-4-Fluorothreonine[70]

Another example for the use of DAST in fluoro dehydroxylation is in the enantioselective synthesis of L-4-fluorothreonine as seen in Fig. 2.2.[70]

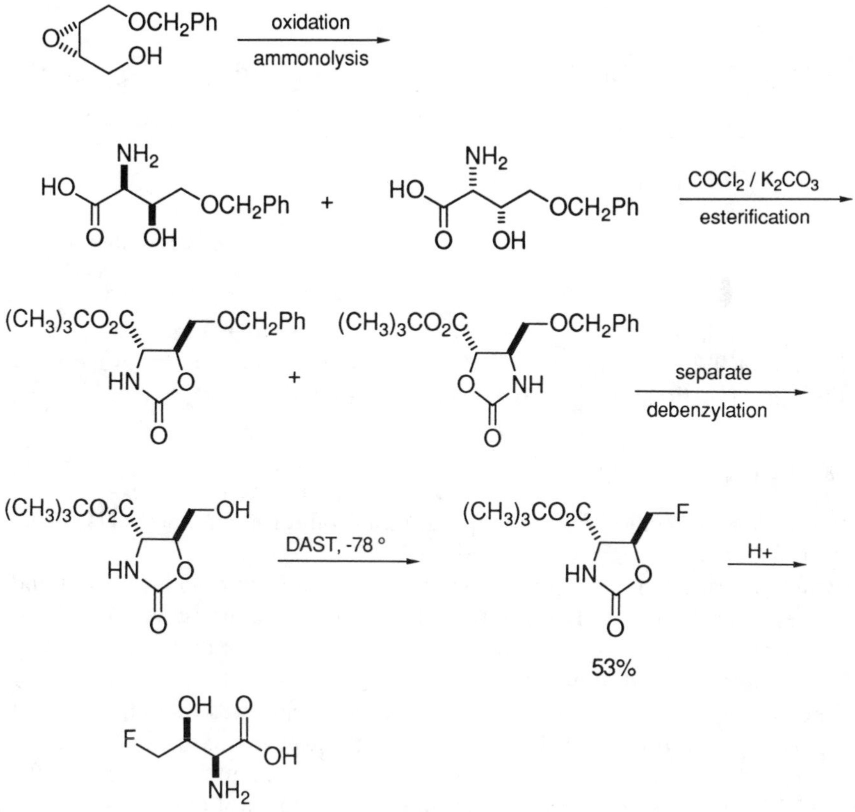

FIGURE 2.2

2.2.3   Yarovenko Reagent

Synthesis of 4-Fluoro glutamic acid

The one-step conversion of a hydroxy derivative to the corresponding fluorocompound (fluorodehydroxylation) by 2-chloro-1,1,2-trifluoroethyldiethyl amine (Yarovenko reagent, $FClCHCF_2NEt_2$)[19] has been used successfully in the synthesis of 4-fluoroglutamic acid (**18**).[20]

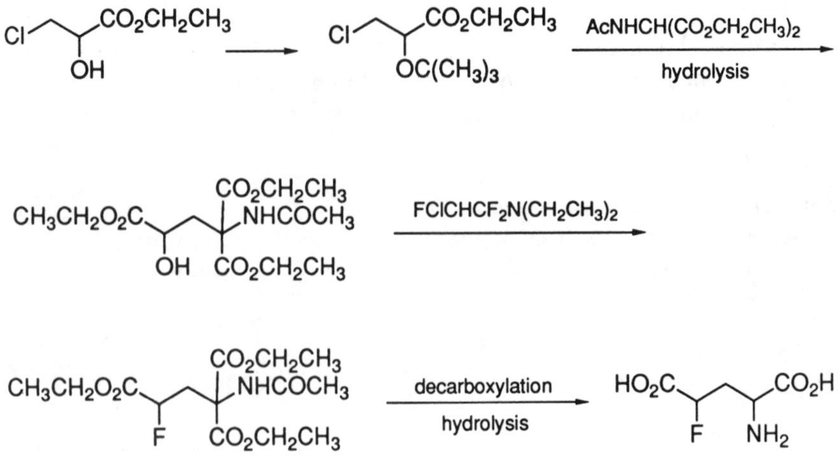

This fluoroalkyl amine reagent is a mild fluorinating agent. A major advantage of 2-chloro-1, 1,2-trifluoroethyldiethylamine is that it is unreactive toward most of the other functional groups like esters and amides.

2.2.4   Hydrogen Fluoride

A simple general method for the preparation of 3-fluoroamino acids (**21**) using hydrogen fluoride is represented in Fig. 2.3.

Bromofluorination of substituted acrylic acids, 2-alkenoic acids (**19**) in liquid hydrogen fluoride, followed by addition of *N*-bromoacetamide gave 2-bromo-3-fluorocarboxylic acids (**20**). On subsequent ammonolysis in liquid NH_3, the 3-fluoroamino acids were obtained.[21]

The bromofluorination of substituted acrylic acids was smooth, without complicating side products. However, the reaction gave mixtures of products with the 2-alkenoic acids, which required repeated recrystallizations for purification. The fluoroamino acids were obtained as mixtures of erythro and threo diastereomers.

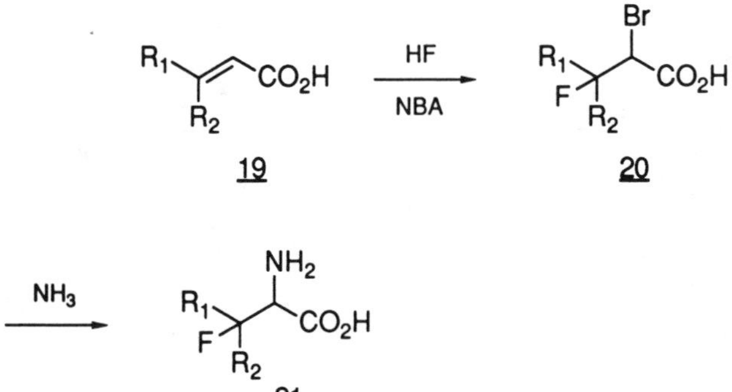

$R_1 = R_2 = H$ (ala)
$R_1 = CH_3$ $R_2 = H$ (BA)
$R_1 = R_2 = CH_3$ (val)
$R_1 = CH_2CH_3$ $R_2 = H$ (norval)
$R_1 = C_3H_7$ $R_2 = H$ (norleu)
$R_1 = C_4H_9$ $R_2 = H$ (2-amino-3-fluoroheptanoic acid)

FIGURE 2.3

Synthesis of β-Fluoroaspartic Acid

β-Fluoroaspartic acid (**23**) is obtained by diazotization of diaminosuccinic acid (**22**) in liquid hydrogen fluoride.[22]

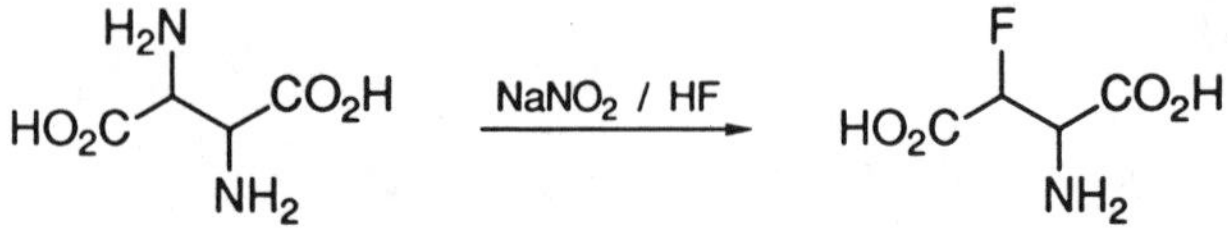

Synthesis of Difluoroborane Analogs of Fluorinated Amino Acids[23]

Difluoroborane analogs, where the boron has replaced the carboxylic acid carbon, can be synthesized from alkyl pinanediol boronates (**24**). On treatment with dichloromethyllithium, chloromethylboronates are obtained which rearrange to form **25**. The amino group is introduced using lithium hexamethyldisilazide. Deprotection of the amino group with very dry tetrabutyl-

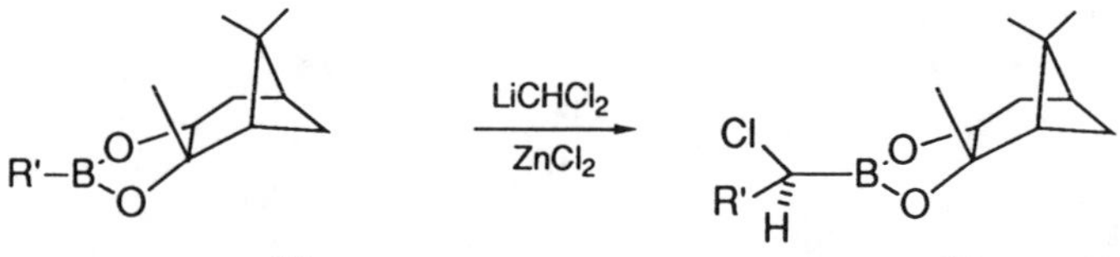

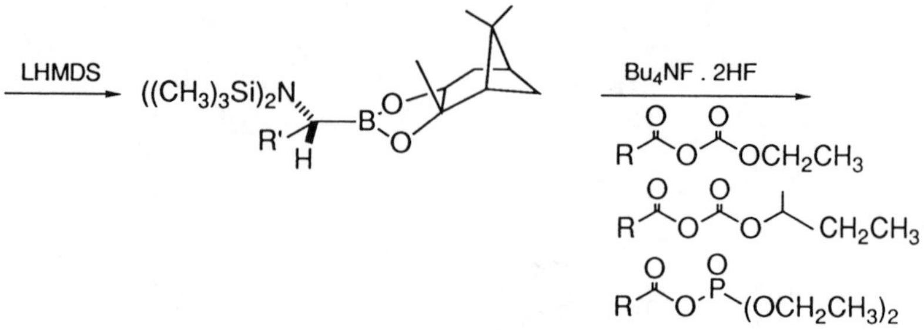

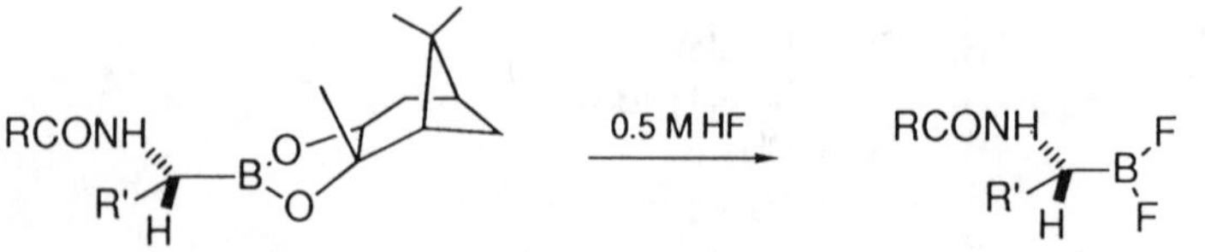

$$R = PhCH_2, Ph, CH_3$$
$$(CH_3)_2CH,$$
$$(CH_3)(CH_3CH_2)CH$$

ammonium fluoride–dihydrogen fluoride complex in the presence of efficient acylating agents gives the boronate esters (**26**), which with boron trichloride followed by 0.5 M aqueous hydrogen fluoride yield the difluoroboranes (**27**).

2.2.5 Hydrogen Fluoride–Pyridine

Polyhydrogen fluoride–pyridine [(HF)$_n$–py] solution is a very versatile reagent for fluorination.[24] Up to 70% by weight of hydrogen fluoride can be dissolved in pyridine to form a stable (up to 55°C), highly concentrated solution of anhydrous hydrogen fluoride in pyridine. This reagent is much easier to handle than anhydrous hydrogen fluoride on a lab scale and permits the reactions to be done at atmospheric pressure.

Ring-opening reactions of aziridines, azirines, or epoxides with hydrogen fluoride–pyridine offer a convenient and effective route of fluorinated amino acids.

Aziridine Ring-Opening Reactions[25-30]

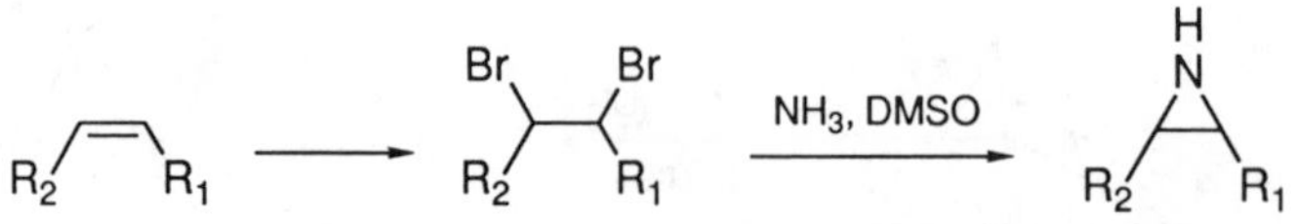

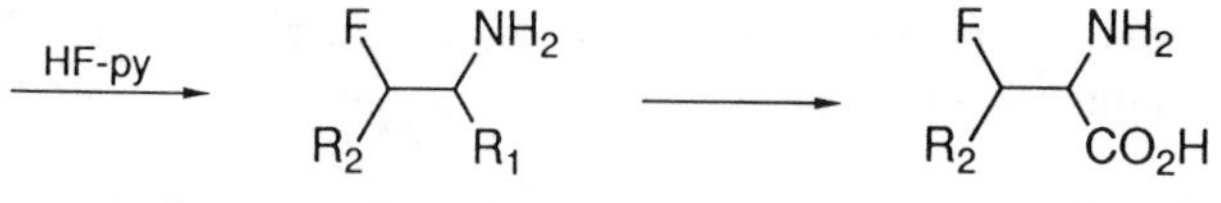

$$\text{29}$$

$$R_1 = CO_2R, \; CN, \; CONH_2$$
$$R_2 = H \; (ala), \; CH_3 \; (BA), \; Ph \; (phe)$$

The ring-opening reaction of **28** with hydrogen fluoride–pyridine is very regioselective. Only the 3-fluoro-2-amino esters (**29**) are obtained. Fluoride attack, in all cases, is directed to the most substituted ring carbon or to the benzylic carbon. The mechanism suggested is an S_N1-type process as seen in Fig. 2.4. Regardless of the cis or trans configuration of the starting aziridine carboxylate, the threo isomer predominates in the product.

The natural configuration of chloramphenicol, the threo isomer (**30**), may be prepared as in Fig. 2.5.[31] Due to the failure of the ring-opening reaction of

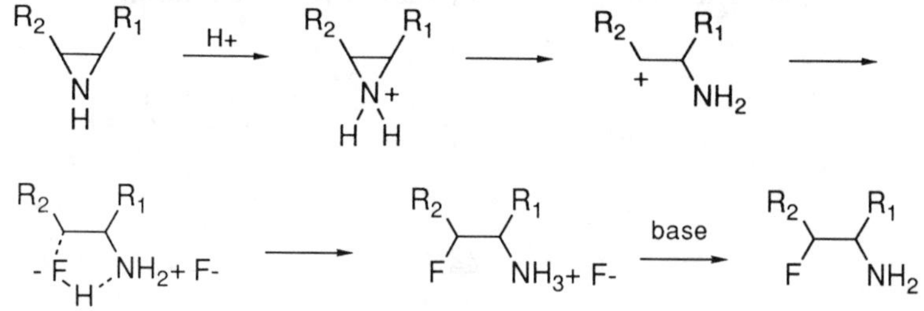

FIGURE 2.4

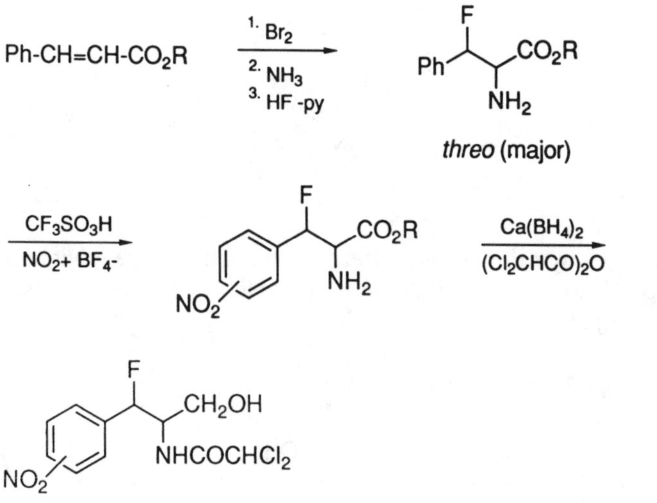

threo (major)

$$\text{30}$$

FIGURE 2.5

a p-nitrophenyl-substituted aziridine, direct nitration of the *threo*-3-fluorophenylalanine is unavoidable.

Azirine Carboxylates with Hydrogen Fluoride–Pyridine[32,33]

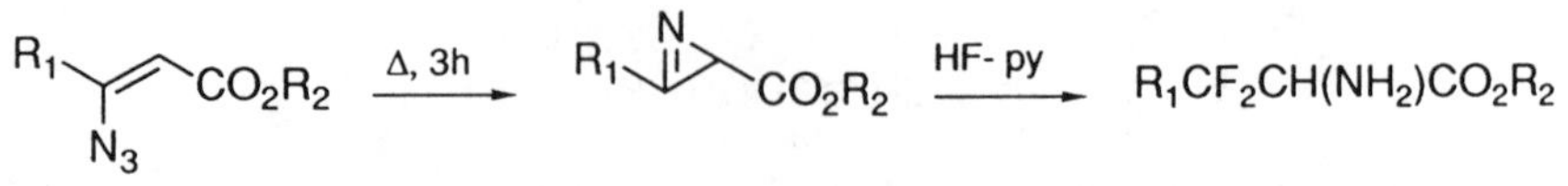

$$R_1 = CH_3, R_2 = CH_3CH_2$$
$$R_1 = Ph, R_2 = CH_3$$

Substituted 1-azirines (**31**) reacted under mild conditions more easily than aziridine homologs with hydrogen fluoride–pyridine. Presumably, the formation of the fluoroaziridine (**32**) is the first step. Two competing ring-opening pathways would be possible, yielding after hydrolysis β,β-difluoro compounds (**33**) or α-keto acids (**34**). The yields of the reaction of hydrogen fluoride–pyridine with

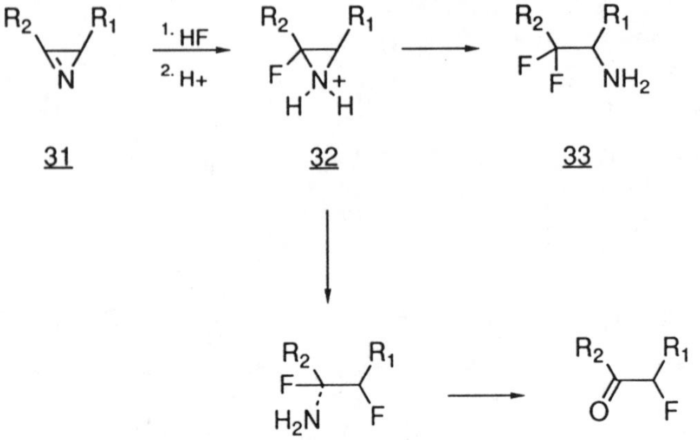

1-azirine were lower than those of the reaction of this reagent with their aziridine homologs. However, 1-azirines were valuable synthons for constructing β,β-difluoro compounds.

Epoxide Ring-Opening Reactions[26,34,35]

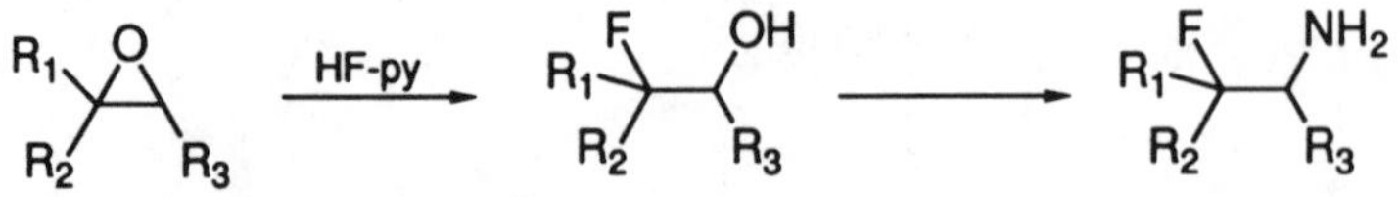

$$\longrightarrow$$

37

$R_1 = R_2 = CH_3$ (val)
$R_1 = CH_3$, $R_2 = CH_3CH_2$ (ile)
$R_3 = CO_2R$, CN

2.2.6 Trifluoromethylhypofluorite

Photofluorination

Photofluorination represents a general and selective method for substitutive fluorination of organic compounds.[36,37] This reaction, represented as follows,

$$R\text{-}H \xrightarrow[\text{liquid phase}]{CF_3OF,\ h\upsilon} R\text{-}F$$

allows the one-step transformation of optically active amino acids into fluoro-amino acids without racemization. Protection of the amino group against the electrophilic CF_3OF is achieved by protonation with the strong acid solvent used (hydrogen fluoride or trifluoroacetic acid).

A radical mechanism is suggested:[36,37]

$$CF_3OF \longrightarrow CF_3O\cdot + F\cdot$$

$$RH + F\cdot \longrightarrow R\cdot + HF$$

$$R\cdot + CF_3OF \longrightarrow RF + CF_3O\cdot$$

$$CF_3O\cdot + RH \longrightarrow R\cdot + COF_2 + HF$$

$CF_3O\cdot$ is thought to be the chain carrier because of the highly selective fluorination observed at C-3 of alanine to form **39**.

$$CH_3\overset{\overset{\displaystyle NH_2}{|}}{-}CHCO_2H \xrightarrow[\text{liquid phase}]{CF_3OF,\ h\upsilon} FCH_2\overset{\overset{\displaystyle NH_2}{|}}{-}CHCO_2H$$

38   **39**

D-alanine[38]	3-fluoro-*D*-alanine (57%)
L-alanine	3-fluoro-*L*-alanine (54%)

Photofluorination of L-isoleucine[36] (**40**) with trifluoromethyl hypofluorite in liquid hydrogen fluoride gave *trans*-3-methyl-L-proline via δ-fluorination and elimination of hydrogen fluoride.

40   41

Fluorodesulfurization

Fluorodesulfurization is the cleavage of C—S bond with concomitant formation of C—F bond:[39]

$$\text{C-SH} \xrightarrow[\text{HF}]{\text{CF}_3\text{OF, h}\upsilon} \text{C-F}$$

Fluoroamino acids can be synthesized from thiolamino acids by this reaction. The amino group is protonated in the highly acidic medium (solvent

42   43

D-penicillamine   3-fluoro-D-valine

hydrogen fluoride) and thus is protected from oxidation by the reagents. The conversion of penicillamine (**42**) to fluorovaline (**43**) has been found to occur equally well in the dark, at −78°C or at 0°C, and also can be effected with chlorine or N-chlorosuccinimide in place of trifluoromethylhypofluorite. This suggests that the solvent is the source of fluorine in the carbon–fluorine bond.

The mechanism proposed[39a] for the reaction involves an electrophilic highly oxidized form of sulfur, such as the difluorosulfonium salt (**44**), which reacts with

$$\text{R-SH} \xrightarrow{\text{CF}_3\text{OF, h}\upsilon} \text{R-SF} \xrightarrow{\text{CF}_3\text{OF, h}\upsilon} \text{R-SF}_2\text{+}$$

44

$$\longrightarrow \text{R}^+ + \text{SF}_2 \xrightarrow{\text{HF}} \text{R-F}$$

45

hydrogen fluoride. The sulfur difluoride (**45**) disproportionates to sulfur and sulfur tetrafluoride, thus accounting for the presence of elemental sulfur in the CF_3OF reaction.

2.2.7   Molecular Fluorine

Fluorination with Replacement of Hydrogen

Replacement of hydrogens attached to carbon by fluorine using molecular fluorine is a practical, useful reaction. Direct fluorination of enol-type 3-substituted pyruvates (**46**) using F_2 proceeds smoothly to give a mixture of keto- and enol-type 3-substituted-3-fluoropyruvates (**47**) in good yields.[40]

$$RCH_2COCO_2R_1 \rightleftharpoons RCH{=}\underset{\underset{\text{\underline{46}\quad OH}}{|}}{C}{-}CO_2R_1 \xrightarrow{F_2} RCHFCOCO_2R_1$$

$$\underline{47}$$

It has been suggested that the rate-determining enolization step precedes the electrophilic reaction with F_2. To obtain a clean reaction, the enolization step should be completed in advance.

Reductive amination of 3-fluorophenylpyruvic acid (**48**) gives *erythro*-3-fluorophenylalanine (**49**) with high stereoselectivity (95:5).[17,27,40] It was

proposed that the stabilizing interactions between fluorine and the neighboring iminium ion favored a single conformation for the reduction leading to the erythro diastereomer (see Fig. 2.6).

Erythro-3-fluoro-*p*-nitrophenylalanine (**50**) obtained by a reductive amination sequence is reduced with diborane and then dichloroacylated to give the unnatural configuration of chloromphenicol the enythro isomer 51.[31] Direct nitration of *erythro*-3-fluorophenylalanine is found to be nonselective in contrast to the exclusive para selectivity observed on nitration of nonfluorinated phenylalanine.

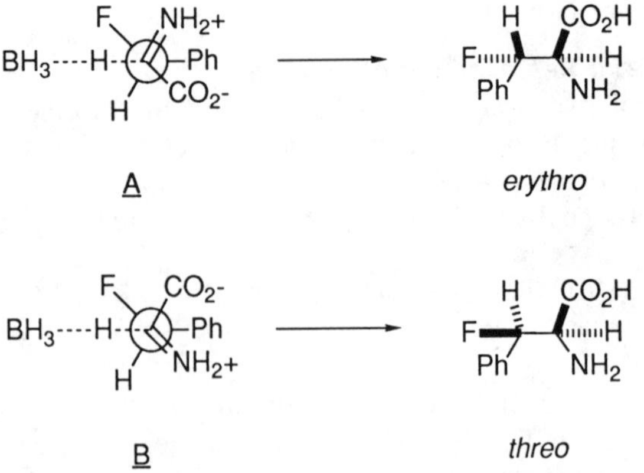

FIGURE 2.6

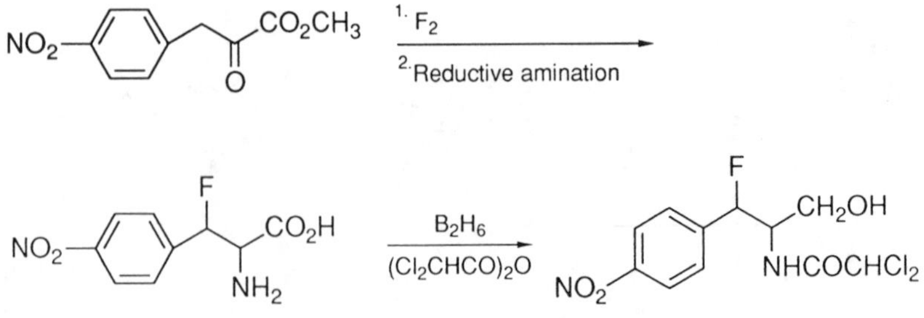

erythro

50

51

Fluorodesulfurization Using Fluorine in Helium[39]

Cysteine (**52**) on fluorodesulfurization with fluorine in helium (1:4 vol/vol) in hydrogen fluoride–fluoroboric acid solvent gave 33% of 3-fluoroalanine (**53**) and 3% of 3,3-difluoroalanine (**54**). Primary thiols like cysteine require more powerful

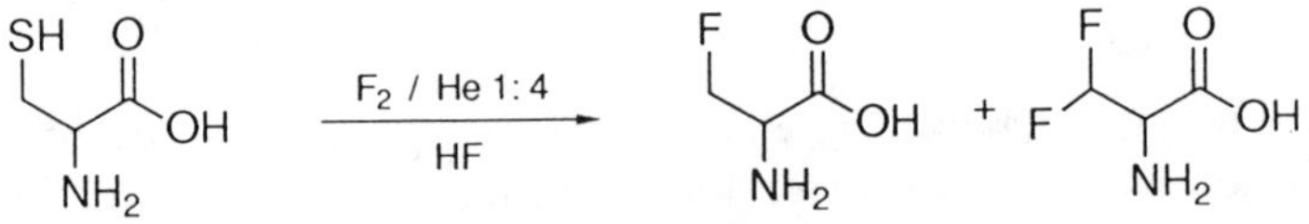

52 53 54

3-fluoro-alanine 3,3-difluoro-alanine

oxidant fluorine since fluorodesulfurization with trifluoromethylhypofluorite, *N*-chlorosuccinimide, or chlorine yielded none of the desired product. This suggests that such systems need a more potent leaving group like a dipositive sulfur species (**55**). This could give rise to a primary carbocation or suffer ready bimolecular displacement by a fluoride ion.

$$\text{R-SH} \xrightarrow{\ 3\ F_2\ } \text{R-SF}_3 \xrightarrow{\ HF\ } \text{R F } + \text{SF}_2 + \text{HF}$$

$$\underline{55}$$

The difluorinated compound can be explained by elimination of hydrogen fluoride to form **56** followed by addition of fluoride to a carbon forming an α-fluorosulfenyl fluoride, which reacts further.

$$\text{RCH}_2\text{-SF}_2+ \xrightarrow{\ -HF\ } \text{RCH=SF+} \xrightarrow{\ +F-\ } \text{RCHFSF}$$

$$\underline{56} \qquad\qquad \underline{57}$$

$$\xrightarrow{\ F_2\ } \text{RCHF-SF}_3 \xrightarrow{\ HF\ } \text{RCHF}_2$$

2.2.8 Perchlorylfluoride

Perchlorylfluoride can be used efficiently for the synthesis of fluoroamino acids.

Synthesis of 4-Fluoroglutamic Acid (58)[41]

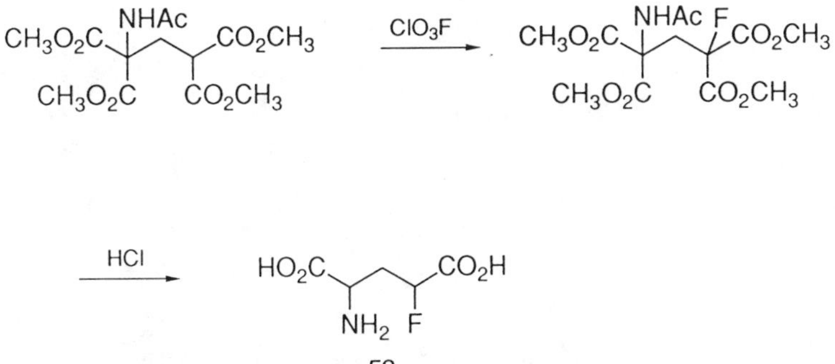

$$\underline{58}$$

Synthesis of β,β-Difluoroaspartic Acid and β,β-Difluoroasparagine[42]

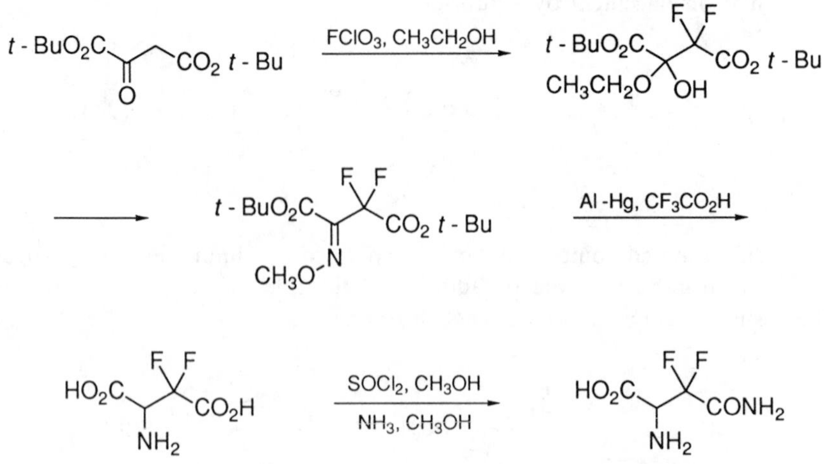

β,β-difluoroaspartic acid

β,β-difluoroasparagine

Synthesis of β-Difluoroaspartic Acid[43]

Dehydrohalogenation of *t*-butyl-β,β-difluoroaspartic acid (**59**) using diazabicy-clononane followed by reduction and hydrolysis affords the β-fluoroaspartic acids (**60**).

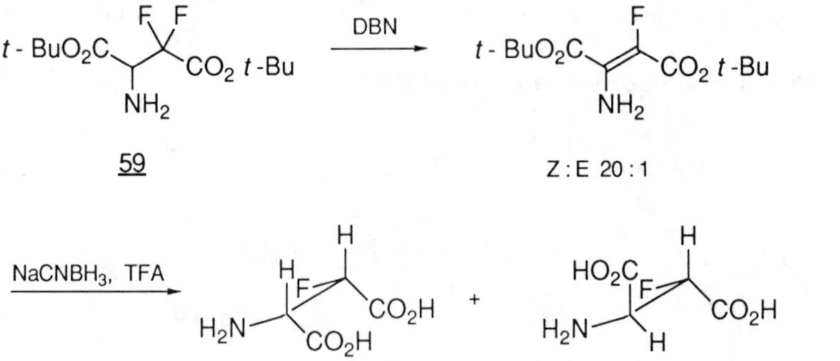

erythro (major)

threo (minor)

60a

60b

Several mechanisms of reaction for FclO₃ have been proposed which include perchlorylfluoride as a source of "positive fluorine." Mesomeric nucleophiles (**61**) form an intermediate (**62**) that can undergo intramolecular transfer of fluoride and form a thermodynamically very stable C—F bond.[19]

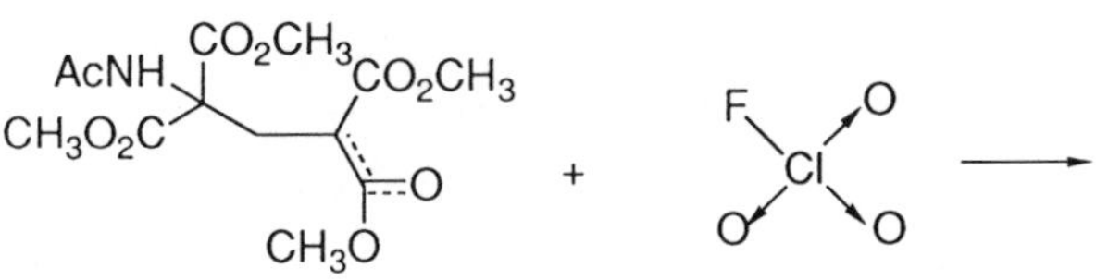

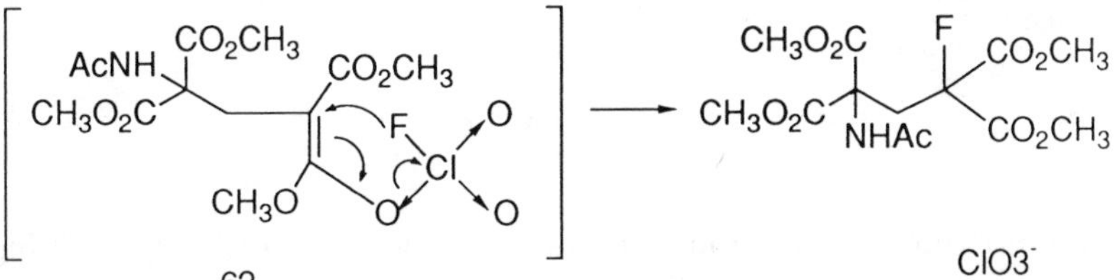

2.2.9 Fluoroboric Acid

Biochemistry and pharmacology of ring-fluorinated aromatic amino acids are of considerable interest. Diazo compounds are known to undergo facile photo-extrusion of nitrogen. Trapping of the reactive intermediates by a fluorine atom would be a direct approach for the synthesis of fluorinated aromatic compounds. This type of photofluorination is indeed possible by doing the photolysis in

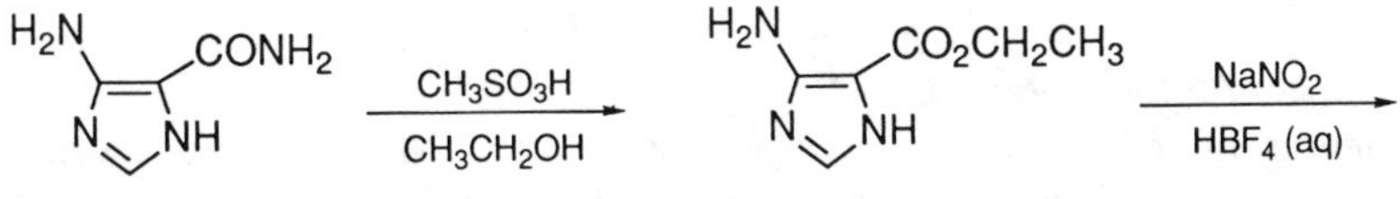

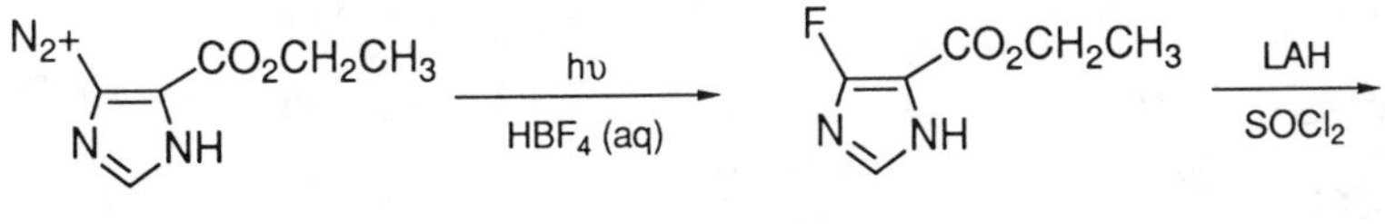

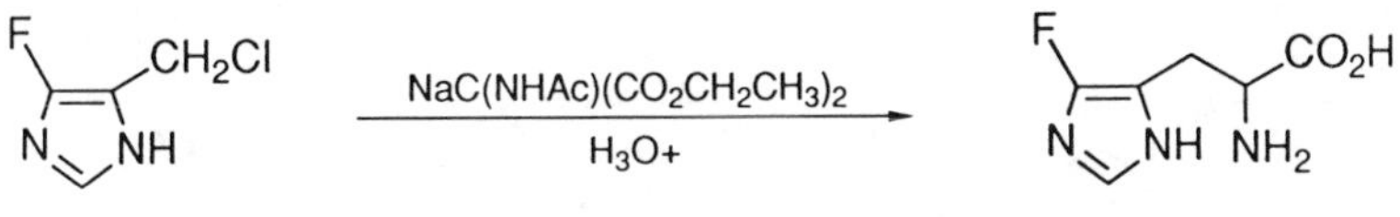

FIGURE 2.7

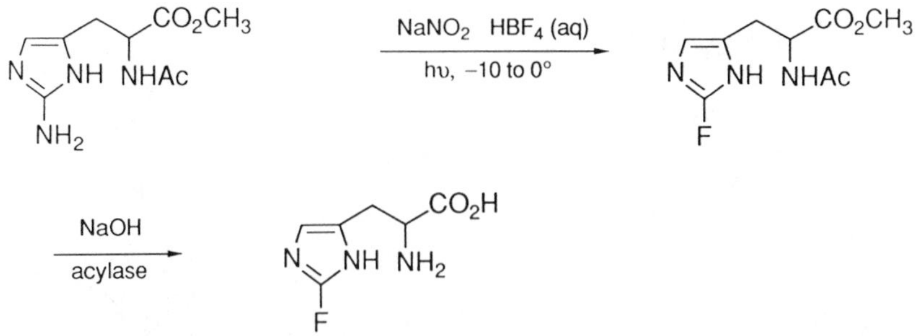

FIGURE 2.8

tetrafluoroboric acid solution. Synthesis of 4-fluoro-DL-histidine (**65**) using this approach is seen in Fig. 2.7.[44] The key step in the synthesis is the photolysis of the imidazole diazonium fluoroborate (**64**) in aqueous fluoroboric acid. It was also shown that isolation of the intermediate diazonium fluoroborate was unnecessary. Thus when a solution of the amine (**63**) in 50% fluoroboric acid was treated with 1.1 equiv wt of concentrated aqueous sodium nitrite and the resulting mixture was photolyzed, good yields of the fluoro compound were obtained. Similar methodology can also be extended to the corresponding 2-fluoro series,[45] as seen in Fig. 2.8.

Photochemical fluorination using fluoroboric acid generally can be applied to the synthesis of a variety of aromatic fluorinated compounds. Synthesis of 3-fluoro-L-tyrosine (**66**) and 3,5-difluoro-L-tyrosine (**67**) aptly demonstrate this case.[46]

This methodology thus demonstrates that it is possible to directly introduce fluorine in an aromatic ring using convenient, simple reactions.

Synthesis of 2-Fluorotyrosine[47,48]

Condensation of acetamidomalonic ester with 2-fluoro-4-methoxybenzyl chloride (**68**), which is obtained as seen in Fig. 2.9,[48] followed by hydrolysis affords 2-fluorotyrosine (**69**).

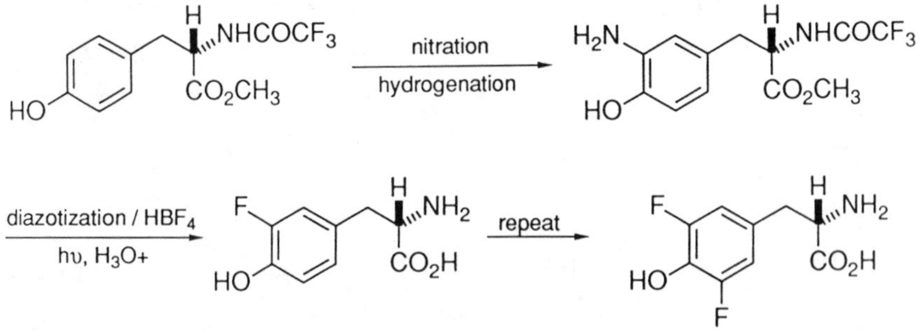

66   67

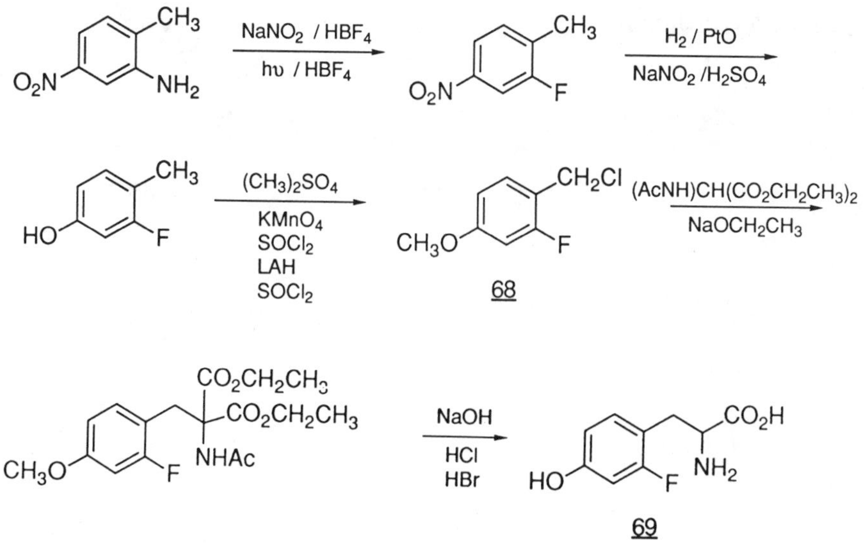

FIGURE 2.9

2.2.10   Metal Fluorides as Fluorinating Agents (KF, AgF, HgF, SbF₃, KHF₂)

Monofluorosubstitution by halogen replacement using metallic or nonmetallic fluorides is fundamentally a simple reaction:[19]

$$RX + MF \longrightarrow RF + MX$$

$$X = Cl,\ Br,\ I$$

Some examples of fluorinations, using the previously listed fluorides in the synthesis of fluoroamino acids, are given in this section.

Synthesis of 3-Fluoroalanine (70)[49] with Potassium Fluoride

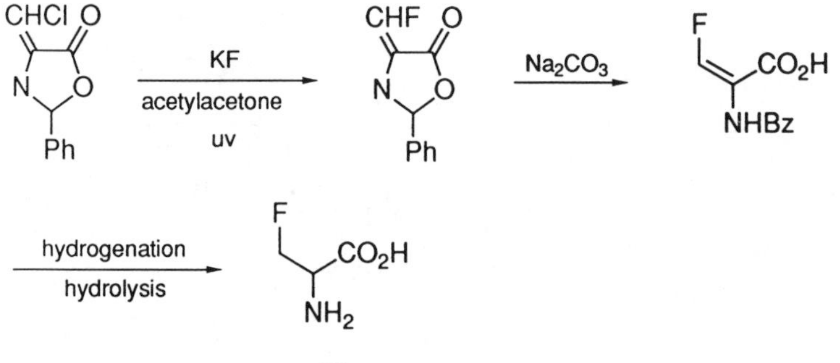

Synthesis of (E)-β-Fluoromethylene Glutamic Acid (71)[50] with Potassium Fluoride

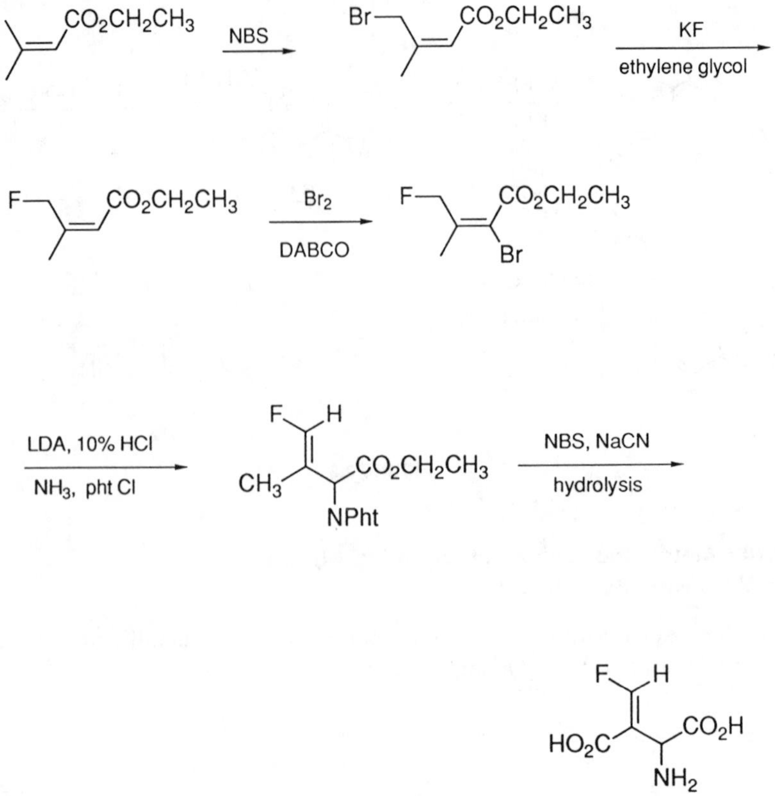

Synthesis of 4,4'-Difluorovaline (72)[51] with Potassium Fluoride

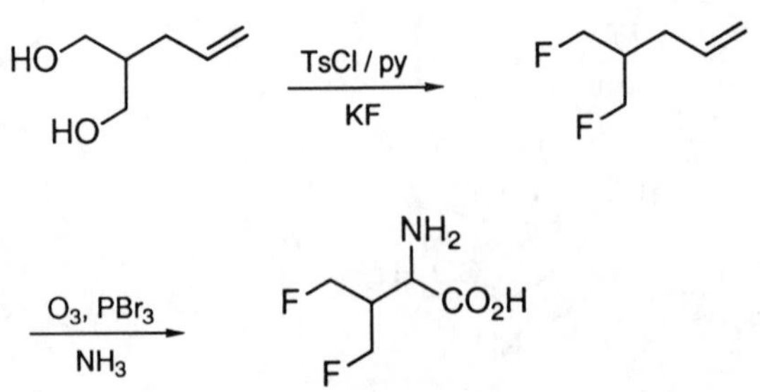

Synthesis of cis- *and* trans- *4-Fluoro-L-proline (73)[52] with Potassium Fluoride*

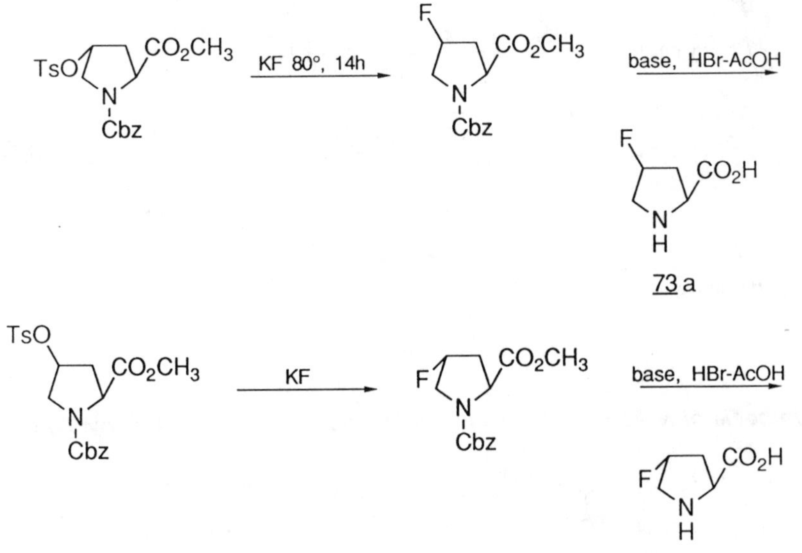

73 a

73 b

Synthesis of 4-Amino-5-Fluoropentanoic Acid (74)[53] with Silver Fluoride

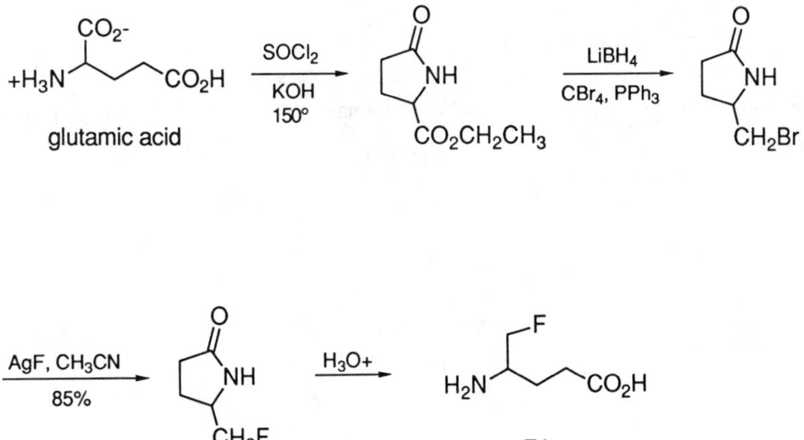

glutamic acid

74

Synthesis of 4-Fluoroglutamic Acid (75)[54] with Mercuric Fluoride

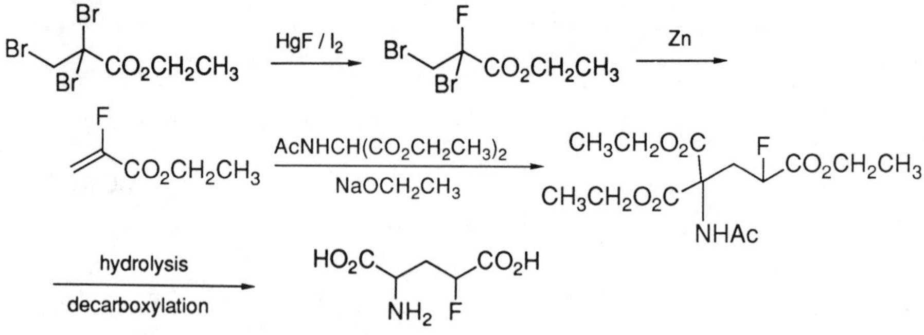

Synthesis of N-Acetyl-4-fluoroglutamic Acid (76)[55] with Antimony Trifluoride

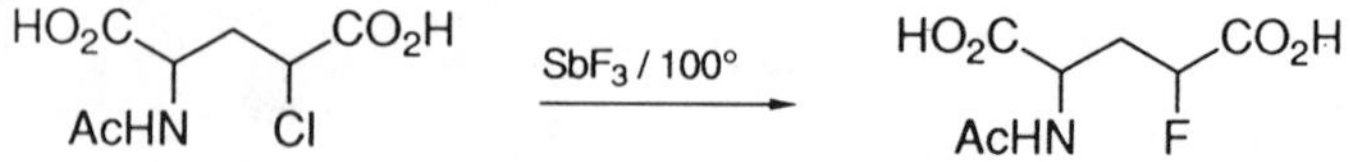

Synthesis of (E)-β-Fluoromethylene-m-tyrosine (77)[56] with Potassium Bifluoride

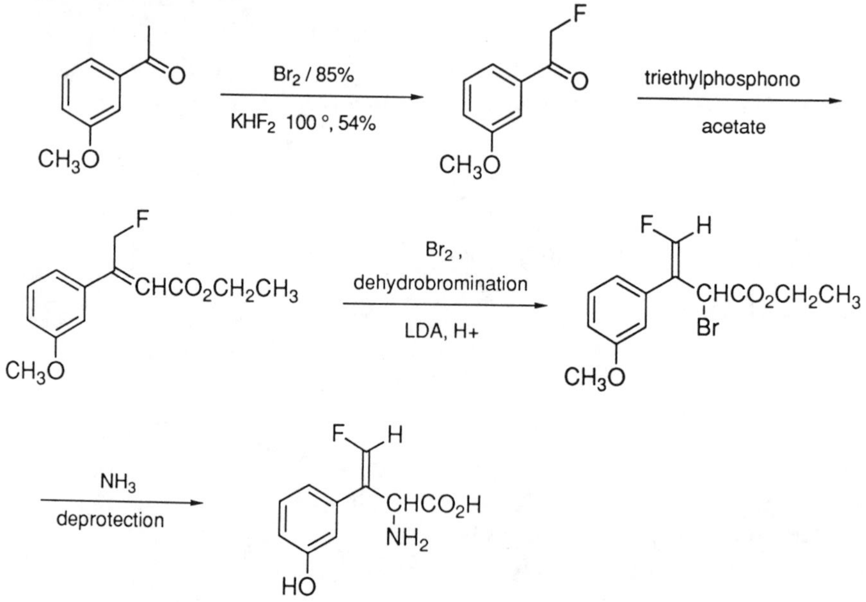

Most substitution fluorinations by potassium fluoride must occur by S_N2 mechanism. Where stereochemistry has been observed, inversion of configuration is usually found. The greater fluorinating power of potassium fluoride is a result of its large ionic size and therefore its enhanced solubility.[19]

2.2.11   Xenon Difluoride

Synthesis of Fluoromethionine (79)[57]

A mild procedure for the fluorination of methionine (**78**) employs the use of xenon difluoride; exclusive fluorination occurs at the methylthio group.

2.2.12   Freons as Building Blocks (CH$_2$FCl, CHFCl$_2$, CHF$_2$Cl, CF$_3$Br)

α-Substituted α-amino acids can be conveniently synthesized via regiospecific alkylation at the α-carbon atom of a derivative of the parent α-amino acid.[58-61] Fluorohalo methylation is an attractive method for the synthesis of fluoroamino acids.

Aminomalonates

Aminomalonates (**80**) give fluoromethylated glycines.[58]

The product β,β-difluoroalanine (**81**) was obtained in good yield (30%, two steps).

Schiff Bases

Schiff base alkyl esters of the parent α-amino acids can be synthesized easily by esterification followed by treatment with benzaldehyde. In the course of this two-step transformation, the other amino and carboxyl, as well as hydroxyl groups also get protected. Deprotonation followed by alkylation of the anion proceeds smoothly. Selective removal of the benzylidine group under mild acidic conditions and subsequent cleavage of the other protecting groups yields the α-

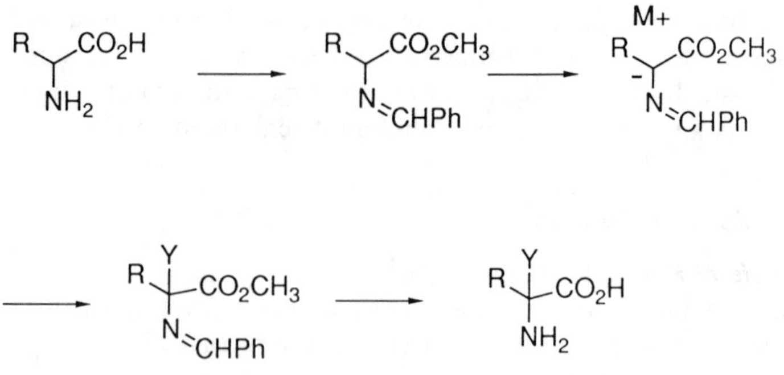

R = CH$_3$ (ala), CH$_2$Ph (phe), CH$_2$PhOH (tyr),
     CH$_2$Ph(OH)$_2$ (DOPA), CH$_3$ S(CH$_2$)$_2$ (met),
     HO$_2$C(CH$_2$)$_2$ (glu), H$_2$N(CH$_2$)$_4$ (lys),
     H$_2$N(CH$_2$)$_3$ (orn), C$_3$H$_3$N$_2$CH$_2$ (his),
     H$_2$NC(=NH)NH(CH$_2$)$_3$ (arg).

Y = CHF$_2$ or CH$_2$F or CF$_3$

FIGURE 2.10

fluoromethyl-α-amino acids (**82**) (Fig. 2.10).[59] The alkylation with difluorochloromethane was instantaneous. It is suggested that the CHF$_2$Cl is first transformed to the electrophilic difluoromethylene (**84**), which then alkylates the anion (**83**). The intermediate (**85**) then reacts with a new molecule of CHF$_2$Cl to afford the difluoromethylated compound and difluorocarbene, allowing a chain process.

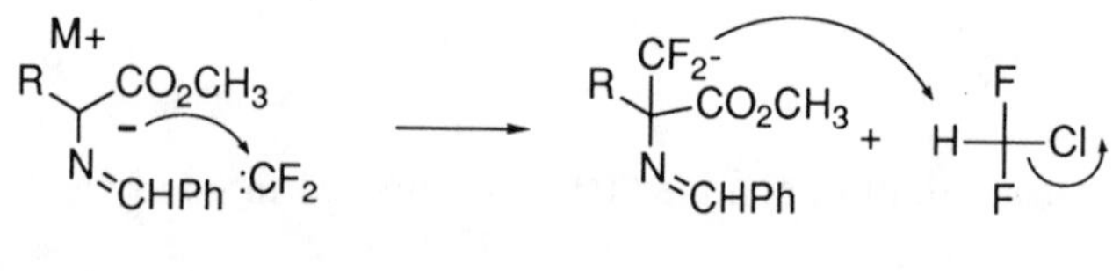

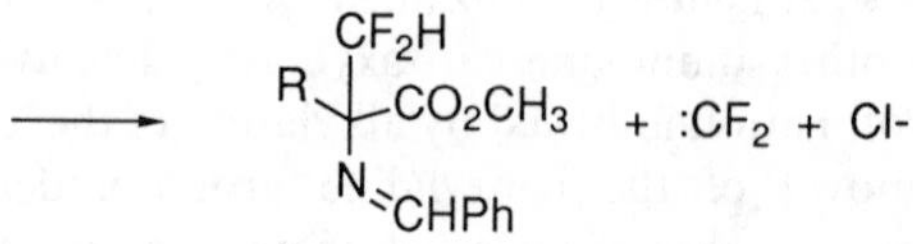

The anion reacted much more slowly with fluorochloromethane, indicating a different mechanism for this alkylation. Considering that polyhalogenomethane derivatives in the presence of nucleophilic reagents can undergo either direct displacement of one halogen atom or α-elimination depending on their substitution pattern, it seems reasonable to propose an S_N2-type mechanism for the reaction with CH_2FCl.

Alkylation of Sodium Enolate

Alkylation of sodium enolate of the readily available monosubstituted malonate diesters with an excess of chlorodifluoromethane gives excellent yields of the difluoromethyl adduct (**87**).[60] Hydrolysis under carefully controlled conditions so as to minimize the defluorination gave the malonic acid hemiester (**88**). Transformation of the carboxylic acid function to an amino group could be achieved under the standard Curtius rearrangement sequence to afford α-difluoromethyl-α-amino acids (**89**).

$$R \stackrel{\displaystyle CO_2R_1}{\underset{\displaystyle CO_2R_2}{\overset{\displaystyle |}{-\!\!\!-\!\!\!-}H}} \quad \xrightarrow[\text{CHF}_2\text{Cl}]{\text{NaH}} \quad R \stackrel{\displaystyle CO_2R_1}{\underset{\displaystyle CO_2R_2}{\overset{\displaystyle |}{-\!\!\!-\!\!\!-}CHF_2}} \quad \xrightarrow{CF_3CO_2H} \quad R \stackrel{\displaystyle CO_2R_1}{\underset{\displaystyle CO_2H}{\overset{\displaystyle |}{-\!\!\!-\!\!\!-}CHF_2}}$$

$$\underline{86} \qquad\qquad\qquad \underline{87} \qquad\qquad\qquad \underline{88}$$

$$\xrightarrow[\text{saponify}]{\substack{\text{SOCl}_2 \\ \text{NaN}_3,\ \Delta, \\ \text{Curtius}}} \quad R \stackrel{\displaystyle CO_2H}{\underset{\displaystyle NH_2}{\overset{\displaystyle |}{-\!\!\!-\!\!\!-}CHF_2}}$$

$$\underline{89}$$

$$R = C_6H_5CH_2 \text{ (phe)}, PhtN(CH_2)_3 \text{ (orn)},$$
$$R_1 = CH_3CH_2, C(CH_3)_3$$

Alkylation of Malonate Esters

Alkylation of anions derived from malonate esters or Schiff base esters of α-amino acids (**90**) with an excess of dichlorofluoromethane gives the α-chlorofluoromethane derivatives (**91**). Reduction of —CHClF group with *tri-n*-butyltin hydride affords α-monofluoromethyl-α-amino acids (**93**).[61] This offers an alternative synthesis of α-monofluoromethyl-α-amino acids without the use of toxic fluoroacetonitrile or less readily available CH_2ClF reagents. The halomethylation of the Schiff base esters was less regiospecific with $CHCl_2F$ than with CHF_2Cl. Also due to the asymmetry of the —CHClF group, mixtures of diastereomers were obtained.

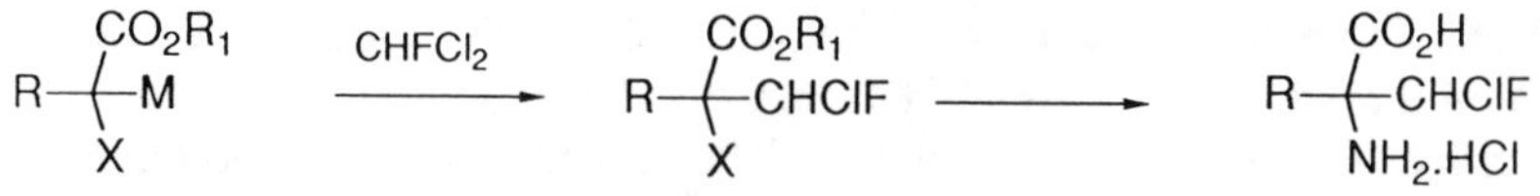

$$R\underset{X}{\overset{CO_2R_1}{\underset{|}{\overset{|}{C}}}}M \xrightarrow{CHFCl_2} R\underset{X}{\overset{CO_2R_1}{\underset{|}{\overset{|}{C}}}}CHClF \longrightarrow R\underset{NH_2.HCl}{\overset{CO_2H}{\underset{|}{\overset{|}{C}}}}CHClF$$

<u>90</u>   <u>91</u>   <u>92</u>

X = CO_2C(CH_3)_3   R_1 = CH_3 or CH_3CH_2
     N=CHPh

R = CH_3 (ala), CH_2Ph (phe), CH_2PhOH (tyr),
     HO_2C(CH_2)_2 (glu), H_2N(CH_2)_3 (orn)

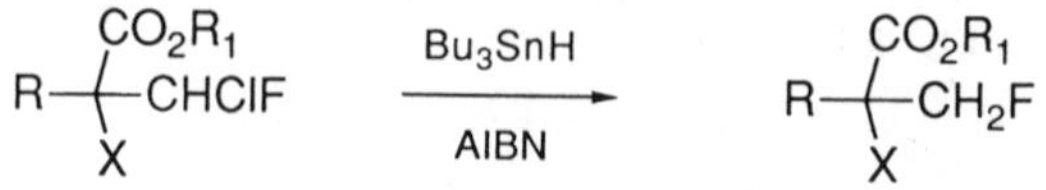

$$R\underset{X}{\overset{CO_2R_1}{\underset{|}{\overset{|}{C}}}}CHClF \xrightarrow[AIBN]{Bu_3SnH} R\underset{X}{\overset{CO_2R_1}{\underset{|}{\overset{|}{C}}}}CH_2F$$

X = CO_2C(CH_3)_3   <u>93</u>
     N_2CO_2R_2

2.2.13 Perfluoroalkyl Iodides

Imidazole and its derivatives undergo facile photochemical trifluoromethylation or perfluoroalkylation with trifluoromethyl iodide or perfluoroalkyl iodide
at room temperature.[62] Both carbon-4 (or carbon-5) and carbon-2 trifluoromethylated isomers are obtained, with carbon-4 or carbon-5 being predominant,
along with trace amounts of bistrifluoromethylated materials. The trifluoromethyl radical generated by ultraviolet radiation of trifluoromethyl iodide in
methanol solution reacts with α-acyl histidine esters (**94**).[62]

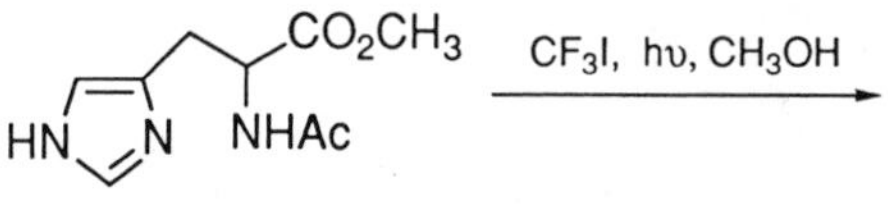

$$\xrightarrow{CF_3I, \ h\upsilon, \ CH_3OH}$$

<u>94</u>

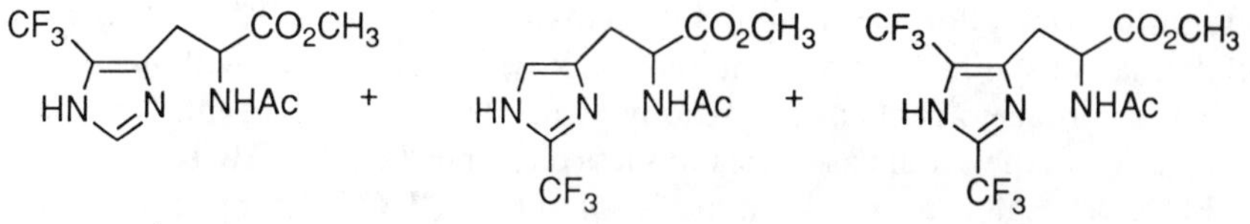

major   +   minor   +   trace

The mechanism suggested involves the formation of a sigma complex with the substrate, such as **95** or **96**, as an initial step. The iodine and not a second trifluoromethyl radical is considered the ultimate hydrogen atom acceptor. The preponderance of attack at carbon-4 (or carbon-5) is consistent with the electrophilic nature of R_f radicals, since carbon-4 (or carbon-5) is found to have higher electron density than carbon-2 (carbon-2 is adjacent to two nitrogen atoms).

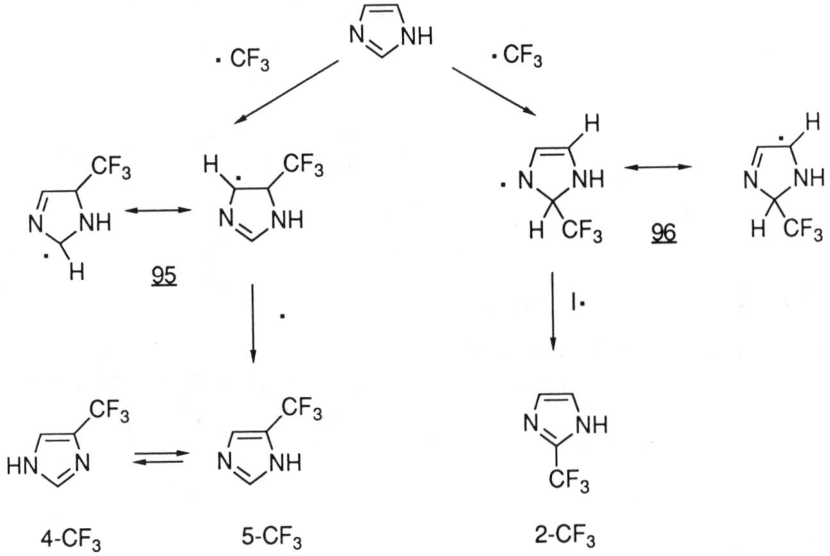

2.2.14 Fluoroacetonitrile

Short and efficient synthesis of fluoroamino acids may employ fluoroacetonitrile.

Synthesis of 3-Amino-4-fluorobutanoic acid (97)[63]

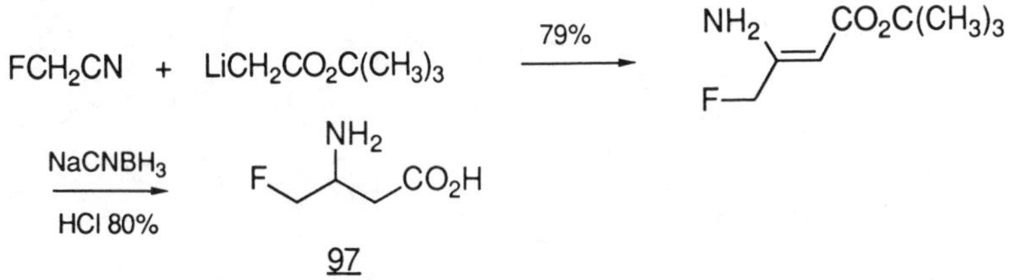

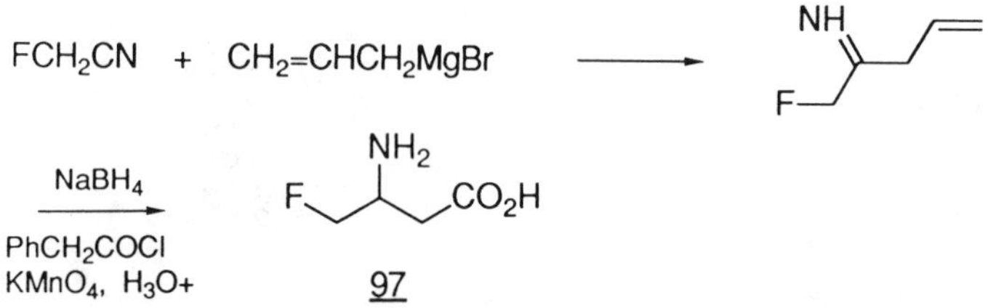

Synthesis of α-Fluoromethyldehydroornithine (98)[64]

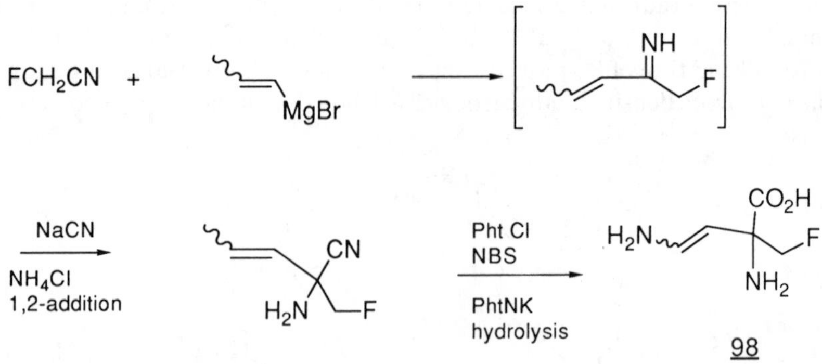
FCH2CN
+
MgBr
NH
F
NaCN
NH4Cl
1,2-addition
CN
H2N
F
Pht Cl
NBS
PhtNK
hydrolysis
CO2H
H2N
F
NH2

98

2.2.15 *N*-Acyl Trifluoroacetaldimines

Synthesis of 3,3,3-Trifluoroalanine

Addition of anhydrous hydrogen cyanide to *N*-acyl trifluoroacetaldimines (**99**) gives 2-acylamino-3,3,3-trifluoropropionitriles (**100**), which can be hydrolyzed to 3,3,3-trifluoroalanine (**101**).[65]

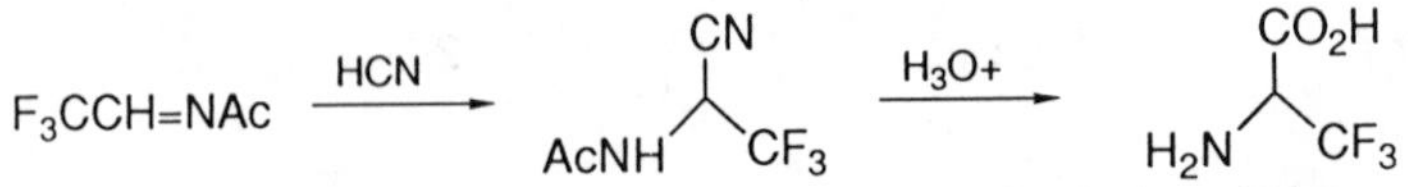
F3CCH=NAc
HCN
CN
AcNH
CF3
H3O+
CO2H
H2N
CF3

99 100 101

Also, addition of vinyl Grignard to **99** followed by oxidation affords 3,3,3-trifluoroalonine (**102**).[84,85]

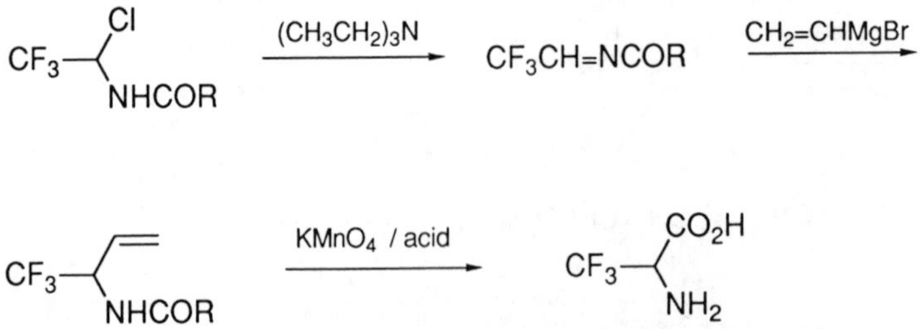
Cl
CF3
NHCOR
(CH3CH2)3N
CF3CH=NCOR
CH2=CHMgBr
CF3
NHCOR
KMnO4 / acid
CO2H
CF3
NH2

102

2.2.16   Ethyl Fluoroacetate

Synthesis of α-Fluoro-β-alanine (103)[66]

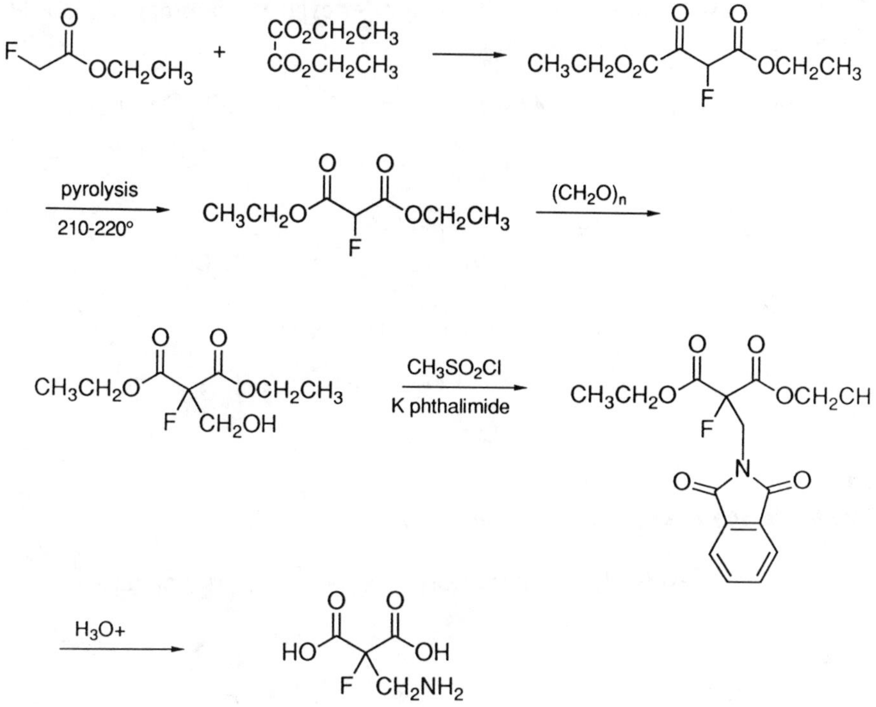

103

Synthesis of 4-Amino-2-fluorocrotonic Acid (104) and 5-Amino-2-fluoropentenoic Acid (105)[67]

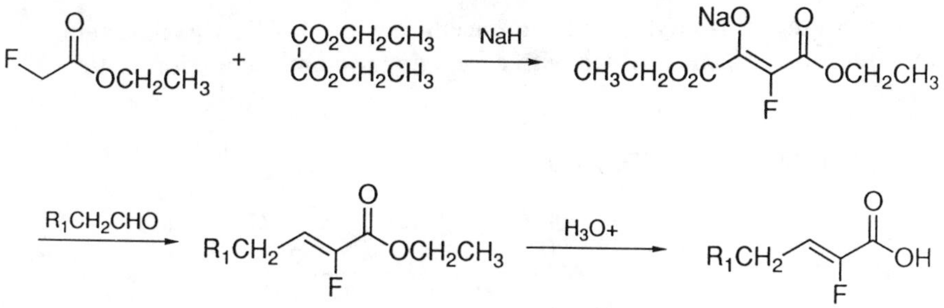

104 R_1 = NH_2 (4-amino-2-fluoro-crotonic acid)
105 R_1 = NH_2CH_2 (5-amino-2-fluoro-pentenoic acid)

Synthesis of γ-Fluoroglutamic Acid (107)[68]

Ethyl fluoroacetate reacts with ethyl chloroformate to give diethyl fluoromalonate (**106**), which undergoes Michael addition with ethyl α-acetamidoacrylate. Hydrolysis and decarboxylation affords the N-acylated γ-fluoroglutamic acid (**107**).[68]

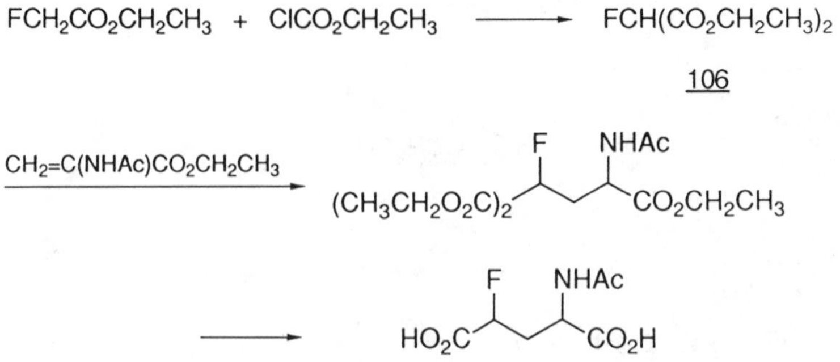

$$FCH_2CO_2CH_2CH_3 \ + \ ClCO_2CH_2CH_3 \ \longrightarrow \ FCH(CO_2CH_2CH_3)_2$$

106

$$\xrightarrow{CH_2=C(NHAc)CO_2CH_2CH_3}$$

(CH₃CH₂O₂C)₂ — F, NHAc — CO₂CH₂CH₃

⟶ HO₂C — F, NHAc — CO₂H

107

Synthesis of 3-Fluoro-ᴅ,ʟ-alanine, 2d (108)[69]

$$FCH_2CO_2CH_2CH_3 \ + \ (CO_2CH_2CH_3)_2 \ \longrightarrow \ FCH_2COCO_2H$$

$$\xrightarrow[NaBD_4]{NH_3}$$

F — D — CO₂H, NH₂

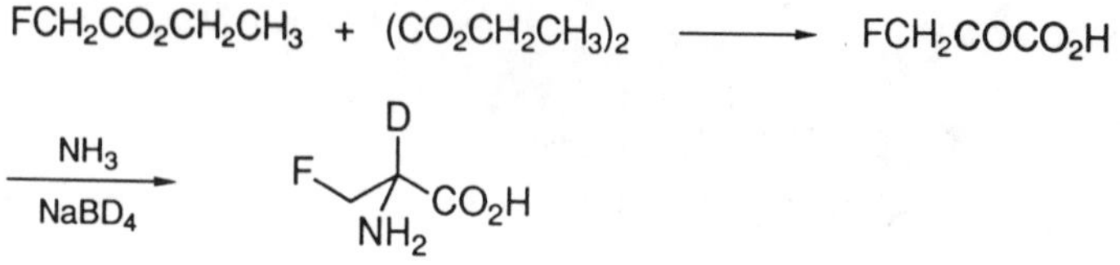

108

2.2.17 t-Butyl Fluoroacetate

Synthesis of erthro- and threo-β-Fluoroaspartic Acids (110, 111) and erythro-β-Fluoroasparagine (113)[43]

Condensation of t-butyl monofluoroacetate with di-t-butyloxalate affords monofluorooxaloacetate, which with excess ammonium acetate forms **109**.

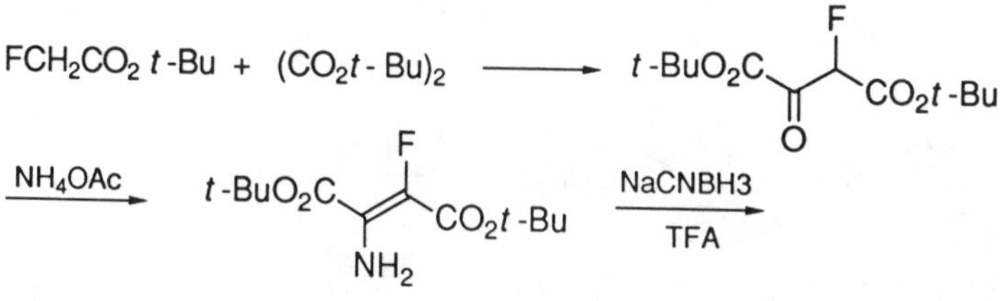

$$FCH_2CO_2 t\text{-}Bu \ + \ (CO_2 t\text{-}Bu)_2 \ \longrightarrow \ t\text{-}BuO_2C \underset{O}{\overset{F}{\diagup}} CO_2 t\text{-}Bu$$

$$\xrightarrow{NH_4OAc} \quad t\text{-}BuO_2C \overset{F}{=}\underset{NH_2}{} CO_2 t\text{-}Bu \quad \xrightarrow[TFA]{NaCNBH3}$$

Z : E   1 : 1

109

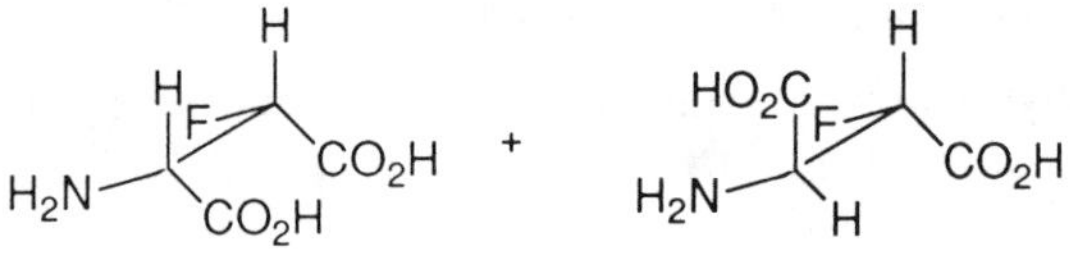

erythro   threo

111

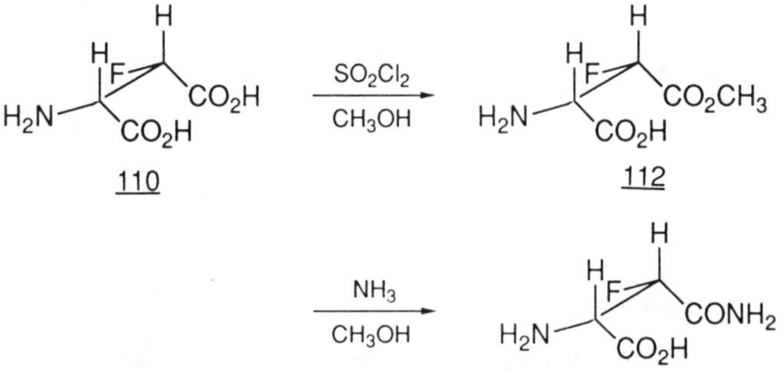

110   **112**

113

2.2.18 Ethyl Bromodifluoroacetate

BOC-L-leucinal (**114**) is condensed with ethyl bromodifluoroacetate under Reformatsky conditions. A mixture of diasteromers (2:1) is obtained under sonicating conditions at room temperature. However, under reflux, only the 3S-hydroxy compound (**115**) is formed. After incorporation into the pentapeptide, the alcohol is oxidized to form the difluoroketone (**116**).

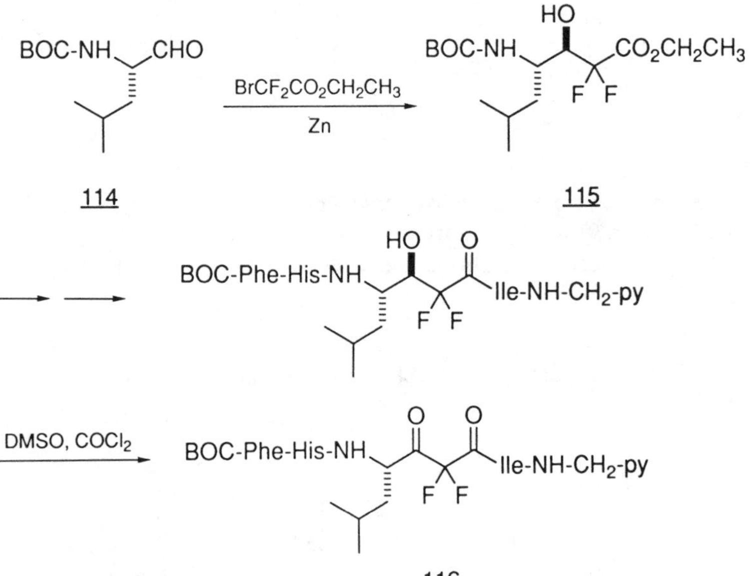

114   **115**

116

2.2.19   Ethyl Trifluoroacetate

Ethyl trifluoroacetate can be used effectively to synthesize 4,4,4-trifluoro-threonines (**117, 118**)[70]

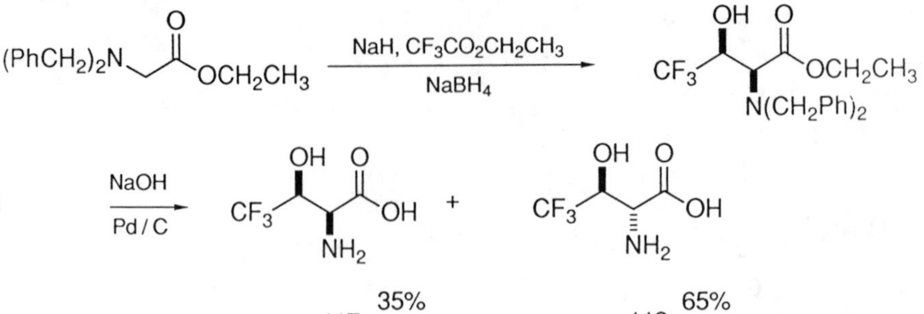

2.2.20   Diethyl Fluoromalonate

Synthesis of 4-Amino-2-fluorobutanoic Acid (120)

Reaction of *N*-aroyl ethylene imine (**119**) with diethyl fluoromalonate followed by hydrolysis affords 4-amino-2-fluorobutanoic acid (**120**).[71]

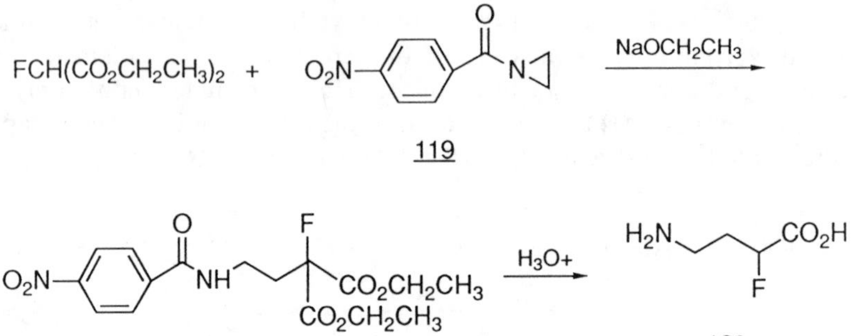

Synthesis of 4-Amino-2-fluorohexanoic Acid

Michael addition of diethyl fluoromalonate to 2-nitro-1-butene (**121**), followed by hydrolysis, decarboxylation, and reduction, affords 4-amino-2-fluoro-hexanoic acid (**122**).[71]

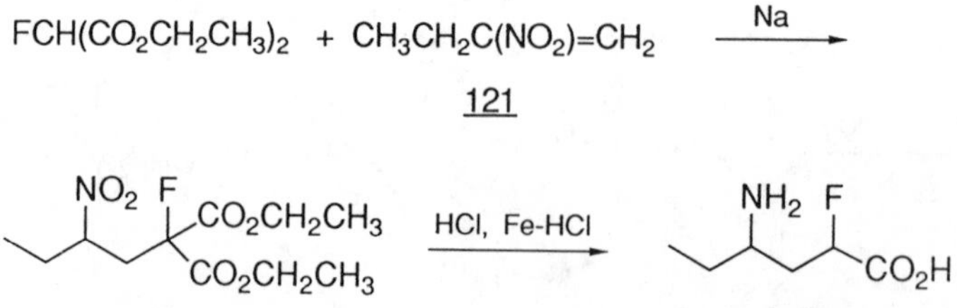

2.2.21   1,1,1-Trifluoro-, 1,1′-Difluoro-, and Hexafluoroacetone

Trifluoromethylated amino acids can be conveniently synthesized using trifluoro-, difluoro-, and hexafluoroacetones.

Synthesis of 3,3,3-Trifluoroalanine (123) Using Hexafluoroacetone[72]

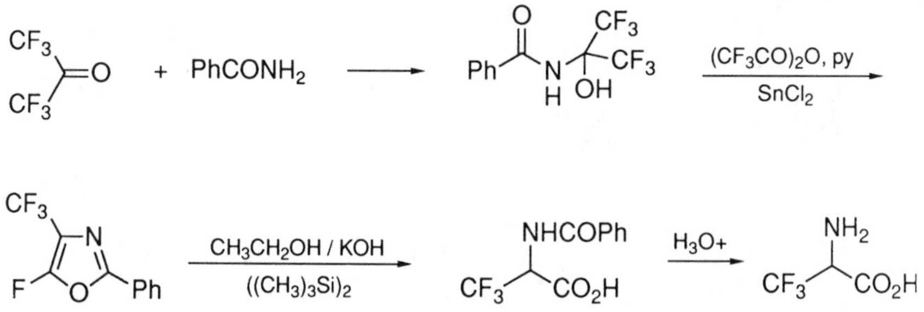

Synthesis of Hexafluorovaline (124) Using Hexafluoroacetone[73]

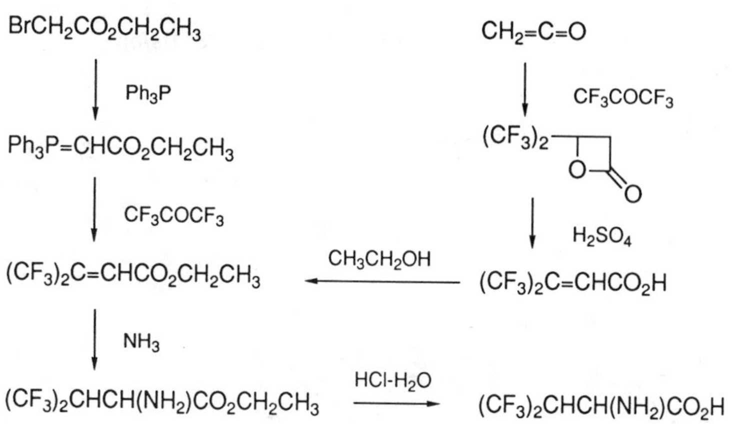

Synthesis of 5,5,5-Trifluoroleucine (125) Using 1,1,1-Trifluoroacetone[74]

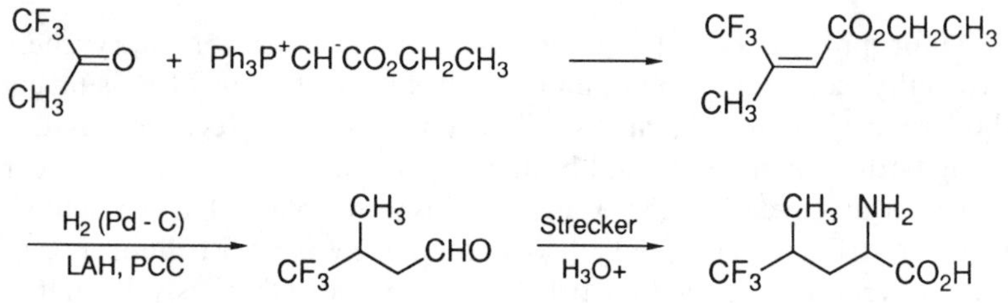

Synthesis of 5-Methyl-6,6,6-trifluoroleucine (126) Using 1,1,1-Trifluoroacetone[7a]

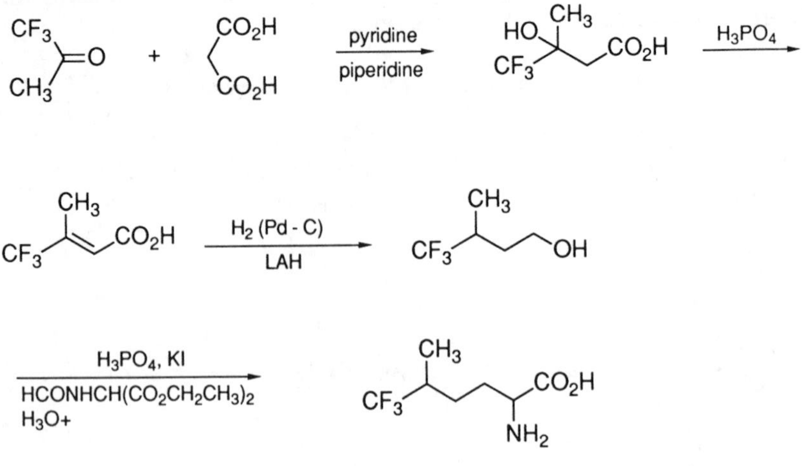

126

Synthesis of α-Amino-β,β-difluoroisobutyric Acid (128) Using 1,1′-Difluoroacetone (127)[75]

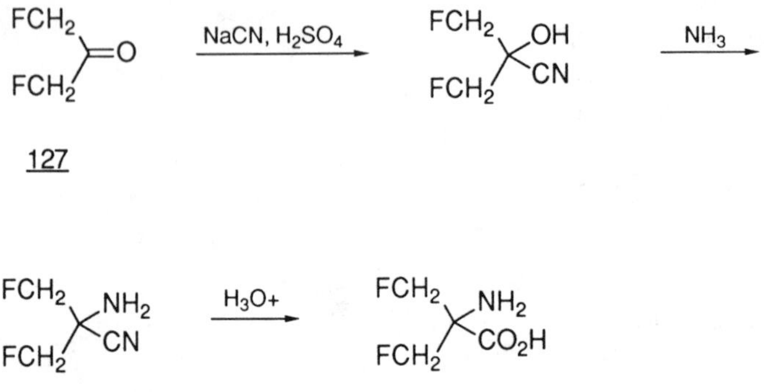

127

128

2.2.22 3,3,3-Trifluoropropene

3,3,3-Trifluoropropene (TFP) is a versatile building block for the synthesis of trifluoromethyl-containing compounds. Hydroformylation of TFP using cobalt and rhodium carbonyls as catalysts affords unusually high regioselectivities for producing both straight chain and branched chain aldehydes. On amidocarbonylation of the aldehydes, N-acyl-α-amino acids were obtained in excellent yields and selectivity (Fig. 2.11).[76-78] Cobalt carbonyl gives 5,5,5-trifluoronorvaline (**129**), whereas rhodium gives 4,4,4-trifluorovaline (**130**). Obviously, the nature of the central metal of the catalyst plays a key role in determining the regioselectivity

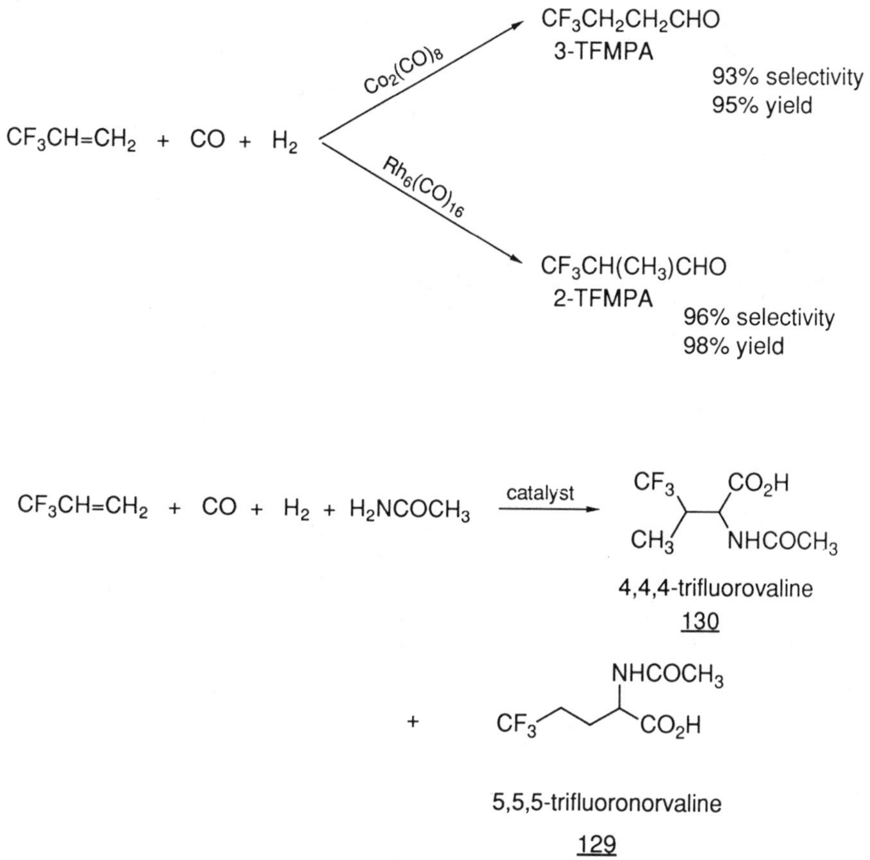

$CF_3CH=CH_2$ + CO + H_2

$Co_2(CO)_8$

$CF_3CH_2CH_2CHO$
3-TFMPA

93% selectivity
95% yield

$Rh_6(CO)_{16}$

$CF_3CH(CH_3)CHO$
2-TFMPA

96% selectivity
98% yield

$CF_3CH=CH_2$ + CO + H_2 + H_2NCOCH_3 →[catalyst]

CF_3 CO_2H

CH_3 $NHCOCH_3$

4,4,4-trifluorovaline
130

+ CF_3 —— NHCOCH_3 / CO_2H

5,5,5-trifluoronorvaline
129

FIGURE 2.11

of the reaction. The suggested mechanism indicates the possibility of two TFP–metal intermediates, **131** and **132**. Since the trifluoromethyl group is strongly

ML_n — CF_3

131

ML_n —— CF_3

132

electron withdrawing, a negative charge on the α-carbon stabilizes the intermediate **131** and a positive charge on the α-carbon stabilizes the intermediate **132**. The cobalt species may cause a fairly large positive charge on the α-carbon, whereas the rhodium species may generate relatively large negative charge on the α-carbon. Thus with cobalt catalyst the unbranched intermediate (**132**) leading to the n-aldehyde is preferred, whereas with rhodium the branched intermediate (**131**) leading to the isoaldehyde predominates.

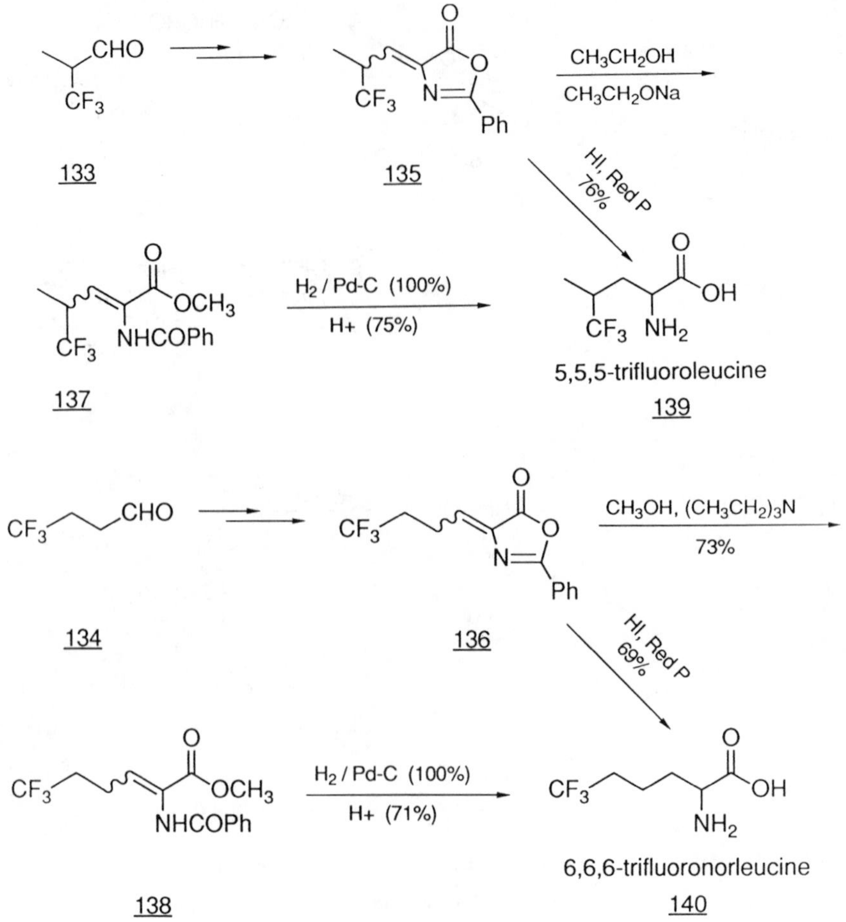

FIGURE 2.12

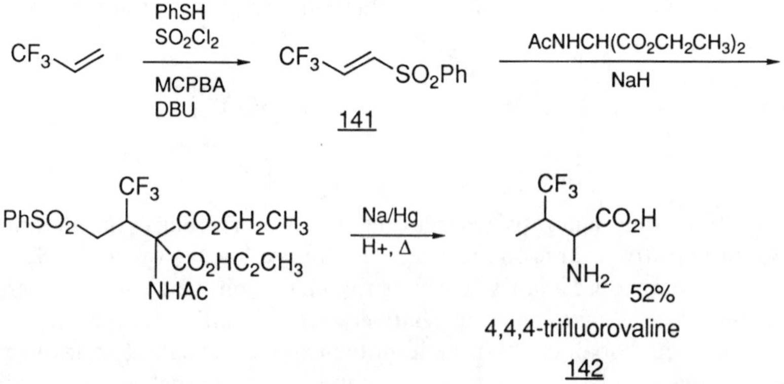

FIGURE 2.13

Trifluoroleucine (**139**) and trifluoronorleucine (**140**) can also be synthesized using 3,3,3-trifluoropropene.[77] The 2- and 3-trifluoromethyl propionaldehydes (**133, 134**) were converted to the azlactones, which with hydrogen iodide and red phosphorus give the amino acids. The azlactones (**135, 136**) can also be converted to dehydroamino acids (**137, 138**), which can be further hydrogenated and on hydrolysis afford the trifluoromethylated amino acids. This is seen in Fig. 2.12.

Another versatile building unit derived from 3,3,3-trifluoropropene is 1-phenylsulfonyl-3,3,3-trifluoropropene (**141**).[79] Due to its high reactivity as a Michael acceptor it is very useful for the construction of trifluoromethylated compounds. Synthesis of 4,4,4-trifluorovaline (**142**) using 1-phenylsulfonyl-3,3,3-trifluoropropene is represented in Fig. 2.13.

2.2.23 Hexafluoropropene

Hexafluoropropene (**143**) is a commercially available perfluoroalkene of low toxicity which can be used efficiently for the synthesis of fluoromalonates (**144**).[80] Conjugate addition of fluoromalonate to ethyl 2-acetamidoacrylate, followed by decarboxylation, affords 4-fluoroglutamic acid (**145**).[81]

$$CF_2=CFCF_3 \xrightarrow[H+\,/\,H_2O]{CH_3CH_2O-\,/CH_3CH_2OH} CH_3CH_2O_2CCHFCF_3$$

<u>143</u>

$$\xrightarrow[-HF]{CH_3CH_2O-} \left[CH_3CH_2O_2CCF=CF_2 \right] \xrightarrow[H+\,/\,H_2O]{CH_3CH_2O-\,/CH_3CH_2OH}$$

<u>144</u>   <u>145</u>

2.2.24 Ethyl α-Fluoroacrylate

Synthesis of 4-Fluoroglutamic Acid (146)[41,68]

Michael addition of α-fluoroacrylate ester and diethyl acetamidomalonate gives the adduct which on hydrolysis and decarboxylation affords 4-fluoroglutamic acid (**146**).

<u>146</u>

2.2.25   3-Fluoro-2-butanone

Synthesis of γ-Fluoroisoleucine[82]

Condensation of 3-fluoro-2-butanone with diethylphosphonoacetate affords **147**, which on hydrogenation followed by bromination gives the ester **148**. The bromoester is converted to the azido ester, which on catalytic hydrogenation and hydrolysis yields γ-fluoroisoleucine (**149**).

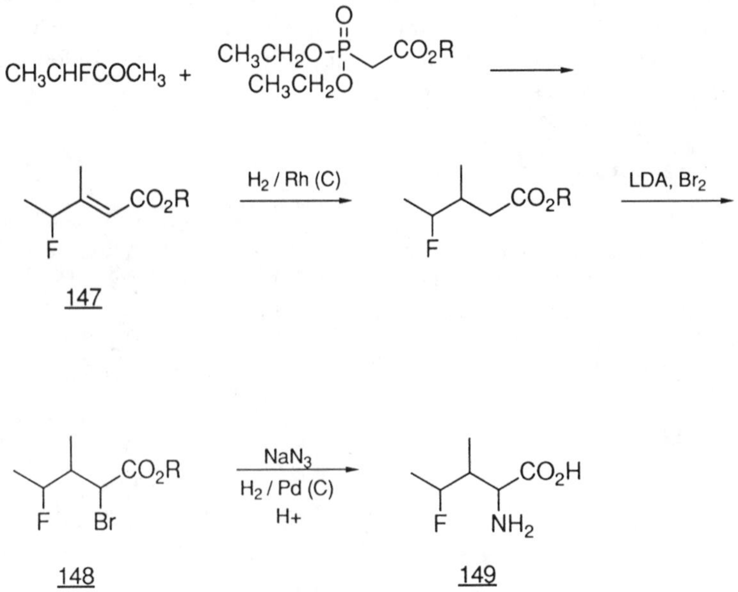

2.2.26   Ethyl α-Fluorocrotonate

In each example in Fig. 2.14, reduction of the double bond was accompanied by simultaneous defluorination. The expected saturated fluorinated compounds were not obtained even in small amounts.

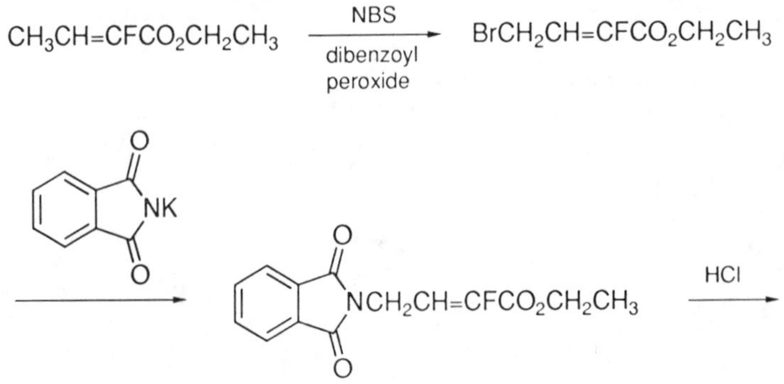

HCl.NH$_2$CH$_2$CH=CFCO$_2$H

2-fluoro-4-amino-crotonic acid

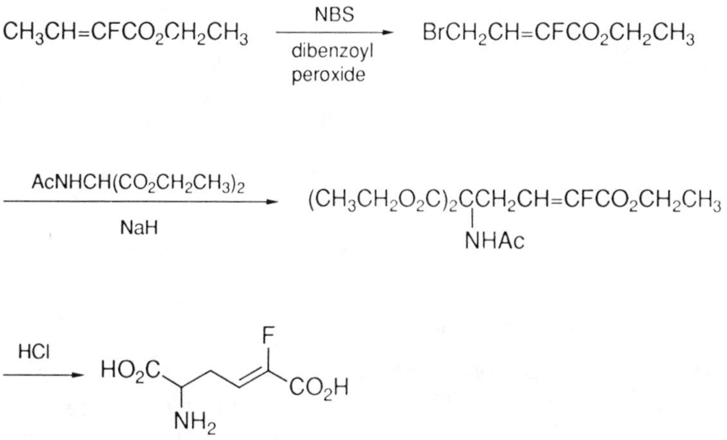

CH$_3$CH=CFCO$_2$CH$_2$CH$_3$ $\xrightarrow[\substack{\text{dibenzoyl} \\ \text{peroxide}}]{\text{NBS}}$ BrCH$_2$CH=CFCO$_2$CH$_2$CH$_3$

$\xrightarrow[\text{NaH}]{\text{AcNHCH(CO}_2\text{CH}_2\text{CH}_3)_2}$ (CH$_3$CH$_2$O$_2$C)$_2$CCH$_2$CH=CFCO$_2$CH$_2$CH$_3$
$\qquad\qquad\qquad\qquad\qquad\qquad\qquad\qquad$ |
$\qquad\qquad\qquad\qquad\qquad\qquad\qquad\qquad$ NHAc

FIGURE 2.14

2.2.27   4,4,4-Trifluorobutanoic Acid

Synthesis of 5,5,5-Trifluoronorvaline (150)[7a]

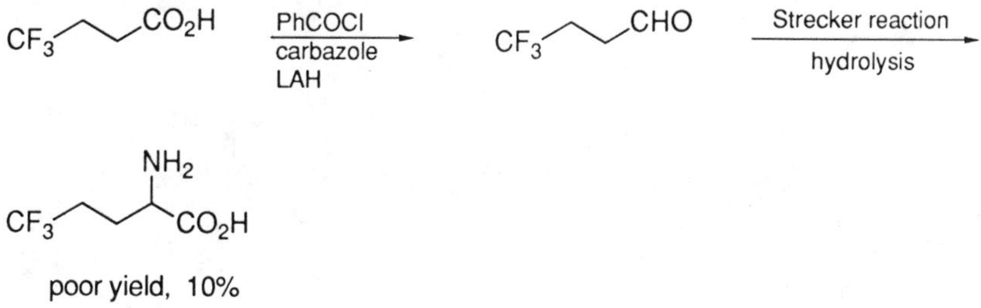

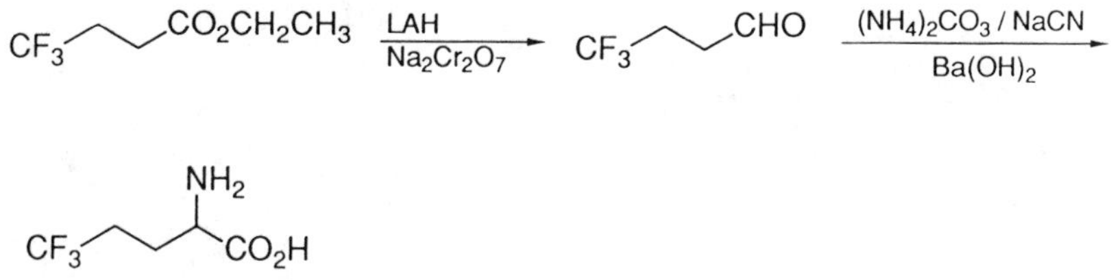

poor yield, 10%

150

Synthesis of 6,6,6-Trifluoronorleucine (151)[7a]

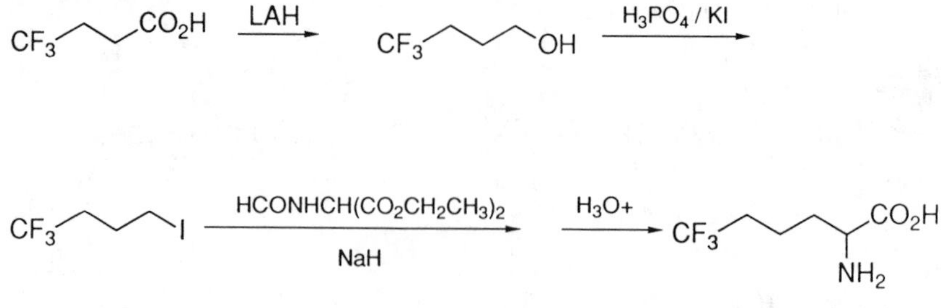

2.2.28 Trifluoroacetic Anhydride

Synthesis of 2-Trifluoromethyl-L-histidine (153)

The synthesis of 2-trifluoromethyl-L-histidine (**153**) using trifluoroacetic anhydride is represented in Fig. 2.15.[83] The ready availability of the starting amino acid (**152**), the short synthetic sequence, the ease of entry into the L-amino and series, and the absence of racemization at any step in the sequence are noteworthy advantages of this approach.

The probable mechanism for the trifluoroacetic anhydride reaction is seen in Fig. 2.16.

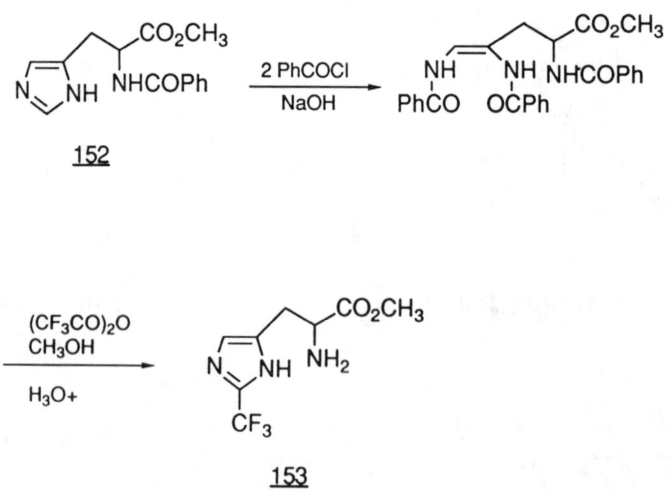

FIGURE 2.15

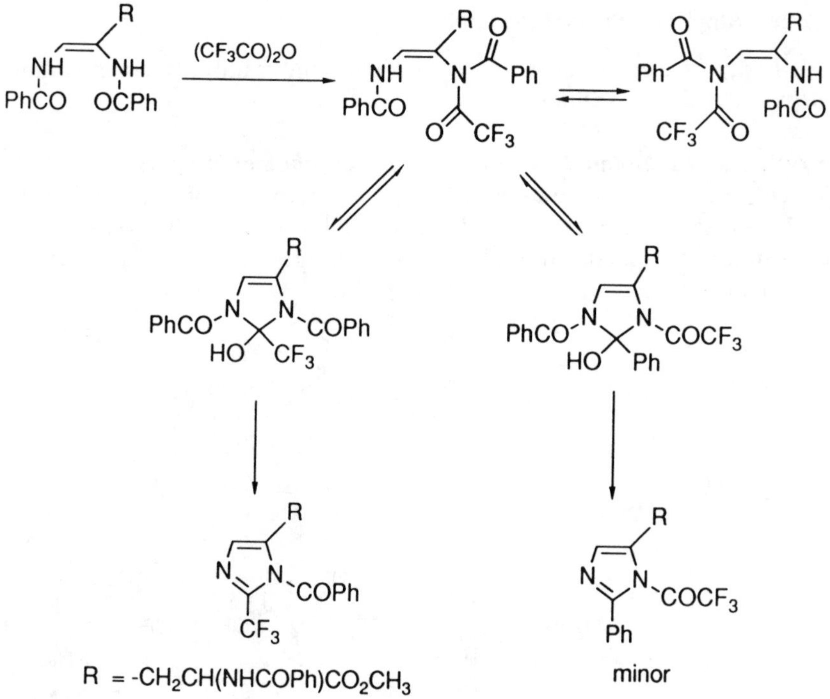
$(CF_3CO)_2O$
R = $-CH_2CH(NHCOPh)CO_2CH_3$
minor

FIGURE 2.16

Synthesis of 3,3,3-Trifluoroalanine (154)[84–86]

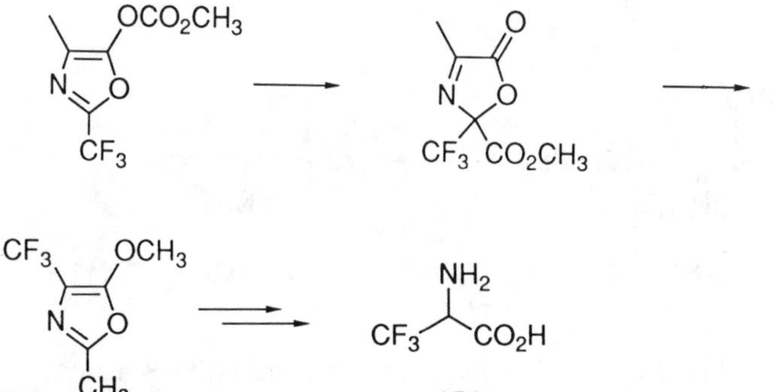

154

2.2.29 Ethyl Trifluoroacetoacetate

Ethyl trifluoroacetoacetate is an easy to handle and readily available synthon for the construction of trifluoromethyl amino acids.

Synthesis of-2-Amino-3-hydroxy 4,4,4-trifluorobutanoic Acids

Threo- and *allo-*2-amino-3-hydroxy-4,4,4-trifluorobutanoic acids (threonines) (**157** and **158**), were synthesized from ethyl trifluoroacetoacetate (**155**) via reduction and saponification of the 4,4,4-trifluoro-3-hydroxy-2-methoxyimino butanoate.[70]

155 156 157 158

trifluorothreonine trifluoro-*allo*-threonine

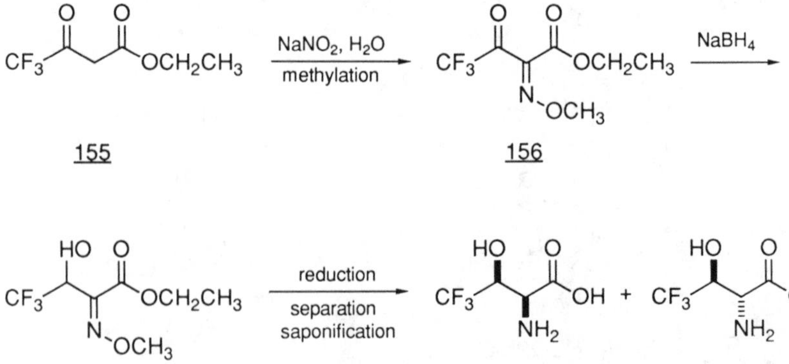

A similar synthesis of 2-amino-3-hydroxy-4,4,4-trifluorobutyric acid (**160**) has been reported by reduction of the phenylhydrazone (**159**).[7b]

159 160

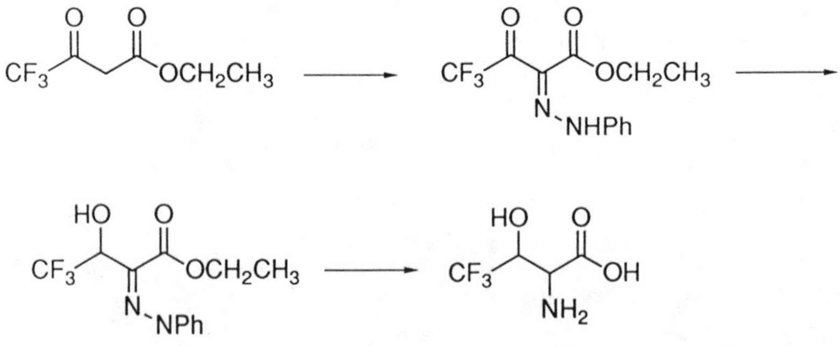

As seen in Fig. 2.17, ethyl trifluoroacetoacetate (**155**) is converted to ethyl 2-chloro-3-keto-4,4,4-trifluorobutyrate (**161**).[7b,7c] Reduction to the chlorohydrin

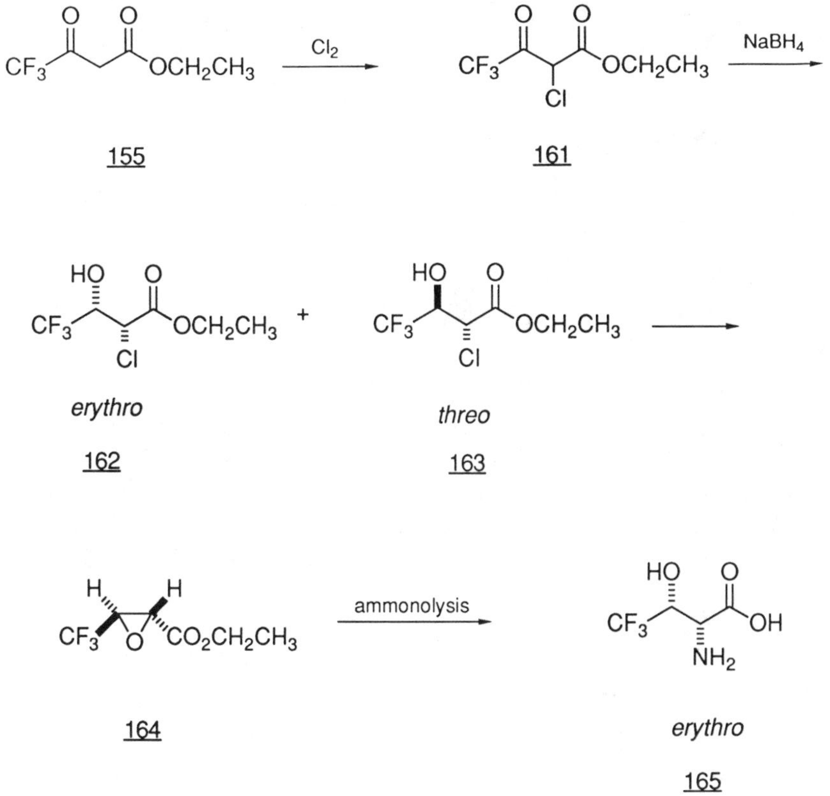

FIGURE 2.17

esters with $NABH_4$ gave a mixture of erythro and threo isomers, **162** and **163**, in the ratio 7:4. The ring closure to the glycidic ester (**164**) with base resulted in the more stable trans compound, which on ammonolysis gave the amino acid (**165**) by a stereospecific trans opening of the epoxide at the α-carbon atom.

It was observed that ammonia and amines attack the β-carbon of glycidic esters to give α-hydroxy-β-amino amides. However, in this case the ammonia attacks the α-carbon exclusively. The strong electronegativity of the trifluoromethyl group may be responsible for the preferential attack at the α-carbon of the fluorinated glycidic ester.

2.2.30   3-Trifluoromethyl-γ-butyrolactone

Synthesis of 5,5,5-trifluoroleucine (168)[87]
Fluorinated chiral building block in enantiomerically pure form is crucial for the synthesis of fluorinated biologically active compounds. Efficient resolution of 3-trifluoromethyl-γ-butyrolactone (**166**) via the amide (**167**) and its conversion to 5,5,5-trifluoroleucine (**168**) is indicated in Fig. 2.18.[87]

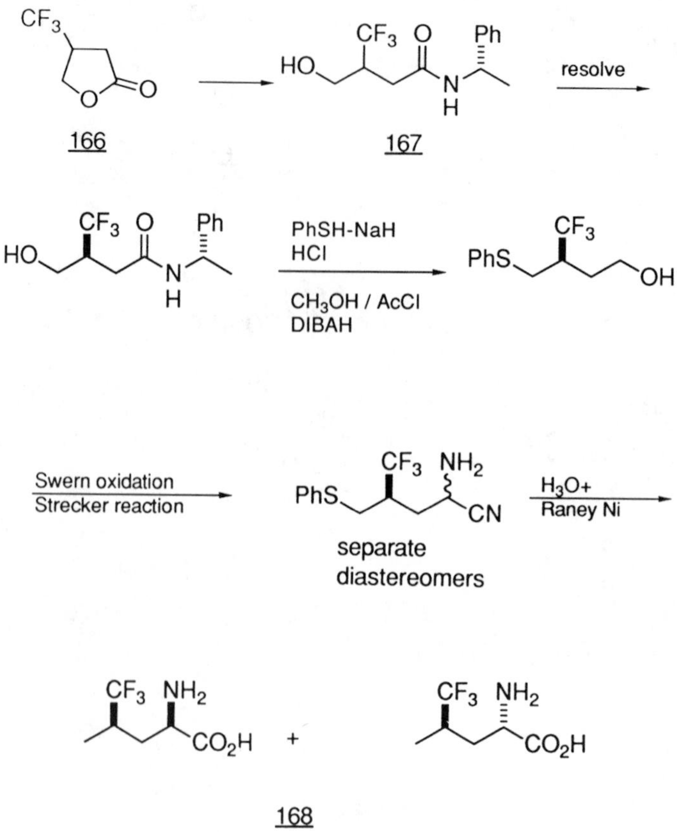

FIGURE 2.18

2.2.31 Pentafluorostyrene

Pentafluorostyrene is an important building block for the synthesis of fluorinated, biologically active amino acids. Hydroformylation of pentafluorostyrene with rhodium catalysts gave the isoaldehyde 2-pentafluorophenylpropionaldehyde with excellent regioselectivity (97–98%) and in a quantitative yield. $CO_2(CO)_8$ gave the *n*-aldehyde 3-pentafluorophenylpropionaldehyde as the major product. The regioselectivity in this case was not as high as in the case of TFP. The 2-pentafluorophenylpropionaldehyde is a convenient synthon for the synthesis of 3-substituted 4,5,6,7-tetrafluoroindoles such as tetrafluorotryptophan (**170**), as seen in Fig. 2.19.[76,77,88]

2.2.32 Fluoroindole

Condensation of 4-, 5-, or 6-fluoroindole (**171**) and diethylpiperidyl methyl formamidomalonate (**172**), followed by hydrolysis and decarboxylation, yields 4-, 5-, or 6-fluorotryptophan (**173**).[89]

FIGURE 2.19

2.3 BIOLOGICAL ACTIVITY

A major justification for developing highly selective enzyme inhibitors is that they have great practical application in clarifying the physiological roles of specific enzymes. Irreversible inactivation of enzymes by fluorinated amino acids may be successful due to the ability of the inactivator to bind at the enzyme-active site and to undergo enzyme-catalyzed conversion by proton abstraction, isomerization, elimination, or oxidation to a reactive species capable of reaction irreversibly with the active site. Since the enzyme becomes inactivated by its own mechanism of action, such inhibitors are referred to as "suicide inhibitors."[3d]

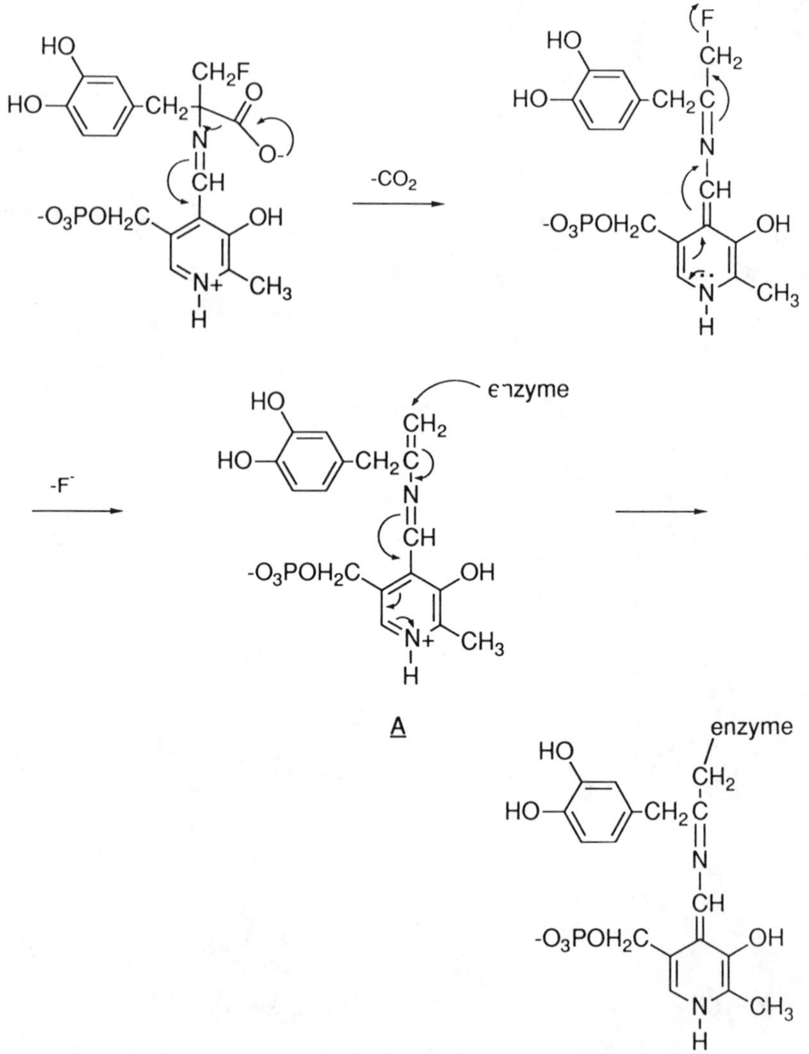

FIGURE 2.20

As mentioned in the introduction, a number of α-monofluoromethyl and difluoromethyl amino acids are potent irreversible inhibitors of the corresponding amino acid decarboxylases.[3d] A proposed general mechanism of inhibition of amino acid decarboxylases by fluoromethyl amino acids is represented in Fig. 2.20. The enzymatic inactivation is thought to be initiated by loss of carbon dioxide and elimination of fluoride ion from the Schiff base intermediate formed between the fluoromethyl amino acid and pyridoxal phosphate. A reactive Michael acceptor (A) that is generated can be alkylated by a nucleophilic functional group in or near the active site to irreversibly bind the enzyme.

Fluoromethyl groups enhance the potency of the α-amino acids. α-Difluoromethyldopa has a selective peripheral activity, whereas α-monofluoromethyldopa inhibits both central and peripheral activities.[3d] α-Monofluoromethyl histidine inhibits gastric acid secretion.[3d] α-Difluoromethyl ornithine[90] has been found to have antigestational, antitrypanosomal, anticoccidial, and antitumor activities.[3d] α-Difluoromethyl ornithine, a specific inhibitor of ornithine decarboxylase, the enzyme that provides the polymines needed for growth and development, is found to be a fungicide with plant protective efficacy.[91] It inhibits cell proliferation.[92] The cytostatic effect of α-difluoromethyl ornithine is due to a decline in intracellular levels of putrescine and spermidine. α-Difluoromethyl dehydroornithine[64] is a more potent suicide inhibitor of ornithine decarboxylase than the corresponding saturated analog. Both trifluoromethyl and monofluoromethyl glutamic acids have been reported to inhibit glutamic acid decarboxylase, the enzyme that catalyzes the formation of the inhibitory hemotransmethyl GABA.[3c,d]

Trifluoroalanine[3a,c] acts as a suicide inhibitor for a number of pyridoxal enzymes such as γ-cystathionase,[93,94] alanine racemase,[95] tryptophanase,[3a] tryptophane synthetase,[3a] β-cystathionase,[3a] pyruvate-glutamate transaminase.[3a] All these enzymes can perform elimination reactions, in addition to their catalytic process. A mechanism similar to that in Fig. 2.20 is proposed (see Fig. 2.21). β-Fluoroalanine,[3c,d,e,96] 3-fluoro-D-alanine-2D,[27,37,69] and β,β-difluoroalanine[3a,95] also inhibit alanine racemase, an enzyme necessary for bacterial cell wall synthesis. Since both D- and L-β-fluoroalanines are active inhibitors of the enzyme, it follows that there might be a common intermediate. Also, it is observed that inactivation of the enzyme by β,β-difluoroalanine is reversible, whereas that by β,β,β-trifluoroalanine is irreversible. In the case of trifluoroalanine,[3a] the initial Michael addition product might hydrolyze to afford a stable acyl derivative (see Fig. 2.22).

GABA transaminase, an enzyme that controls the catabolism of GABA, requires pyridoxal phosphate as the coenzyme. Fluoroamino acids such as trans-4-amino-2-fluorocrotonic acid,[3d] 3-amino-4-fluorobutanoic acid,[63] and 4-amino-5-fluoropentanoic acid[53,97] are found to block GABA transaminase irreversibly. A covalent attachment of the inactivator to the enzyme-active site is suggested. A mechanism-based inactivation is proposed,[97] as seen in Fig. 2.23 Both erythro and threo isomers of 3-fluoro-L-glutamic acids[12] are potent suicide inhibitors of pyridoxal phosphate-dependent enzymes such as transaminases,

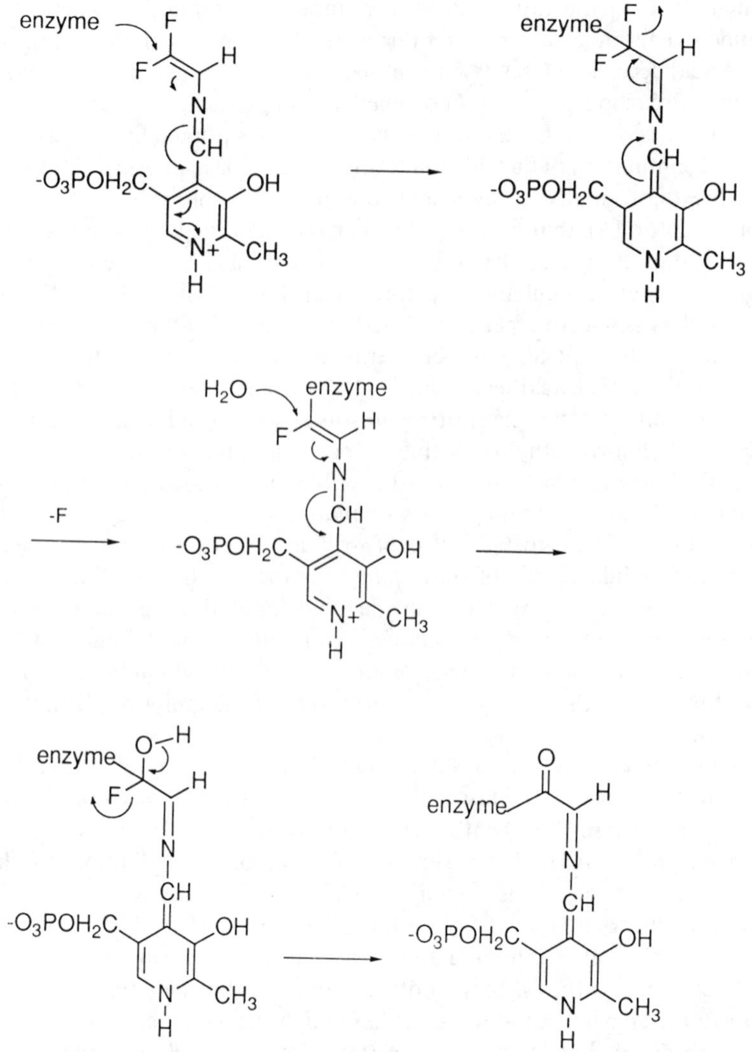

FIGURE 2.21

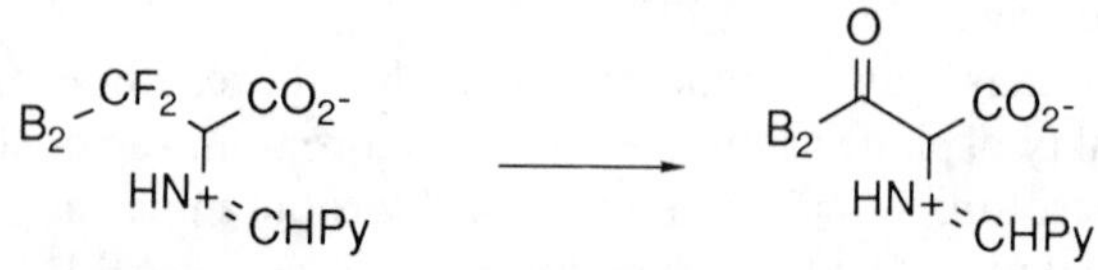

B$_2$ is a nucleophlic residue
on the enzyme

FIGURE 2.22

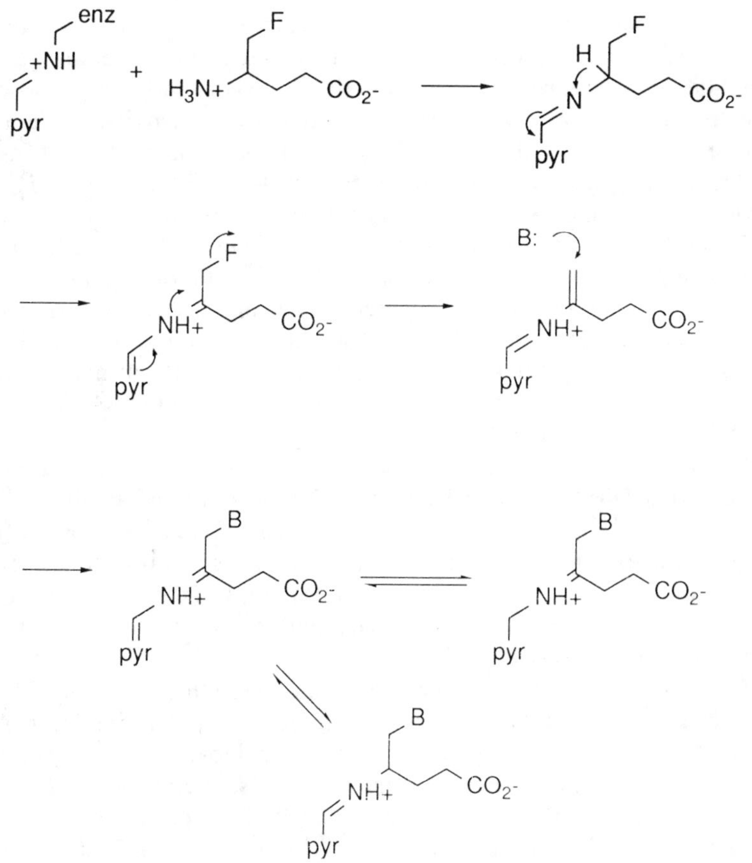

FIGURE 2.23

glutamate racemase, and glutamate decarboxylase, which use glutamic acid as substrate. D,L-Threo-4-fluoroglutamate is an effective concentration-dependent inhibitor of polyglutamylation of both tetrahydrofolate and methotrexate, while the erythro isomer is weakly inhibitory.[98]

γ-Fluoromethotrexate (**174**), a methotrexate analog with a fluorine at the γ-carbon of the glutamate moiety, is reported to be a poor substrate for folypoly (γ-glutamate) synthetase and a poor inhibitor of dihydrofolate reductase than methotrexate.[99]

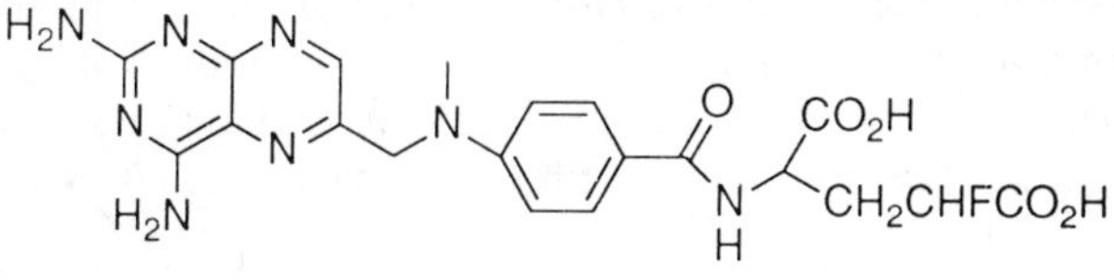

174

Several fluoroamino acids exhibiting antimetabolite or inhibitory activities have been reported. 3-Fluoro-D-alanine[37] is a potent antibacterial agent. 3-Fluoro-2-deuterio-D-alanine[36,69] and a derivative of the antibiotic cycloserine exhibit an unusually broad spectrum of antimicrobial activity. α-Fluoro-β-alanine, a metabolic product of 5-fluorouracil, has potential value as an antimetabolite of β-alanine and its derivatives.[66] 4-Fluoroglutamic acid[41] and 5,5,5-trifluoronorvaline[7a] show inhibitory activity on the growth of *Escherichia coli*. 5,5,5-Trifluoroleucine can replace leucine in proteins and peptides with little effect on biological activity of mutants of *E. coli*.[14,100] γ-Fluoroglutamic acid, an antimetabolite of glutamic acid, has a potential of being effective against tubercle bacillus.[68] 5-Fluorolysine and 4-fluoroarginine strongly inhibit the growth of *E. coli*, whereas 4-fluoroornithine and 4-fluorocitrulline are relatively inactive under the same conditions. 4-Amino-2-fluorobutanoic acid, fluoroGABA, is found to be an antimetabolite lf GABA.[71]

2-Fluorohistidine has antimetabolic and antiviral properties, although 4-fluorohistidine fails to show such effects.[101] 2-Fluorohistidine inhibits cellular protein synthesis.[101] 4-Fluorohistidine[44] was also found to produce mild soporific and anesthetic effects. It parallels histidine as a substrate for several enzymes and serves as a competitive inhibitor for histidine-ammonia lyase. Histidine-ammonia lyase effects α-β elimination of ammonia from L-4-fluorohistidine to generate *trans*-4-fluorourocanic acid.[102]

4-, 5-, or 6-Fluorotryptophan supports limited growth of a strain of *E. coli* which is incapable of synthesizing tryptophan and is a potent inhibitor of the wild type *E. coli*, which is capable of tryptophan biosynthesis.[103] The fluorotryptophan is incorporated into bacterial proteins, altering their properties. Such substituted proteins may be used for the study of protein structure and function by ^{19}F NMR and other techniques.[104] Induction of inducible enzymes at the time of addition of the fluoroanalog offers an attractive approach for incorporating large amounts of analog into a specific protein. 4-, 5-, and 6-Fluorotryptophans have been used to study the heat stability and turn-over rate of tyrosine aminotransferase.[105] 5-Fluorotryptophan binds to plasma albumin, facilitating ^{19}F NMR studies of this substance.[106] 4,5,6,7-Tetrafluorotryptophan[88] exhibits strong activities in the inhibition of both tryptophanyl hydroxamate and amino acyl *t*-RNA formation. Mono- and trifluorothreonines and allothreonines[70] are possible precursors for the synthesis of Aztreonam analogs which possess excellent activity against gram-negative bacteria.

3-Fluorophenylalanine[28] has antibacterial activity similar to that of chloramphenicol and tetracycline. Peptides containing *m*-fluorophenylalanine inhibit the growth of the yeast *Candida albicans* with minimum inhibitor concentrations.[107] These peptides may be transported into the yeast cells via peptide permeases followed by release of the inhibitory *m*-fluorophenylalanine inside the cell by the intracellular peptidases.

The fluorine-containing amino acid analogs of the well-known antibiotic chloramphenicol have been synthesized and tested for antimicrobial activity.[31]

Unfortunately, neither the threo (natural) nor the erythro (unnatural) compounds (**175**) demonstrated appreciable antibacterial or antifungal activities, indicating that substitution of the secondary hydroxyl by fluorine destroys the antibacterial activity of the parent compound. This is in contrast to the substitution of the 3-hydroxy by fluorine (**176**),[106] which enhances the antibac-

175 *threo*
      erythro

176

terial activity against many chloramphenicol-resistant strains. 2- and 3-Fluorotyrosines are effective growth inhibitors of the microorganisms *E. coli*, *Streptococcus faecalis*, and *Lactobacillum plantarum*.[48] 3-Fluorotyrosines were also found to be toxic to rats and mice.[108]

The fluoroprolines can be incorporated into the proteins of *E. coli*.[52] *Trans*-4-hydroxyproline is formed from *trans*-4-fluoroproline by enzymatic hydroxylation during collagen biosynthesis. 4,4-Difluoro-L-proline, once activated by prolyl-*t*-RNA ligase, on incorporation into procollagen inhibits peptidylproline hydroxylase due to the presence of the electronegative difluoro group at the site of hydroxylation.[14]

Agents that selectively control collagen biosynthesis may be of therapeutic utility. 5,5-Difluorolysine is a potential inhibitor of collagen biosynthesis.[13] The 5,5-geminal fluorines block the critical hydroxylation at the site by peptidyl lysine hydroxylase, resulting in the formation of abnormal precollagen. The geminal difluoro group β to the ε-amino group drastically decreases the pK_a of the amino group and alters the ability of 5,5-difluorolysine to act as a substrate for ATP–PPi exchange reactions in the presence of lysyl-*t*-RNA ligase.[13]

β-Fluoroaspartates show selective cytotoxicity against various mammalian cells.[11] The introduction of fluorine at the β position in aspartate causes facile elimination on β-carbon (dehydrofluorination) and no transamination is observed with aspartate transaminase.[110] The effect of fluorine on transamination has been studied using ^{19}F NMR.[109]

β,β-Difluoroaspartic acid and β,β-difluoroasparagine are potential carcinostatics.[42] β,β-Difluoroaspartic acid is a potential inhibitor of various aspartate-utilizing enzymes. Racemic β,β-difluoroaspartic acid competitively inhibits glutamate-oxaloacetate transaminase (aspartate aminotransferase), which is responsible for conversion of aspartate into oxaloacetate.[42] β,β-Difluoroasparagine may be tested for activity in asparagine-dependent leukemias.

β-Fluoroaspartic acid is found to inhibit adenylosuccinate synthetase and lyase.[43] D,L-Threo-β-fluoroasparagine shows in vitro antileukemic activity to murine and human cells.[111] The threo isomer, but not the erythro, was found to be an effective inhibitor of glycosylation and to cause partial inhibition of protein synthesis. Fluoroasparagine is a useful complement to tunicamycin as it inhibits glycosylation by a different mechanism.

Aspartyl protease renin cleaves the protein substrate angiotensinogen into the decapeptide angiotensin I, which in turn is cleaved into the pressor octapeptide angiotensinogen II.[112] Highly potent competitive inhibitors of renin have been reported to contain statine (**177**) incorporated into renin substrate analogs. A

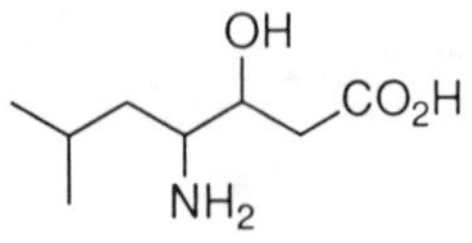

177

difluorostatine-containing peptide is also a potent renin inhibitor, more potent than the nonfluorinated statine-containing peptide. The greater electrophilicity of the carbonyl in difluorostatone over statine facilitates the addition of water to form the tetrahedral species similar to that formed during the enzyme-catalyzed hydrolysis of a peptidic bond. Although the ketone is a less effective inhibitor of pepsin than pepstatin, it does exhibit high specificity for renin, which is desirable for a potentially useful therapeutic agent for control of renin-associated hypertension. Substitution of hexafluorovaline into angiotensin II results in analogs that resemble the parent peptide in biological activity.[73] These enable the investigation of possible conformational differences between angiotensin II agonists and antagonists by ^{19}F NMR.

Acylaminoboronic acids are potential transition state analog inhibitors of the serine proteases chymotrypsin and elastase.[23] The phenylalanine and phenylgly-

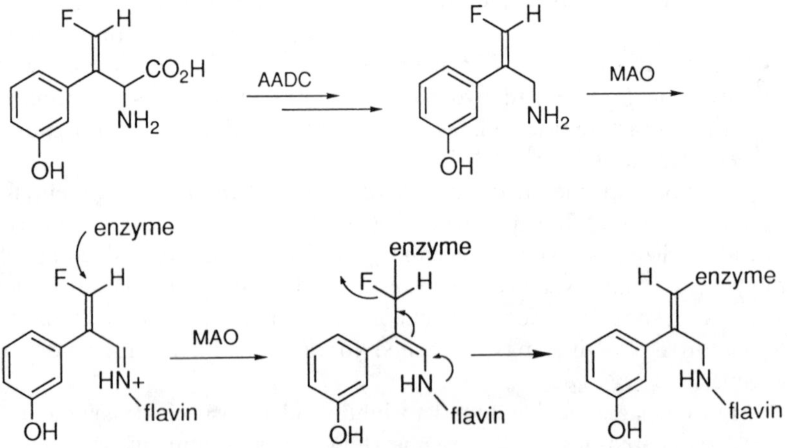

FIGURE 2.24

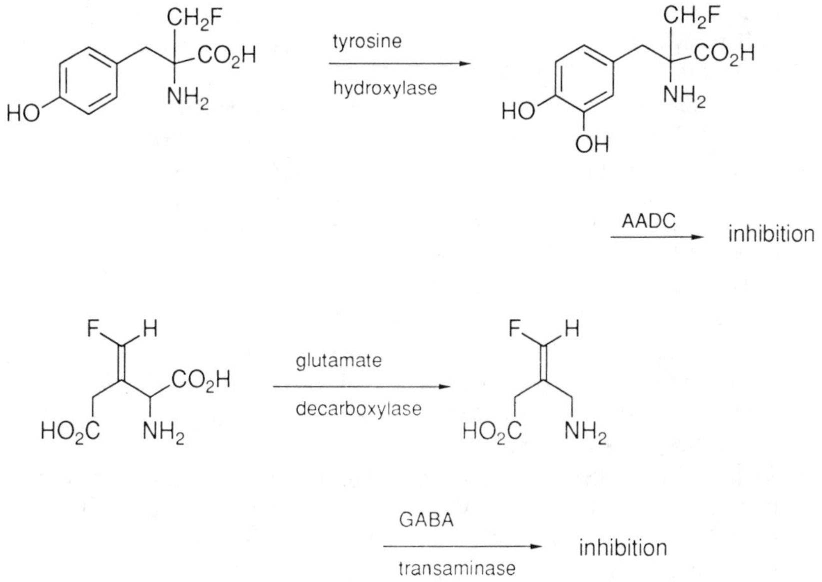

FIGURE 2.25

cine analogs are good competitive inhibitors of α-chymotrypsin, and the alanine, valine, and isoleucine analogs are good inhibitors of elastase.

The dual enzyme-activated approach to the design of enzyme inhibitors has led to the synthesis of inhibitors possessing target enzyme specificity as well as site selectivity. An example of such advantageous dual specificity is the inhibition of monoamine oxidase.[3e,50,56]

(E)-β-Fluoromethylene-m-tyrosine[3e,56] requires activation by the metabolic enzyme aromatic amino acid decarboxylase (AADC) to form the fluoroalkyl amine (Fig. 2.24), which is a potent enzyme-activated inhibitor of the catabolic enzyme monoamine oxidase (MAO). The predominant neuronal location of AADC results in selective MAO inhibition in nerve endings. Other examples of dual enzyme-activated inhibitors are seen in Fig. 2.25.[50]

REFERENCES

1. (a) P. Goldman, *Science* **164**, 1123, 1969; (b) F. A. Smith, *Chem. Tech.* **1973**, 422; (c) R. Filler, *Chem. Tech.* **1974**, 752; (d) *Ciba Foundation Symposium, Carbon–Fluorine Compounds, Chemistry, Biochemistry and Biological Activities*, Elsevier, New York, 1972; (e) R. Filler, in *Organofluorine Chemicals and Their Industrial Applications* (ed. R. E. Banks), Halsted, New York, 1979; (f) R. Filler and Y. Kobayashi, eds., *Biomedical Aspects of Fluorine Chemistry*, Kodansha, Tokyo, 1982;

(g) R. Filler, ed., *Biochemistry Involving Carbon–Fluorine Bonds*, American Chemical Society, Washington, D.C., 1976; (h) M. R. C. Gerstenberger and A. Haas, *Angew. Chem. Int. Ed.* **20**, 647, 1981.

2. J. Kollonitsch, L. M. Perkins, A. A. Patchett, G. A. Doldouras, S. Marburg, D. E. Duggan, A. L. Maycock, and S. D. Aster, *Nature* **274**, 906, 1978.

3. (a) R. H. Abeles and A. L. Maycock, *Acc. Chem. Res.* **9**, 313–319, 1976; (b) R. R. Rando, *Enzyme Inhibitors* **8**, 281–288, 1975; (c) C. Walsh, *Tetrahedron* **38**, 871–909, 1982; (d) B. W. Metcalf, *Ann. Rep. Med. Chem.* **16**, 289–297, 1981; (e) P. Bey, *Ann. Chim. Fr.* **9**, 695–702, 1984.

4. D. H. G. Versteeg, M. Palkouts, J. Van der Gugten, H. J. L. M. Wijnen, G. W. M. Smeets, and W. DeJong, *Prog. Brain Res.* **47**, 111, 1977.

5. W. W. Douglass, in *The Pharmacological Basis of Therapeutics*, 5th ed. (ed. L. S. Goodman and A. Gilman), Macmillan, New York, 1975, pp. 590–629.

6. D. H. Russell, ed., *Polyamines in Normal and Neoplastic Growth* (ed. D. H. Russell), Raven, New York, 1973, pp. 1–13.

7. (a) H. M. Walborsky, M. Baum, and D. F. Loncrini, *J. Am. Chem. Soc.* **77**, 3637–3640, 1955; (b) H. M. Walborsky and M. E. Baum, *J. Am. Chem. Soc.* **80**, 187–192, 1958; (c) H. M. Hill, E. B. Towne, and J. B. Dickey, *J. Am. Chem. Soc.* **72**, 3289, 1950.

8. S. Loy and M. Hudlicky, *J. Fluorine Chem.* **7**, 421–426, 1976.

9. J. Kollonitsch, S. Marburg, and L. M. Perkins, *J. Org. Chem.* **40**, 3808–3809, 1975.

10. J. Kollonitsch, S. Marburg, and L. M. Perkins, *J. Org. Chem.* **44**, 771–777, 1979.

11. A. M. Stern, B. M. Foxman, A. H. Tashjian, Jr., and R. H. Abeles, *J. Med. Chem.* **25**, 544–550, 1982.

12. (a) A. Vidal-Cros, M. Gaudry, and A. Marquet, *J. Org. Chem.* **50**, 3163–3167, 1985; (b) R. M. Babb and F. W. Bollinger, *J. Org. Chem.* **35**, 1438–1440, 1970.

13. F. N. Shirota, H. T. Nagasawa, and J. A. Elberling, *J. Med. Chem.* **20**, 1623–1627, 1977.

14. F. N. Shirota, H. T. Nagasawa, and J. A. Elberling, *J. Med. Chem.* **20**, 1176–1181, 1977.

15. W. J. Middleton, *J. Org. Chem.* **40**, 574–578, 1975.

16. T. Tsushima, T. Sato, and T. Tsuji, *Tetrahedron Lett.* **21**, 3591–3592, 1980.

17. T. Tsushima, J. Nishikawa, T. Sato, H. Tanida, K. Tori, T. Tsuji, S. Misaki, and M. Suefuji, *Tetrahedron Lett.* **21**, 3593–3594, 1980.

18. U. Groth and U. Schollkopf, *Synthesis* **1983**, 673–675.

19. G. A. Boswell, Jr., W. C. Ripka, R. M. Scribner, and C. W. Tullock, *Org. React.* **21**, 1–124, 1974.

20. E. D. Bergmann and L. Chun-Hsu, *Synthesis* **1973**, 44–46.

21. H. Gershon, M. W. McNeil, and E. D. Bergmann, *J. Med. Chem.* **16**, 1407–1409, 1973.

22. K. Matsumoto, Y. Ozaki, T. Iwasaki, H. Horikawa, and M. Miyoshi, *Experientia* **35**, 850–851, 1978.

23. D. H. Kinder and J. A. Katzenellenbogen, *J. Med. Chem.* **28**, 1917–1925, 1985.

24. G. Olah, J. T. Welch, Y. D. Vankar, M. Nojima, I. Kerekes, and J. A. Olah, *J. Org. Chem.* **44**, 3872–3881, 1979.

25. A. I. Ayi, M. Remli, and R. Guedj, *J. Fluorine Chem.* **18**, 93–96, 1981.

26. R. Guedj, A. I. Ayi, and M. Remli, *Ann. Chim. Fr.* **9**, 691–694, 1984.

27. T. Tsushima, K. Kawada, J. Nishikawa, T. Sato, K. Tori, T. Tsuji, and S. Misaki, *J. Org. Chem.* **49**, 1163–1169, 1984.

28. T. N. Wade, F. Gaymard, and R. Guedj, *Tetrahedron Lett.* **1979**, 2681–2682.

29. T. N. Wade, *J. Org. Chem.* **45**, 5328–5333, 1980.

30. T. N. Wade and R. Kheribet, *J. Chem. Res.* (S) **1980**, 210–211.

31. T. Tsushima, K. Kawada, T. Tsuji, and K. Tawara, *J. Med. Chem.* **28**, 253–256, 1985.

32. T. N. Wade and R. Guedj, *Tetrahedron Lett.* **1979**, 3953–3954.

33. T. N. Wade and R. Kheribet, *J. Org. Chem.* **45**, 5333–5335, 1980.

34. A. I. Ayi, M. Remli, and R. Guedj, *Tetrahedron Lett.* **1981**, 1505–1508.

35. A. Ourari, R. Condom, and R. Guedj, *Can. J. Chem.* **60**, 2707–2710, 1982.

36. J. Kollonitsch, L. Barash, and G. A. Doldouras, *J. Am. Chem. Soc.* **92**, 7494–7495, 1970.

37. J. Kollonitsch and L. Barash, *J. Am. Chem. Soc.* **98**, 5591–5593, 1976.

38. J. Kollonitsch, L. Barash, F. M. Kahan, and H. Kropp, *Nature* **243**, 346–347, 1973.

39. (a) J. Kollonitsch, S. Marburg, and L. M. Perkins, *J. Org. Chem.* **41**, 3107–3111, 1976; (b) Merck, U. S. Pat. 4049708 (1977).

40. T. Tsushima, K. Kawada, and T. Tsuji, *J. Org. Chem.* **47**, 1107–1110, 1982.

41. (a) V. Tolman and K. Veres, *Collect. Czech. Chem. Commun.* **32**, 4460–4469, 1967; (b) V. Tolman and K. Veres, *Tetrahedron Lett.* **1964**, 1967–1969; (c) V. Tolman and K. Veres, *Tetrahedron Lett.* **1966**, 3909–3912.

42. J. J. M. Hageman, M. J. Wanner, G. Koomen, and U. K. Pandit, *J. Med. Chem.* **20**, 1677–1679, 1977.

43. M. J. Wanner, J. J. M. Hageman, G. Koomen, and U. K. Pandit, *J. Med. Chem.* **23**, 85–87, 1980.

44. K. L. Kirk and L. A. Cohen, *J. Am. Chem. Soc.* **95**, 4619–4624, 1973.

45. K. L. Kirk, W. Nagai, and L. A. Cohen, *J. Am. Chem. Soc.* **95**, 8389–8392, 1973.

46. K. L. Kirk, *J. Org. Chem.* **45**, 2015–2016, 1980.

47. T. J. McCord, D. R. Smith, D. W. Winters, J. F. Grimes, K. L. Hulme, L. Q. Robinson, L. D. Gage, and A. L. Davis, *J. Med. Chem.* **18**, 26–29, 1975.

48. E. L. Bennett and C. Niemann, *J. Am. Chem. Soc.* **72**, 1806–1807, 1950.

49. C.-Y. Yuan, C.-N. Chang, and I.-F. Yeh, *Chem. Abstr.* **54**, 12096(i), 1960.

50. I. A. McDonald, M. G. Palfreyman, M. Jung, and P. Bey, *Tetrahedron Lett.* **26**, 4091–4092, 1985.

51. H. Lettre and U. Wolcke, *Liebigs Ann. Chem.* **708**, 75–85, 1967.

52. A. A. Gottlieb, Y. Fujita, S. Udenfriend, and B. Witkop, *Biochemistry* **4**, 2507–2513, 1965.

53. R. B. Silverman and M. A. Levy, *J. Org. Chem.* **45**, 815–818, 1980.

54. M. Hudlicky, *Tetrahedron Lett.* **1960**, 21–22.

55. L. V. Alekseeva, N. L. Burde, and B. N. Lundin, *Chem. Abstr.* **75**, 36618, 1971.

56. (a) I. A. McDonald, J. M. Lacoste, P. Bey, J. Wagner, M. Zreika, and M. G. Palfreyman, *J. Am. Chem. Soc.* **106**, 3354–3356, 1984; (b) I. A. McDonald, J. M. Lacoste, P. Bey, J. Wagner, M. Zreika, and M. G. Palfreyman, *Bioorg. Chem.* **14**, 103–118, 1986.

57. A. F. Janzen, P. M. C. Wang, and A. E. Lemire, *J. Fluorine Chem.* **22**, 557, 1983.

58. T. Tsushima and K. Kawada, *Tetrahedron Lett.* **26**, 2445–2448, 1985.

59. (a) P. Bey and J. P. Vevert, *Tetrahedron Lett.* **1978**, 1215–1218; (b) P. Bey, J. P. Vevert, V. Van Dorsselaer, and M. Kolb, *J. Org. Chem.* **44**, 2732–2742, 1979; (c) B. W. Metcalf, P. Bey, C. Danzin, M. J. Jung, P. Casara, and J. P. Vevert, *J. Am. Chem. Soc.* **100**, 2551–2553, 1978.

60. P. Bey and D. Schirlin, *Tetrahedron Lett.* **1978**, 5225–5228.

61. P. Bey, J. B. Ducep, and D. Schirlin, *Tetrahedron Lett.* 1978, 5225–5228.

62. (a) H. Kimoto, S. Fujii, and L. A. Cohen, *J. Org. Chem.* **49**, 1060–1064, 1984; (b) H. Kimoto, S. Fujii, and L. A. Cohen, *J. Org. Chem.* **47**, 2867–2872, 1982.

63. J. Mathew, B. J. Invergo, and R. B. Silverman, *Synth. Commun.* **15**, 377–383, 1985.

64. P. Bey, F. Gerhart, V. Van Dorsselaer, and C. Danzin, *J. Med. Chem.* **26**, 1551–1556, 1983.

65. F. Weygand, W. Steglich, and F. Fraunberger, *Angew Chem. Int. Ed.* **6**, 808, 1967.

66. (a) E. D. Bergmann and S. Cohen, *J. Chem. Soc.* **1961**, 4669–4671; (b) E. D. Bergmann, S. Cohen, and I. Shahak, *J. Chem. Soc.* **1959**, 3286–3288.

67. E. D. Bergmann and A. Cohen, *Tetrahedron Lett.* **1965**, 2085–2086.

68. R. L. Buchanan, F. H. Dean, and F. L. M. Pattison, *Can. J. Chem.* **40**, 1571, 1962.

69. (a) G. Gal, J. M. Chemerda, D. F. Reinhold, and R. M. Purick, *J. Org. Chem.* **42**, 142–143, 1977; (b) U. H. Dolling, A. W. Douglas, E. J. J. Grabowski, E. F. Schoenewaldt, P. Sohar, and M. Sletzinger, *J. Org. Chem.* **43**, 1634–1640, 1978.

70. C. Scolastico, E. Conca, L. Prati, G. Guanti, L. Banfi, A. Berti, P. Farina, and U. Valcavi, *Synthesis* **1985**, 850–855.

71. R. L. Buchanan and F. L. M. Pattison, *Can. J. Chem.* **43**, 3466, 1965.

72. K. Burger, D. Hubl, and P. Gertitschke, *J. Fluorine Chem.* **27**, 327–332, 1985.

73. W. H. Vine, K. Hsieh, and G. R. Marshall, *J. Med. Chem.* **24**, 1043–1047, 1981.

74. K. Weinges and E. Kromm, *Leibigs Ann. Chem.* **1985**, 90–102.

75. E. D. Bergmann and A. Shani, *J. Chem. Soc.* **1963**, 3462–3463.

76. T. Fuchikami and I. Ojima, *J. Am. Chem. Soc.* **104**, 3527–3529, 1982.

77. I. Ojima, *Third Regular Meeting of Soviet-Japanese Fluorine Chemists*, Tokyo **1983**, pp. 81–100.

78. I. Ojima, *J. Organomet. Chem.* **279**, 203–214, 1985.

79. T. Taguchi, G. Tomizawa, M. Nakajima, and Y. Kobayashi, *Chem. Pharm. Bull.* **33**, 4077–4080, 1985.

80. N. Ishikawa and A. Takaoka, *Chem. Lett.* **1981**, 107–110.

81. T. Tsushima, K. Kawada, O. Shiratori, and N. Uchida, *Heterocycles* **23**, 45–49, 1985.

82. D. Butina and M. Hudlicky, *J. Fluorine Chem.* **16**, 301–323, 1980.

83. H. Kimoto, K. L. Kirk, and L. A. Cohen, *J. Org. Chem.* **43**, 3403–3405, 1978.

84. F. Weygand, W. Steglich, and W. Oettmeier, *Chem. Ber.* **103**, 818–826, 1970.

85. F. Weygand, W. Steglich, and W. Oettmeier, *Chem. Ber.* **103**, 1655–1663, 1970.

86. G. Hofle and W. Steglich, *Chem. Ber.* **104**, 1408–1419, 1971.

87. T. Taguchi, A. Kawara, S. Watanabe, Y. Oki, H. Fukushima, Y. Kobayashi, M. Okada, K. Ohta, and Y. Iitaka, *Tetrahedron Lett.* **27**, 5117–5120, 1986.

88. M. Fujita and I. Ojima, *Tetrahedron Lett.* **24**, 4573–4576, 1983.

89. M. Bentov and C. Roffman, *Isr. J. Chem.* **7**, 835–837, 1969.

90. C. J. Bacchi, H. C. Nathan, S. H. Hutner, P. P. McCann, and A. Sjoerdsma, *Science* **210**, 332–334, 1980.

91. M. V. Rajam, L. H. Weinstein, and A. W. Galston, *Proc. Natl. Acad. Sci. USA* **82**, 6874–6878, 1985.

92. J. Seidenfeld and L. J. Marton, *Biochim. Biophys. Res. Commun.* **86**, 1192–1198, 1979.

93. C. W. Fearon, J. A. Rodkey, and R. H. Abeles, *Biochemistry* **21**, 3790–3794, 1982.

94. R. B. Silverman and R. H. Abeles, *Biochemistry* **16**, 5515, 1977.

95. E. A. Wang and C. Walsh, *Biochemistry* **20**, 7539–7546, 1981.

96. D. Roise, K. Soda, T. Yagi, and C. T. Walsh, *Biochemistry* **23**, 5195–5201, 1984

97. R. B. Silverman and M. A. Levy, *Biochim. Biophys. Res. Commun.* **95**, 250–255, 1980.

98. J. J. McGuire and J. K. Coward, *J. Biol. Chem.* **260**, 6747–6754, 1985.

99. J. Galivan, J. Inglese, J. J. McGuire, Z. Nimec, and J. K. Coward, *Proc. Natl. Acad. Sci. USA* **82**, 2598–2602, 1985.

100. N. H. Pardanani and N. Muller, *Org. Prep. Proc. Int.* **10**, 279–283, 1978.

101. E. De Clercq, A. Billiau, V. G. Edy, K. L. Kirk, and L. A. Cohen, *Biochim. Biophys. Res. Commun.* **82**, 840–846, 1978.

102. C. B. Klee, K. L. Kirk, L. A. Cohen, and P. McPhie, *J. Biol. Chem.* **250**, 5033–5040, 1975.

103. D. T. Browne, G. L. Kenyon, and G. D. Hegeman, *Biochim. Biophys. Res. Commun.* **39**, 13–19, 1970.

104. E. A. Pratt and C. Ho, *Biochemistry* **14**, 3035, 1975.

105. R. W. Johnson and F. T. Kenney, *J. Biol. Chem.* **248**, 4528–4531, 1973.

106. T. L. Nagabhushan, D. Kandasamy, H. Tsai, W. N. Turner, and G. H. Miller, *Current Chemotherapy Proceedings of 11th International Congress on Chemotherapy*, American Society of Microbiology, Washington, D.C., 1980, pp. 442–443.

107. W. D. Kingsbury, J. C. Boehm, R. J. Mehta, and S. F. Grappel, *J. Med. Chem.* **26**, 1725–1729, 1983.

108. C. Niemann, and M. M. Rapport, *J. Am. Chem. Soc.* **68**, 1671–1672, 1946.

109. M. C. Salon, S. Hamman, and C. G. Beguin, *Org. Magn. Res.* **21**, 265–270, 1983.

110. M. C. Salon, S. Hamman, and C. G. Beguin, *J. Fluorine Chem.* **27**, 361–370, 1985.

111. (a) A. W. Stern, R. H. Abeles, and A. H. Tashjian, Jr., *Cancer Res.* **44**, 5614–5618, 1984; (b) M. A. Phillips, A. W. Stern, R. H. Abeles, and A. H. Tashjian, Jr., *J. Pharmacol. Exp. Ther.* **226**, 276–281, 1983; (c) G. Hortin, A. W. Sterm, B. Miller, R. H. Abeles, and I. Boime, *J. Biol. Chem.* **258**, 4047–4050, 1983.

112. (a) S. Thaisrivongs, D. T. Pals, W. M. Kati, S. R. Turner, and L. M. Thomasco, *J. Med. Chem.* **28**, 1553–1555, 1985; (b) S. Thaisrivongs, D. T. Pals, W. M. Kati, S. R. Turner, L. M. Thomasco, and W. Watt, *J. Med. Chem.* **29**, 2080–2087, 1986.

3

FLUORINATED AMINES

3.1 INTRODUCTION

The synthesis of fluorinated analogs of biologically important amines has been explored extensively. Their direct preparation is difficult since most of the common fluorination agents can also react with the amino group. However, by the use of appropriate solvents, like liquid hydrogen fluoride, which protects the amine functionality by protonation, successful fluorinations have been achieved.[1] β-Fluorinated amines are important targets in the design of antimetabolites and drugs since fluorine causes minimal structural changes and maximal shift in electron distribution.

3.2 SYNTHESIS

As in the case of fluorinated amino acids, the synthesis of fluoroamines is organized based on synthon size.

3.2.1 Sulfur Tetrafluoride

The alcoholic hydroxyl groups in hydroxyamines such as **1** can be selectively replaced by fluorine (fluorodehydroxylation) using SF_4 in hydrogen fluoride at low temperature ($-78°C$) and atmospheric pressure.[1,2] Since the hydroxy compounds are easily accessible, many of them in the optically active form and with established stereochemistry, this is a very useful transformation for the synthesis of fluoroamines. A wide variety of alcohols containing basic nitrogen can serve as substrates for fluorodehydroxylation with the SF_4–HF system. The reaction requires more stringent conditions ($0°C$) when the alcohol is flanked by two positive charges as in the case of β-hydroxy histamine (**1**).[1] Both ephedrine (**3**) and pseudoephedrine (**4**) afford the same 2:1 mixture of fluoro isomers (**5**) on reaction with sulfur tetrafluoride–hydrogen fluoride, suggesting an S_N1 mechanism since the developing positive charge can be stabilized very well in this case.[1]

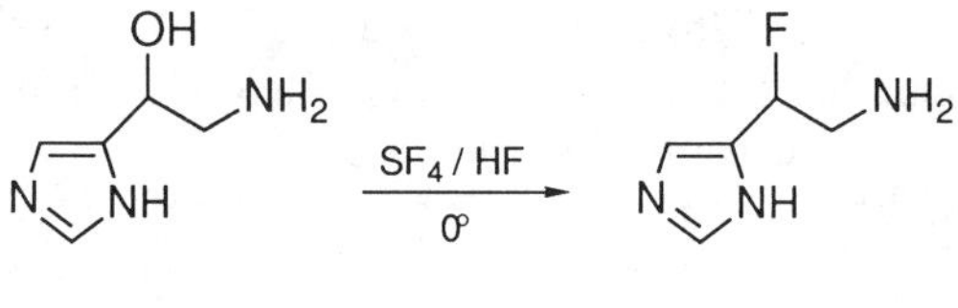

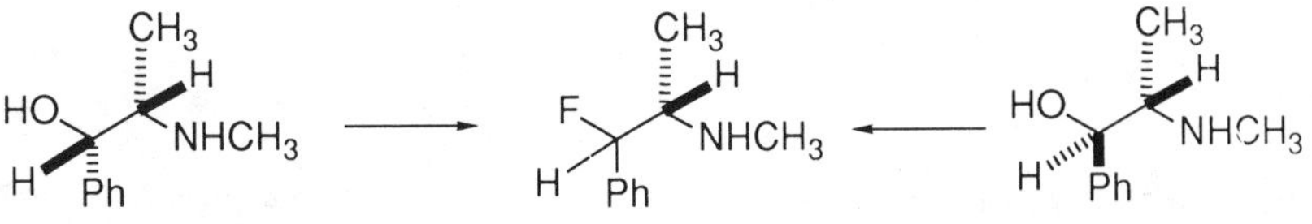

3.2.2 Diethylaminosulfur Trifluoride

Synthesis of Monofluoroputrescine[26]

Fluorination of 2-hydroxy-1, 4-dibromobutanone (**6**) with DAST gives fluoro compound **7**, which is converted to fluorodiamine (**8**) through the diphthalimide.

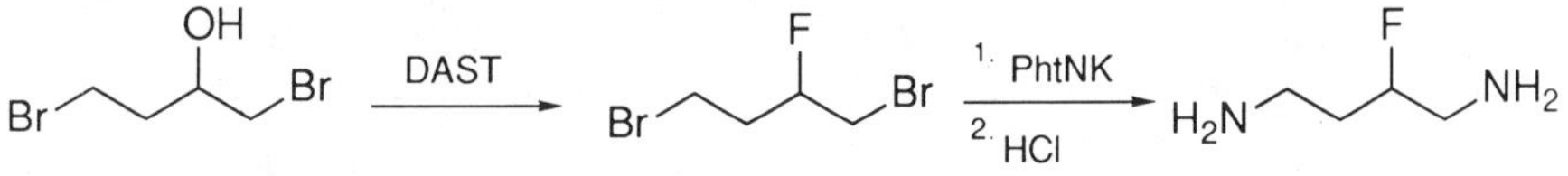

3.2.3 Hydrogen Fluoride–Pyridine

Hydrogen fluoride–pyridine is a very useful reagent for the synthesis of fluoroamines.[3] It can be handled more easily than sulfur tetrafluoride and anhydrous hydrogen fluoride.

Ring-Opening Reactions of Aziridines[3-5]

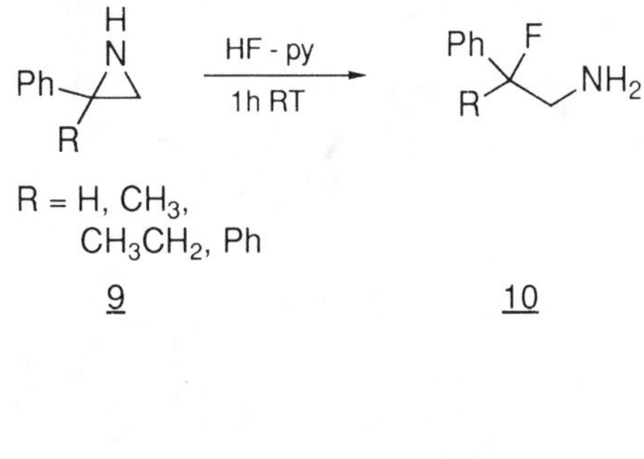

A convenient method for synthesizing β-fluoroamines (**10**) is by way of ring-opening reactions of aziridines (**9**) with hydrogen fluoride–pyridine. Aziridines with a variety of functional groups are easily accessible. The regioselectivity of the reaction is excellent, fluoride attack being directed to the most substituted ring carbon or the benzylic carbon in most cases.[3]

Both the cis and the trans isomers of 2-phenyl-3-methyl aziridine (**11**) and 2,3-diphenyl aziridine (**12**) gave predominantly the threo fluoroamine (**13**).[3]

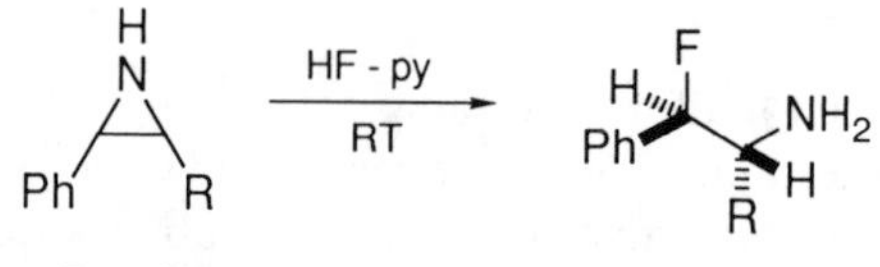

$\underline{11}$ R = CH$_3$,
$\underline{12}$ R = Ph
cis and trans

$\underline{13}$
threo

A possible mechanism is represented in Fig. 3.1.[3] The cis isomer reacts much more slowly than trans (70°C vs. RT). This could be due to the fact that the cis substituent makes the phenyl group less effective in delocalizing the developing positive charge on the aziridinyl carbon (steric inhibition of resonance).

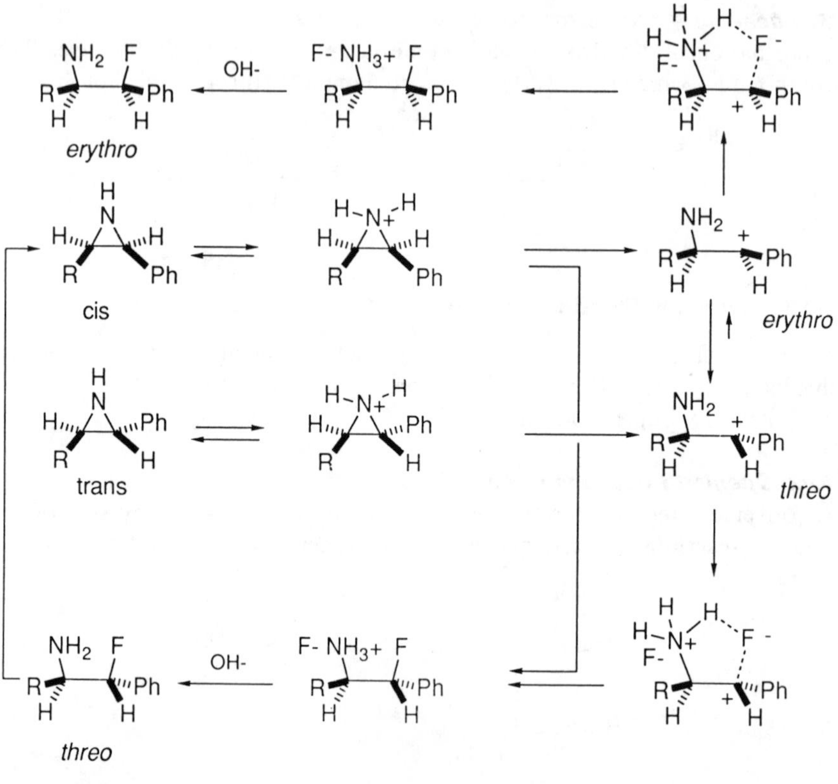

FIGURE 3.1

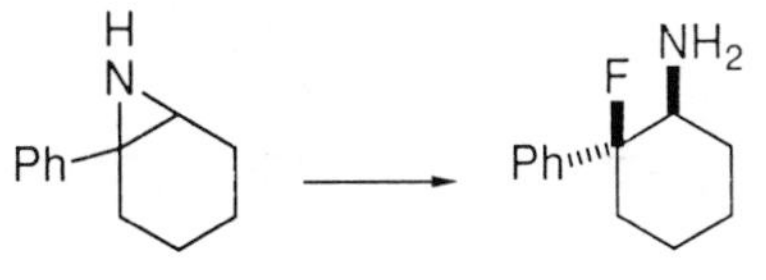

In 2-fluoro-2-phenyl-cyclohexylamine (**15**) the –F and –NH$_2$ are syn to each other, indicating cis opening of the aziridine ring. An SN$_i$-type mechanism is suggested, as seen in Fig. 3.2.[3] However, in 2-fluorocyclohexyl amine (**17**) the –F and –NH$_2$ are anti to each other, suggesting a trans ring opening of cyclohexenimine by hydrogen fluoride–pyridine.

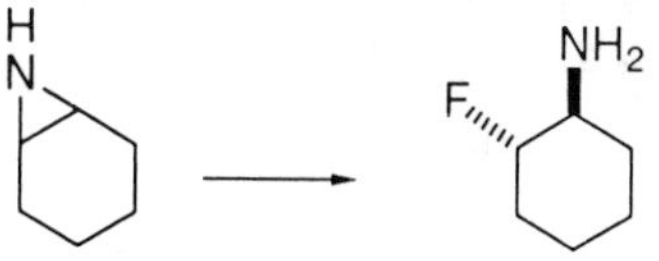

Hydroxyaziridines such as **18** and **20** react with hydrogen fluoride–pyridine to form fluorohydroxyamines (**19**) or difluoroamines (**21**).[6]

A general method for the synthesis of β-fluoroamino amide (**22**) via an aziridine ring-opening reaction by hydrogen fluoride–pyridine has been developed (Fig. 3.3).[4]

Addition of Hydrogen Fluoride–Pyridine to 1-Azirines[7]

The addition of hydrogen fluoride–pyridine to the unsaturated rings of substituted 1-azirines (**23**) is a convenient synthetic route to β,β-difluoroamines (**24**). Presumably, the formation of fluoroaziridine (**26**) is the first step. The fluoroaziridine then adds another molecule of hydrogen fluoride to give the β,β-difluoro compound (**27**). 1-Azirines add a proton easily on their nitrogen in the

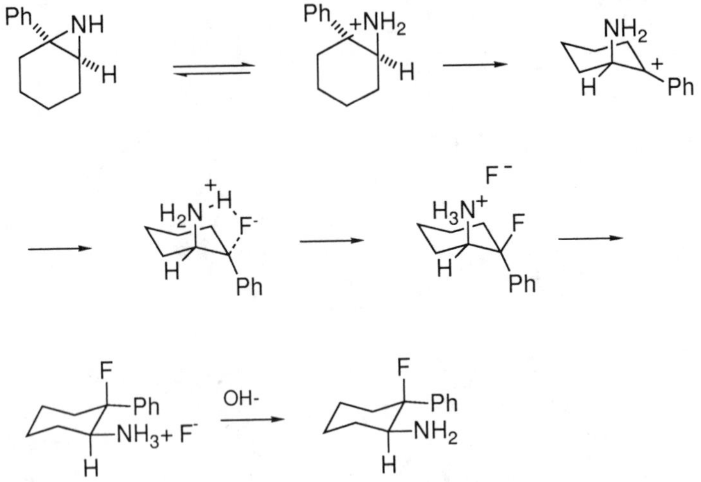

FIGURE 3.2

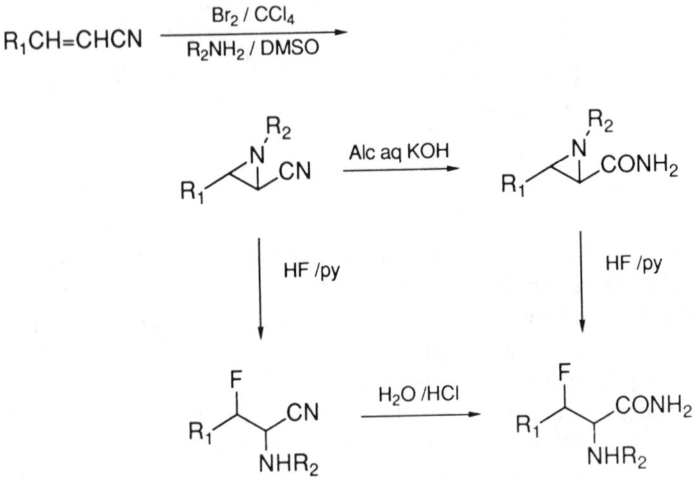

FIGURE 3.3

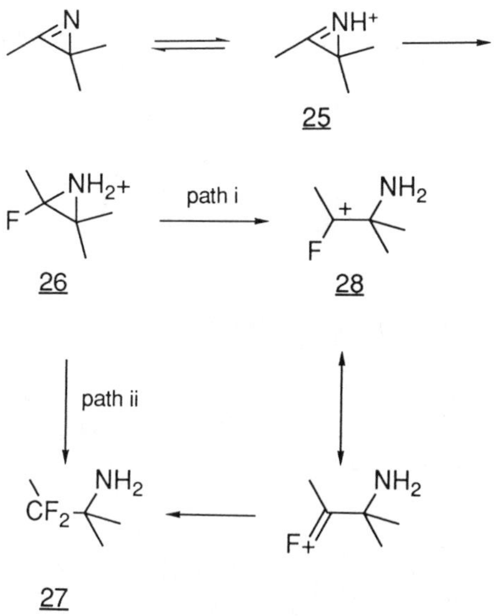

25

26   path i   28

path ii

27

FIGURE 3.4

$$\text{Ph} \overset{N}{\diagup} R \longrightarrow PhCF_2CH(NH_2)R + PhCOCHFR + \text{pyrazine}$$

R = CH$_3$	20%	5%	54%
R = Ph	-	37%	-
		29	30

presence of an acid-forming azirinium ion (**25**). Two competing pathways are possible, one leading to the β,β-difluoroamines, the other, to the α-fluoroketones and pyrazines. The formation of β,β-difluoroamine is represented in Fig. 3.4.[7] The ring opening can proceed via the stabilized carbonium ion (**28**) (path i) or by the direct displacement of the amine group by a fluoride ion (path ii).

Another competing pathway yields α-fluoroketone (**29**) and pyrazine (**30**).[7] The probable pathways for α-fluoroketone are seen in Fig. 3.5. The formation of the pyrazine can be represented as in Fig. 3.6. The yields of the reaction of hydrogen fluoride–pyridine with azirines are lower than those with aziridines. However, 1-azirines are valuable synthons for β,β-difluoroamines.

Epoxide Ring Opening with Hydrogen Fluoride–Pyridine[4]

The regiospecificity of epoxide ring-opening reactions with hydrogen fluoride–pyridine is excellent. Synthesis by this methodology is seen in Fig. 3.7.

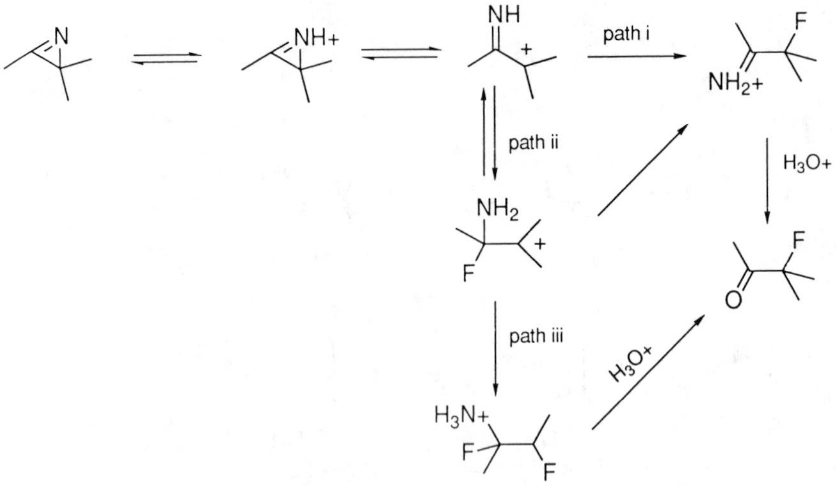

FIGURE 3.5

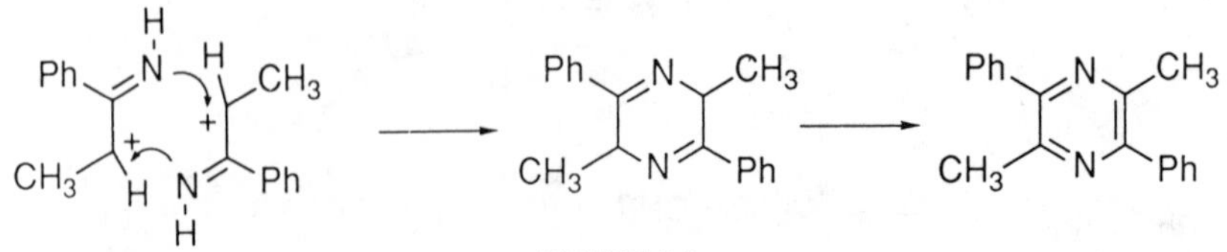

FIGURE 3.6

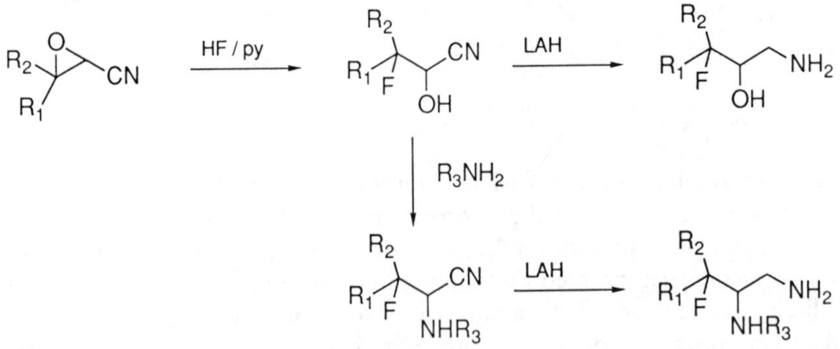

FIGURE 3.7

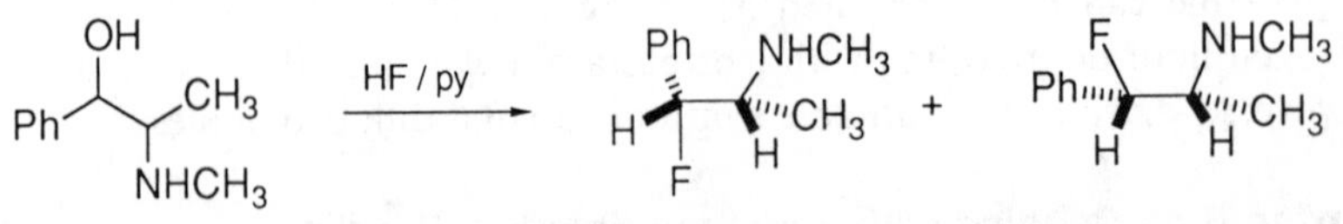

ephedrine (*erythro*) <u>31</u>
pseudoephedrine (*threo*) <u>32</u>

threo 70%
<u>33</u>

erythro 30%
<u>34</u>

Fluorodehydroxylation Using Hydrogen Fluoride–Pyridine[6]

Ephedrine (**31**) and pseudoephedrine (**32**) undergo fluorodehydroxylation with hydrogen fluoride–pyridine to form the fluoroisomers **33** and **34** in a 70:30 ratio. The rationale for the predominant formation of the threo isomer is seen in Fig. 3.8.

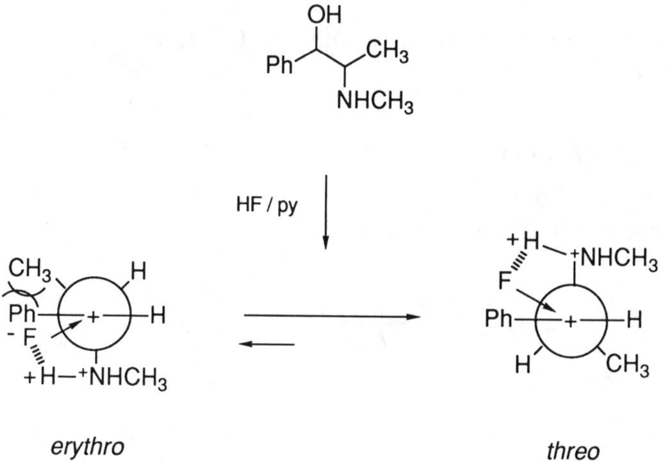

FIGURE 3.8

Fluorodediazoniation with Hydrogen Fluoride–Pyridine

Hydrogen fluoride–pyridine is used as the fluorinating agent in the conversion of the diazoketone to fluoroketone for the synthesis of α-fluoromethyl putrescine (**36**) from 4-phthalimido-1-butyroyl chloride (**35**).[8]

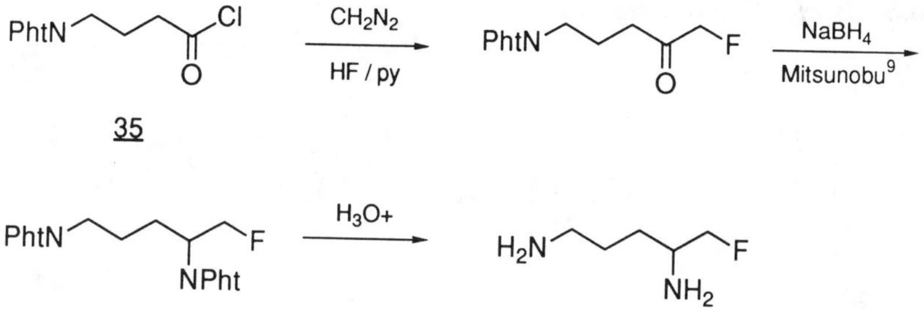

3.2.4   Trifluoromethylhypofluorite

Both ephedrine and pseudoephedrine (**37**), on reaction with hydrogen sulfide in liquid hydrogen fluoride, afford *threo*-2-methylamino-1-phenyl-propanethiol

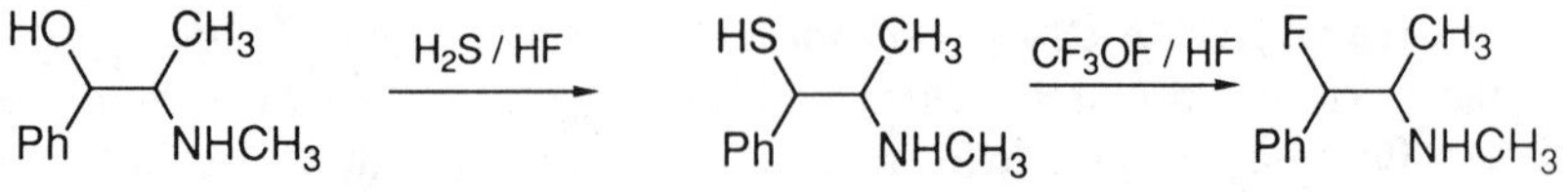

37 → 38 → 39

(**38**), which on fluorodesulfurization with CF_3OF forms 3-fluoro-2-methyl-amino-3-phenyl-propane (**39**).[10]

3.2.5 Molecular Fluorine

N-substituted 2-amino thiols (**40**) undergo fluorodesulfurization with fluorine in helium to give aminoalkyl fluorides (**41, 42**).[10]

$$(CH_3CH_2)_2NCH_2CH_2SH \xrightarrow[\text{HF / HBF}_4]{\text{F}_2/\text{He (1:4)}} (CH_3CH_2)_2NCH_2CH_2F \ +$$

40   41

$$(CH_3CH_2)_2NCH_2CF_2H$$

42

3.2.6 Fluoroboric Acid

Diazotization of amino imidazoles such as **43** in fluoroboric acid affords imidazole diazonium ions, which on in situ irradiation decompose to form fluoroimidazoles (**44**).

Synthesis of 4-Fluorohistamine (47)[11]

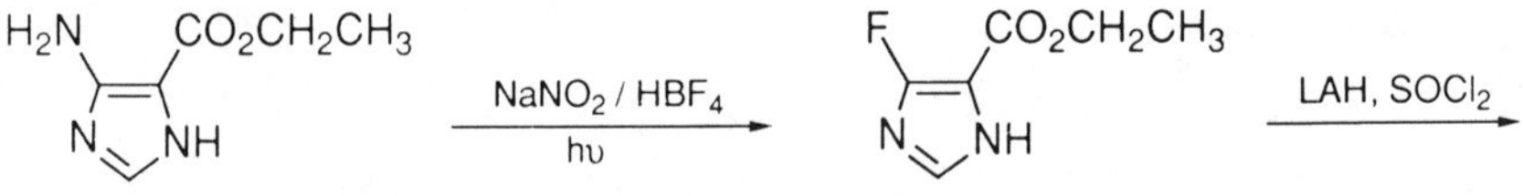

43 → 44

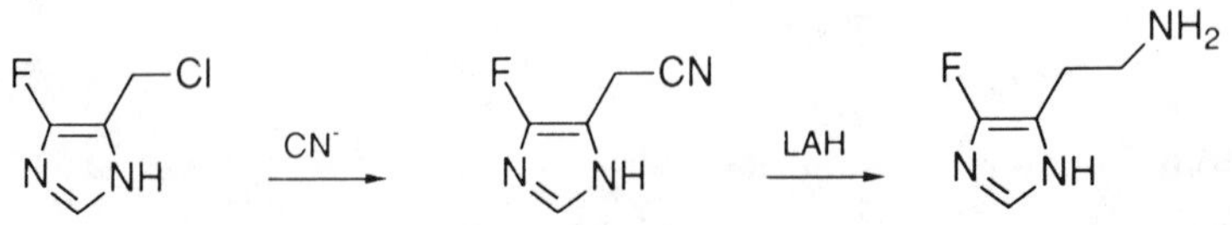

45 → 46 → 47

This synthesis is not practically useful because the yield of the cyanomethyl compound (**46**) is poor and there are problems associated with the lithium aluminum hydride reduction.

Synthesis of 2-Fluorohistamine(48)[12]

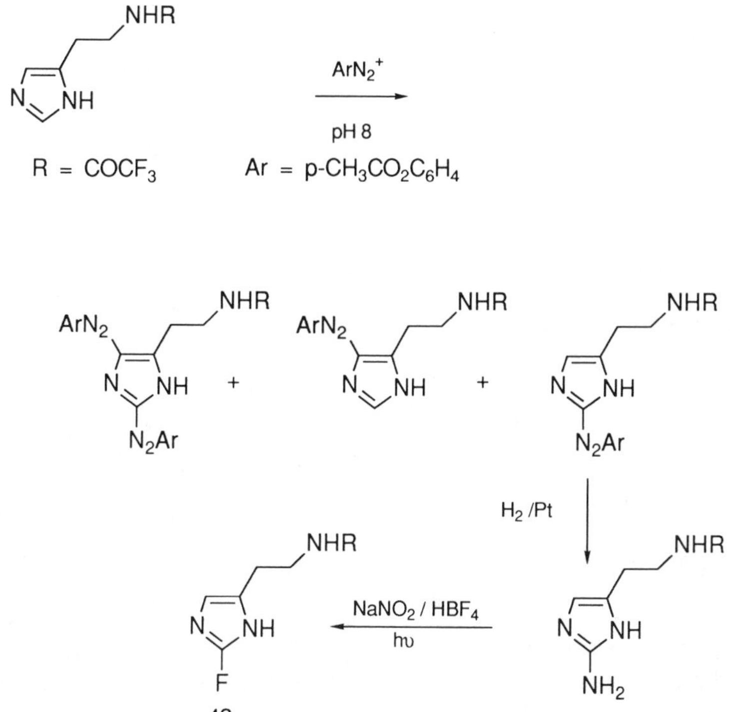

R = COCF$_3$ Ar = p-CH$_3$CO$_2$C$_6$H$_4$

3-Fluorotyramine (**49**), 3,5-difluorotyramine (**50**), and 6-fluorodopamine (**53**) are synthesized from protected phenethylamine precursors by nitration, catalytic reduction, diazotization, photochemical decomposition, and finally deprotection.[13]

Synthesis of 3-Fluorotyramine[13]

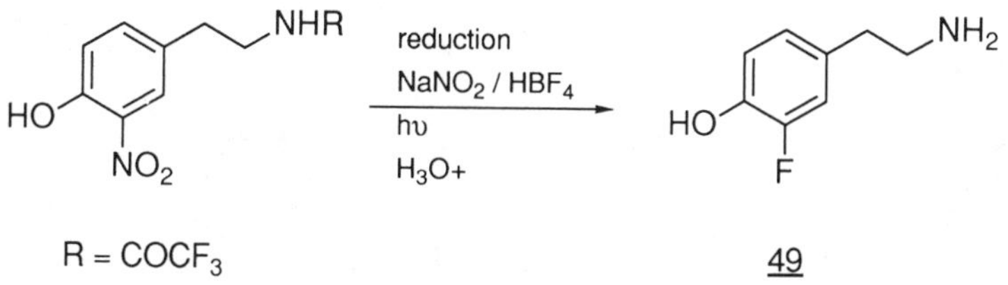

R = COCF$_3$

Synthesis of 3,5-Difluorotyramine (50)[13]

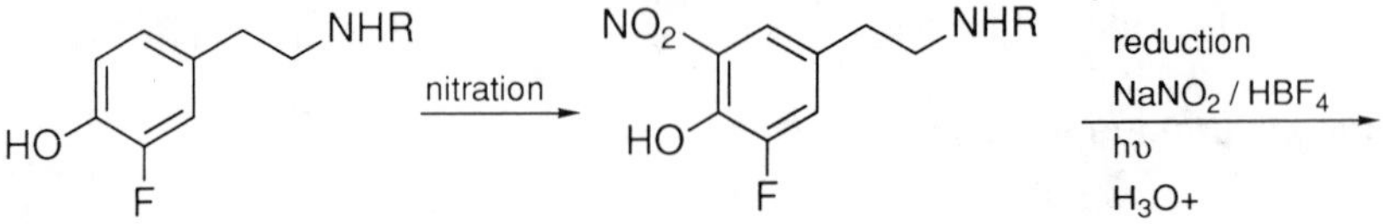

R = COCF$_3$

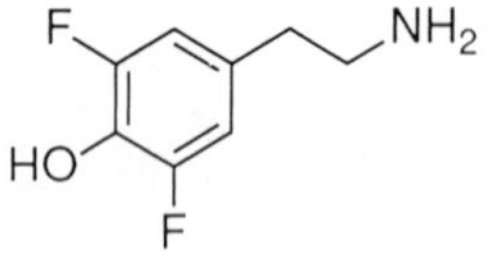

<u>50</u>

The hydroxylated compound **51** is obtained as a by-product of the irradiation procedure.

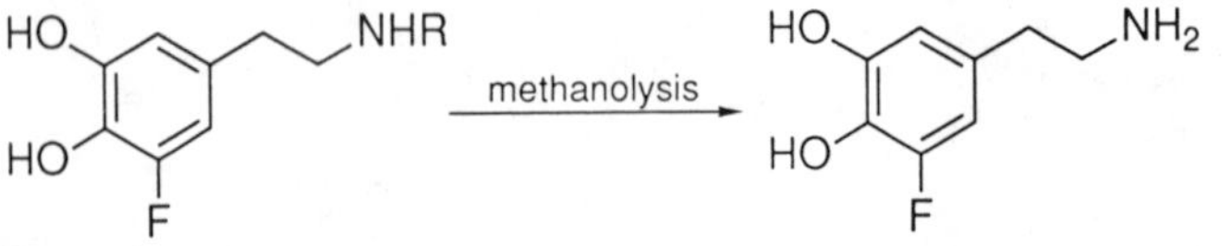

<u>51</u>   5-fluorodopamine

Synthesis of 6-Fluorodopamine (53)[13] and 2-Fluorodopamine (55)[13]

This is the most readily accessible isomer because mononitration of **52** occurs exclusively at the 6 position. There is no obvious direct route to 2-flu-orodopamine (**55**). The benzaldehyde derivative (**54**) is used as the starting material.[13]

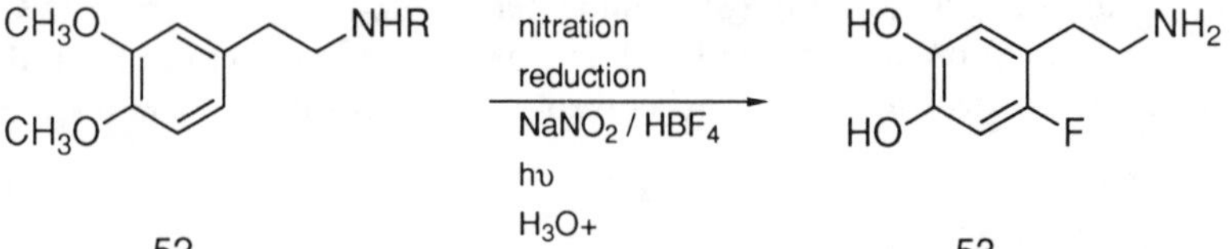

R = COCF$_3$

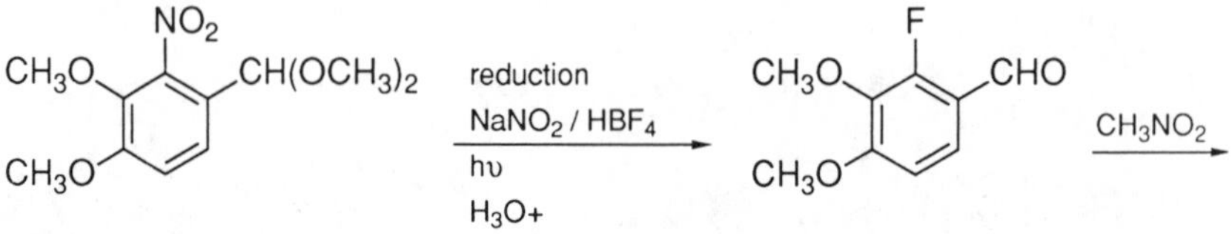

<u>54</u>

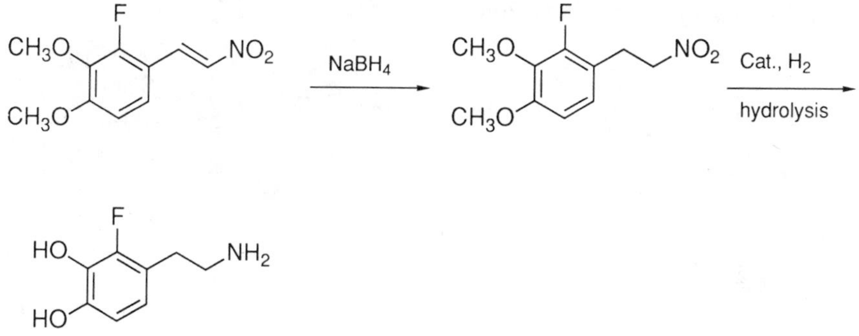

55

3.2.7   Chlorodifluoromethane

Regiospecific alkylation of the anion derived from substituted malonate esters with CHF_2Cl is a very versatile reaction, which may be used for the introduction of difluoromethyl group. A number of α-difluoromethyl amines may be synthesized by this methodology.[14-17]

Synthesis of α-Difluoromethyl Dopamine (56)[14]

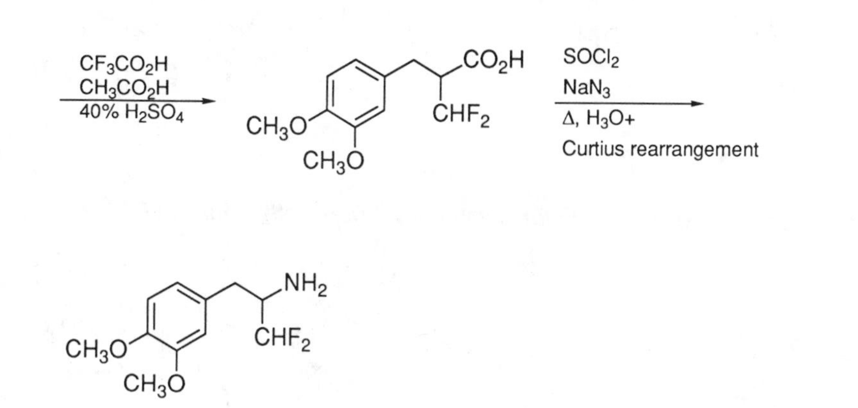

56

Synthesis of α-Difluoromethyl Putrescine (57a)[14,15]

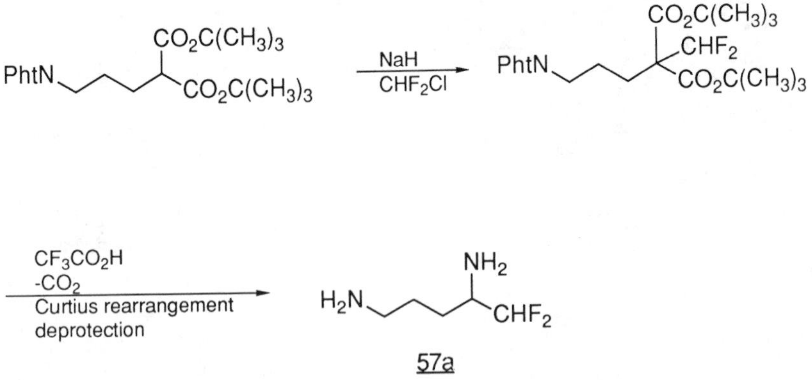

Synthesis of α-Difluoromethyl Dehydroputrescine (57b)[17]

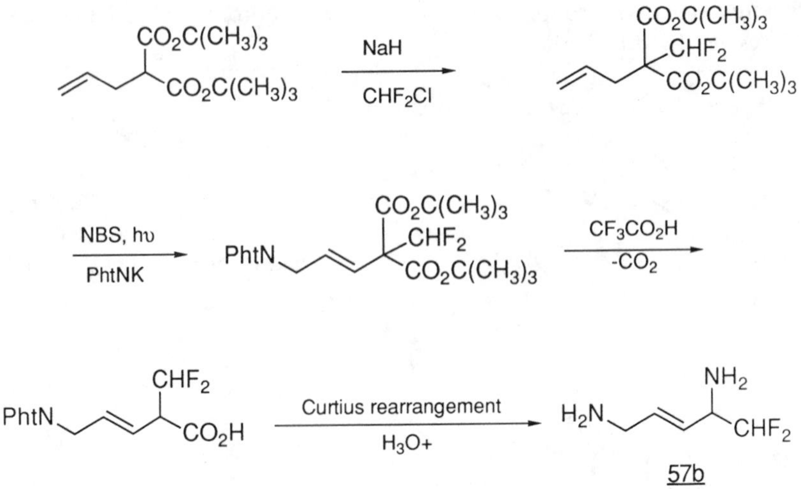

Synthesis of (E)-2-(3,4-Dihydroxyphenyl)-3-fluoroallylamine (58)[16,27]

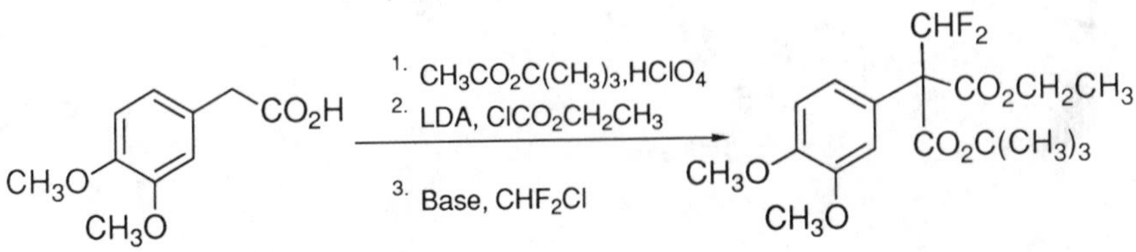

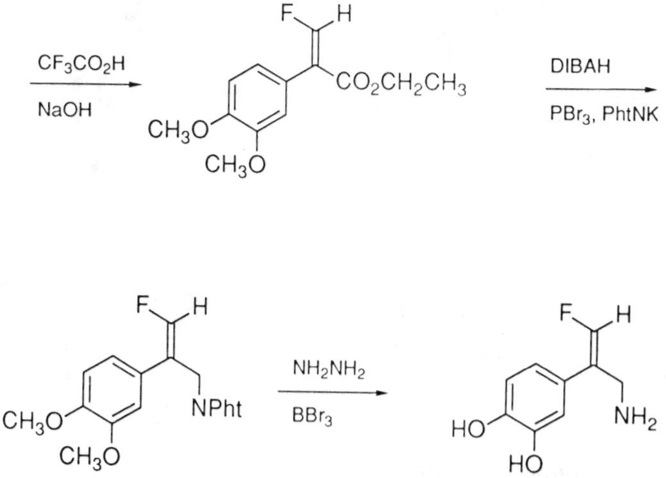

Synthesis of (Z)-2-Phenyl-3-fluoroallylamine (61)[27]

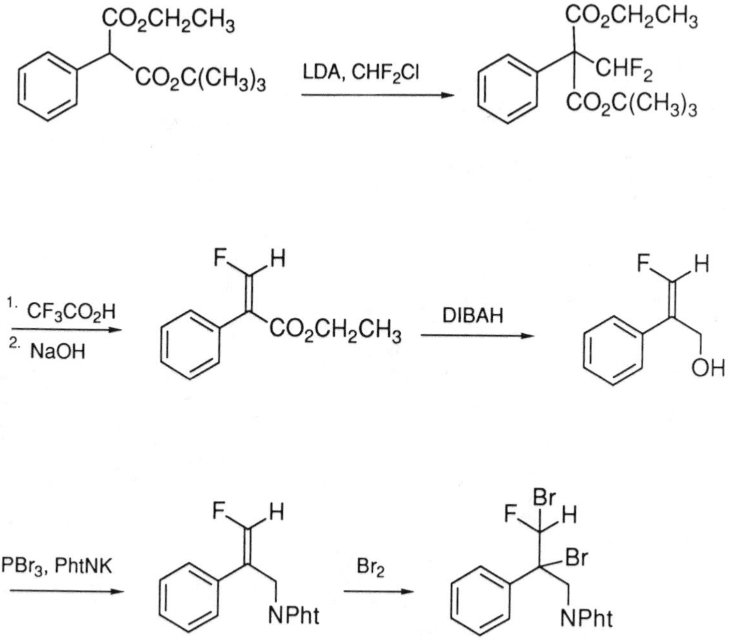

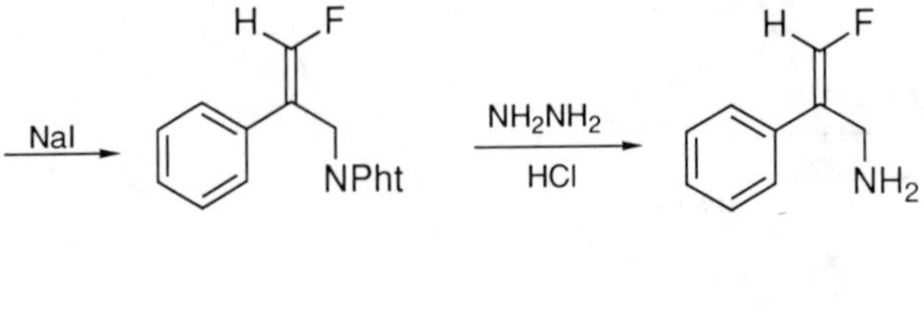

60   61

The synthesis depends on the cis addition of bromine to the double bond of the phthalimide derivative **59** followed by sodium iodide–mediated trans debromination to form **60**.

3.2.8   Dibromodifluoromethane

Synthesis of 2-Phenyl-3,3-difluoroallylamine (62)[27]

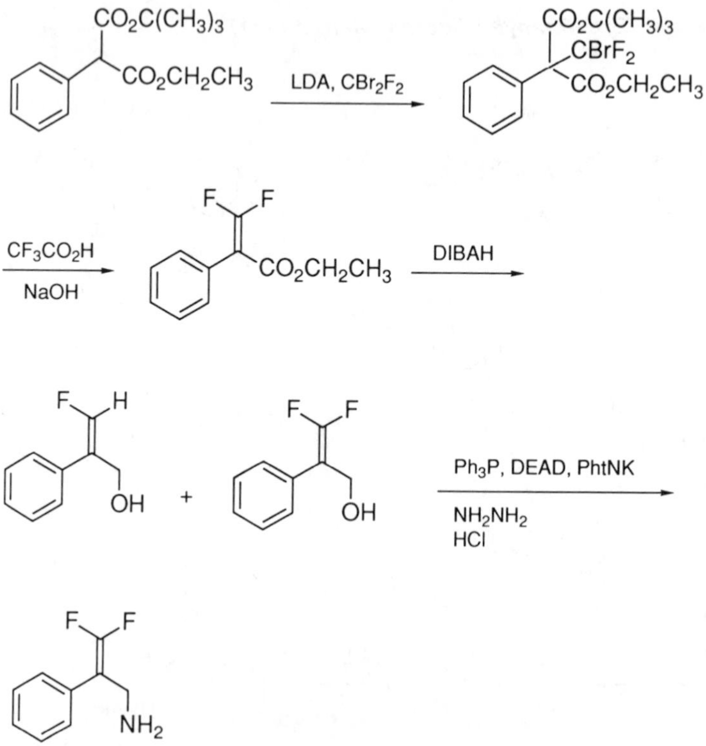

62

3.2.9   Trifluoromethyl Iodide

Synthesis of 2- and 4-Trifluoromethyl Histamine[18]

N-acyl histamines (**63**) can be trifluoromethylated at C-2 to form **64** and at C-4 to form **65** using CF_3I in methanol solution. A small amount of 2,4-bistrifluorom-ethyl product (**66**) is also formed. The electrophilic CF_3 generated by ultraviolet irradiation shows different degrees of preference for attack at C-4 and C-2.

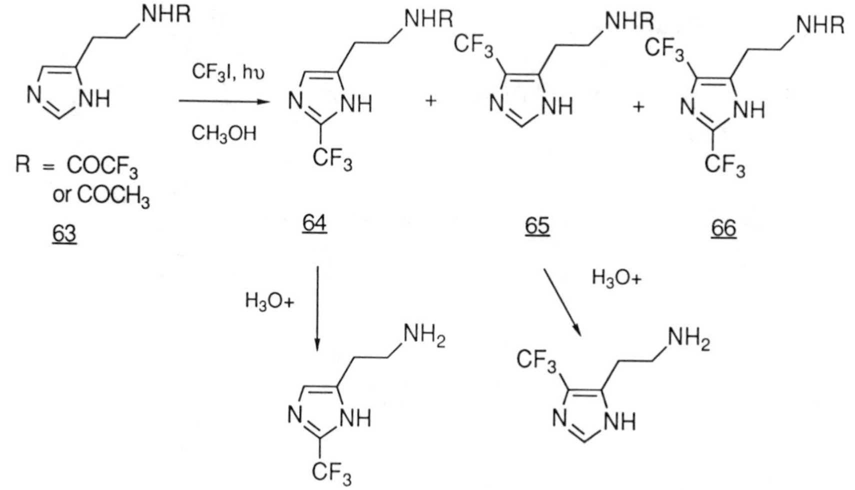

Synthesis of Trifluoromethylurocanic Acids[18]

Due to the electron-withdrawing effect of the acrylic ester moiety as well as the high reactivity of trifluoromethyl radical, low yields of complex mixtures of **67**, **68**, **69**, and **70** are obtained.

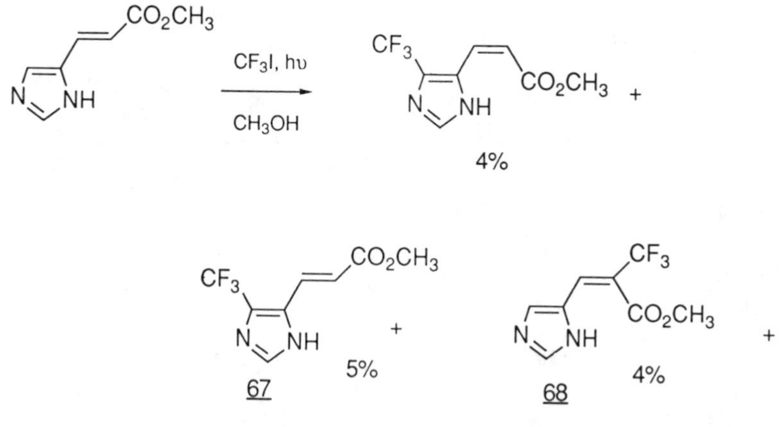

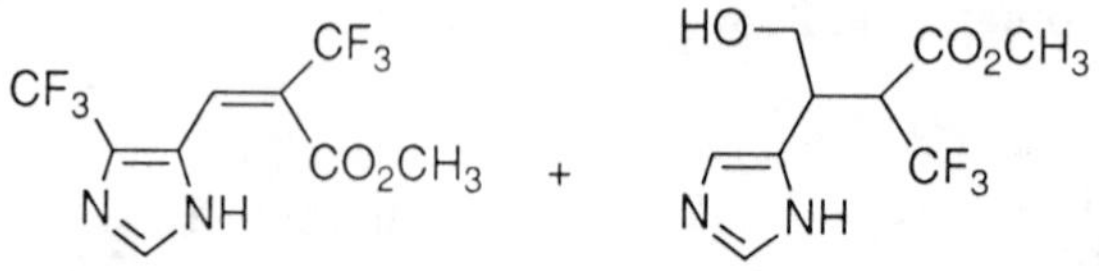

69   trace

70   trace

3.2.10   Fluoroacetonitrile

Synthesis of α-Fluoromethyl Dehydroputrescine (71)[17]

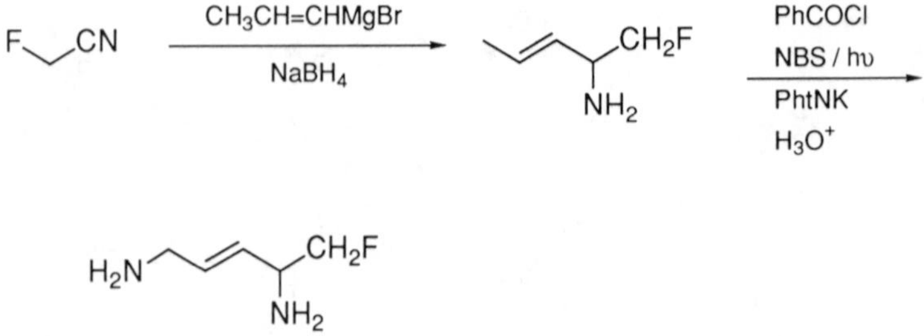

71

3.2.11   Trifluoroacetic Anhydride

Synthesis of 2-trifluoromethyl histamine (**72**) using $(CF_3CO)_2O$ as the trifluoromethylating agent is shown below.[19]

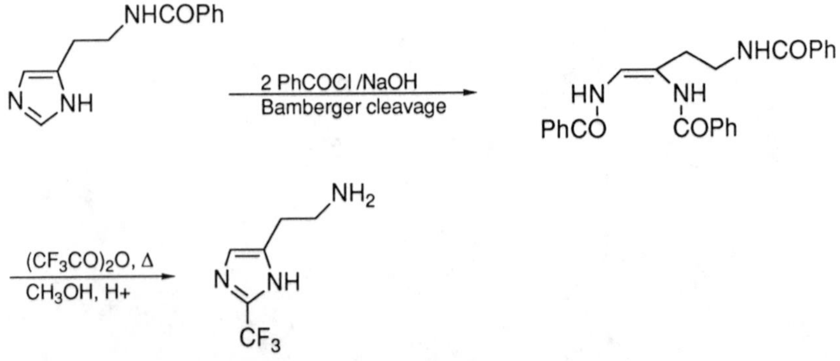

72

3.2.12 Perfluorosuccinamide

Tetrafluoroputrescine (**74**) is obtained when commercially available perfluorosuccinamide (**73**) is reduced with borane in tetrahydrofuran.[26]

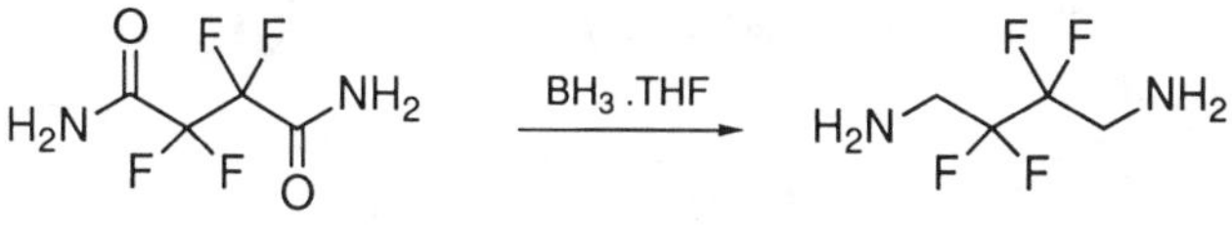

73   **74**

3.3 ENZYMATIC SYNTHESIS

3.3.1 Synthesis of 4-Fluoro-L-Histamine[11]

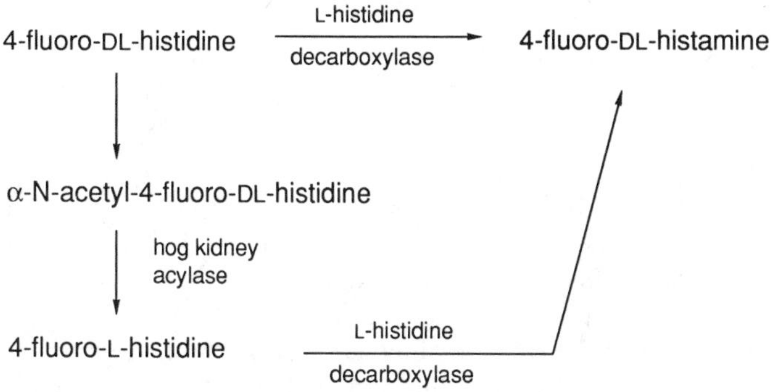

3.3.2 Synthesis of (E)-β-Fluoromethylene-*meta*-Tyramine (75)[20–22]

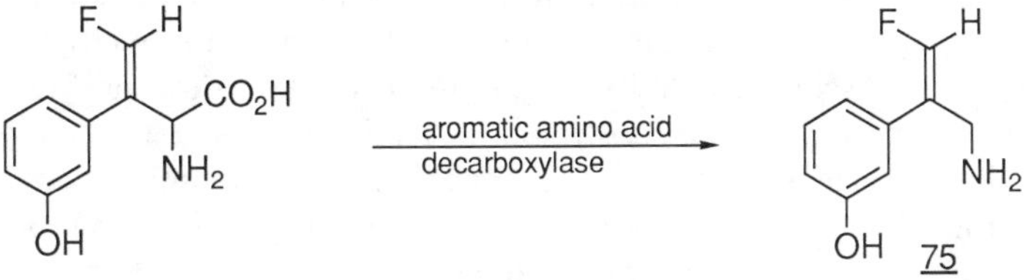

3.3.3 Synthesis of (E)-β-Fluoromethylene GABA (76)[20]

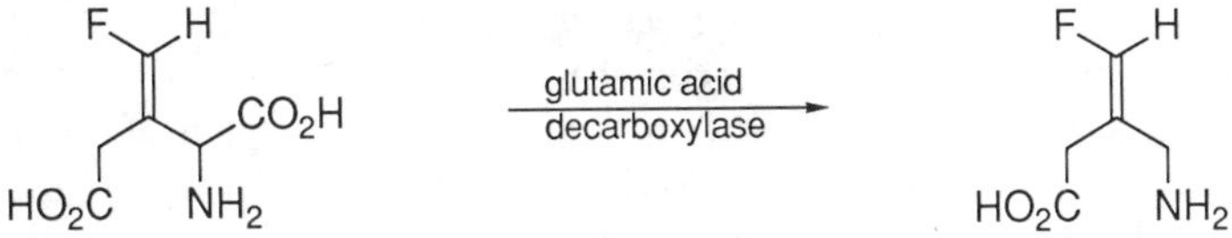

76

3.3.4 Synthesis of 4-Fluorourocanic Acid (77)[11]

3.3.5 Synthesis of 2-Fluorourocanic Acid (78)[23]

3.4 BIOLOGICAL ACTIVITY

Synthesis of β-fluoroamines has been an active area of research as a result of their potential biological and pharmacological properties. The diamine putrescine and the polyamines spermidine and spermine have been known to regulate growth processes.[15] α-Fluoromethyl and difluoromethyl amines are found to be potent enzyme-activated irreversible inhibitors of corresponding carboxylases.[14] Thus α-difluoromethyl dopamine specifically inhibits L-aromatic-α-amino acid decarboxylase, and α-difluoromethyl and monofluoromethyl putrescines irreversibly inhibit rat liver ornithine decarboxylase.[8,15] Both the fluoromethyl and difluoromethyl putrescines activate S-adenosyl-L-methionine decarboxylase.[8] The difluoromethyl derivative is a substrate for diamine oxidase and the monofluoromethyl putrescine is a substrate for mitochondrial monoamine oxidase. The monofluoromethyl compound is also a potent inhibitor of $E.$ $coli$–ornithine decarboxylase.[24] If used in combination with α-difluoromethyl arginine, both ornithine and arginine decarboxylase activities are inhibited. The fluorinated putrescines labeled with ^{18}F may be used in tumor localization in conjunction with positron emission tomography.[26] The (E)-dehydro analogs of α-fluoromethyl and difluoromethyl putrescines are more potent irreversible inhibitors of rat liver ornithine decarboxylase than are their saturated analogs.[17]

A new approach to enzyme inhibition leading to the design of a number of therapeutically useful compounds depends on both site selectivity and target enzyme specificity.[20–22] A biosynthetic enzyme (metabolic) generates in situ an irreversible enzyme-activated inhibitor of a catabolic enzyme in the same pathway.

Thus (E)-β-fluoromethylene-meta-tyrosine is decarboxylated by the metabolic enzyme aromatic L-amino acid decarboxylase (AADC), an enzyme predominantly located in monoamine nerves, to generate (E)-β-fluoromethylene-meta-tyramine, which is a potent inhibitor of the catabolic enzyme monoamine oxidase (MAO).[20,27] Also, β-fluoromethylene dopamine is a highly selective inhibitor of MAO type B, with therapeutic possibilities as an adjunct to the L-dopa treatment of Parkinsonism.[16,27] Similarly, (E)-β-fluoromethylene glutamic acid is decarboxylated by glutamic acid decarboxylase (GAD) to produce (E)-β-fluoromethylene GABA, which inhibits GABA transminase.[20]

Imidazole ring plays a key role in biological structure and function. Fluoro analogs of histamine[11,12] are found to be valuable tools in studying enzyme-receptor mechanims.

Ring-hydroxylated phenethylamines like tyramine and dopamine have important biological and pharmacological functions.[13] Introduction of fluorine in the ring alters properties like pK_a and oxidation potential of the phenolic group, due to the strong electronic effect of fluorination. 3-Fluoro- and 3,5-difluorotyramines and 2-, 5-, and 6-fluorodopamines were studied.[13] The phenolic groups of these fluorinated phenethyl amines are significantly ionized at physiological pH, which might affect the interaction of these compounds with biomembranes and receptor sites.

2-Fluorourocanic acid is reported to have interesting and potentially important biological properties.[18,19,23] It is found to be an especially potent irreversible inhibitor of urocanase, the second enzyme in the pathway for histidine catabolism.[25]

REFERENCES

1. J. Kollonitsch, S. Marburg, and L. M. Perkins, J. Org. Chem. **44**, 771–777, 1979.

2. J. Kollonitsch, S. Marburg, and L. M. Perkins, J. Org. Chem. **40**, 3808–3809, 1975.

3. T. N. Wade, J. Org. Chem. **45**, 5328–5333, 1980.

4. R. Guedj and A. M. Remli, Ann. Chim. Fr. **9**, 691–694, 1984.

5. T. N. Wade and R. Guedj, Tetrahedron Lett. **1978**, 3247–3250.

6. G. Alvernhe, S. Lacombe, A. Laurent, and C. Rousset, J. Chem. Res. (S) **1983**, 246–247.

7. T. N. Wade and R. Kheribet, J. Org. Chem. **45**, 5333–5335, 1980.

8. C. Danzin, P. Bey, D. Schirlin, and N. Claverie, Biochem. Pharmacol. **31**, 3871–3878, 1982.

9. O. Mitsunobu, M. Wada, and T. Sano, J. Am. Chem. Soc. **94**, 679–680, 1972.

10. J. Kollonitsch, S. Marburg, and L. M. Perkins, J. Org. Chem. **41**, 3107–3111, 1976.

11. K. L. Kirk and L. A. Cohen, J. Am. Chem. Soc. **95**, 4619–4624, 1973.

12. K. L. Kirk, W. Nagai, and L. A. Cohen, J. Am. Chem. Soc. **95**, 8389–8392, 1973.

13. K. L. Kirk, J. Org. Chem. **41**, 2373–2376, 1976.

14. P. Bey and D. Schirlin, Tetrahedron Lett. **1978**, 5225–5228.

15. B. W. Metcalf, P. Bey, C. Danzin, M. J. Jung, P. Casara, and J. P. Vevert, *J. Am. Chem. Soc.* **100**, 2551–2553, 1978.

16. P. Bey, J. Fozard, J. M. Lacoste, I. A. McDonald, M. Zreika, and M. G. Palfreyman, *J. Med. Chem.* **27**, 9–10, 1984.

17. P. Bey, F. Gerhart, V. VanDorsselaer, and C. Danzin, *J. Med. Chem.* **26**, 1551–1556, 1983.

18. H. Kimoto, S. Fujii, and L. A. Cohen, *J. Org. Chem.* **49**, 1060–1064, 1984.

19. H. Kimoto, K. L. Kirk, and L. A. Cohen, *J. Org. Chem.* **43**, 3403–3405, 1978.

20. I. A. McDonald, M. G. Palfreyman, M. Jung, and P. Bey, *Tetrahedron Lett.* **26**, 4091–4092, 1985.

21. I. A. McDonald, J. M. Lacoste, P. Bey, J. Wagner, M. Zreika, and M. G. Palfreyman, *Bioorg. Chem* **14**, 103–118, 1986.

22. I. A. McDonald, J. M. Lacoste, P. Bey, J. Wagner, M. Zreika, and M. G. Palfreyman, *J. Am. Chem. Soc.* **106**, 3354–3356, 1984.

23. J. T. Welch, *Tetrahedron* **43**, 3123–3197, 1987.

24. A. Kallio, P. P. McCann, and P. Bey, *Biochem. J.* **204**, 771–775, 1982.

25. (a) K. L. Kirk and C. R. Creveling, *Med. Res. Rev.* **4**, 189, 1984; (b) C. B. Klee, L. E. LaJohn, K. L. Kirk, and L. A. Cohen, *Biochim. Biophys. Res. Commun.* **75**, 674, 1977.

26. K. Saito, G. A. Digenis, A. A. Hawi, and J. Chaney, *J. Fluorine Chem.* **35**, 663–667, 1987.

27. I. A. McDonald, J. M. Lacoste, P. Bey, M. G. Palfreyman, and M. Zreika, *J. Med. Chem.* **28**, 186–193, 1985.

4

FLUORINATED ACIDS AND ESTERS

α-Fluorocarboxylic acids and esters are of considerable biological interest as enzymatic blocking agents and also as synthetic intermediates. They have been widely used as biological probes.[1]

4.1 SYNTHESIS BY FLUORINATION REACTIONS

4.1.1 Diethylaminosulfur Trifluoride

The reaction of DAST with alcohols to replace hydroxyls by fluorine is a broadly general, very useful reaction.[2] Primary, secondary, and tertiary alcohols react to give high yields of unrearranged products. The reactions are conducted under very mild conditions so that groups like ester and other halogens can also be present. Figure 4.1 is one example.

4.1.2 Hydrogen Fluoride

A general approach to the preparation of 2-bromo-3-fluorocarboxylic acids (2) involves bromofluorination of unsaturated acids (1) in liquid HF, followed

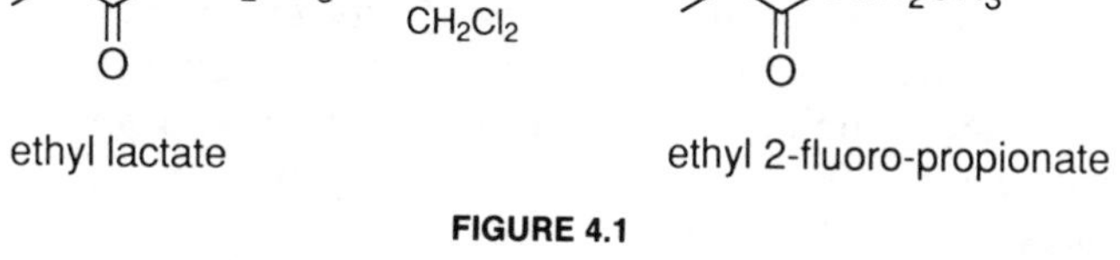

ethyl lactate   ethyl 2-fluoro-propionate

FIGURE 4.1

by addition of *N*-bromoacetamide (NBA).[3] The reaction proceeded smoothly for acrylic, crotonic, and 3-methylcrotonic acids. However, for 2-pentenoic, 2-hexenoic, and 2-heptenoic acids, mixtures of products were obtained.

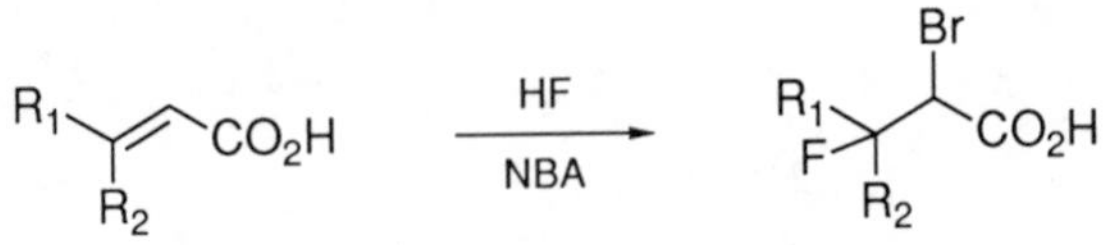

$$R_1 = R_2 = H$$
$$R_1 = CH_3 \quad R_2 = H$$
$$R_1 = R_2 = CH_3$$
$$R_1 = CH_2CH_3 \quad R_2 = H$$
$$R_1 = CH_3(CH_2)_2 \quad R_2 = H$$
$$R_1 = CH_3(CH_2)_3 \quad R_2 = H$$

4.1.3 Hydrogen Fluoride–Pyridine

Ring Opening of Glycidic Esters[4,5]

The opening of glycidic esters (**3**) with hydrogen fluoride–pyridine (70% w/w) followed by oxidation with Jones reagent gives derivatives of fluoropyruvic esters (**5**) in good yields (Fig. 4.2).

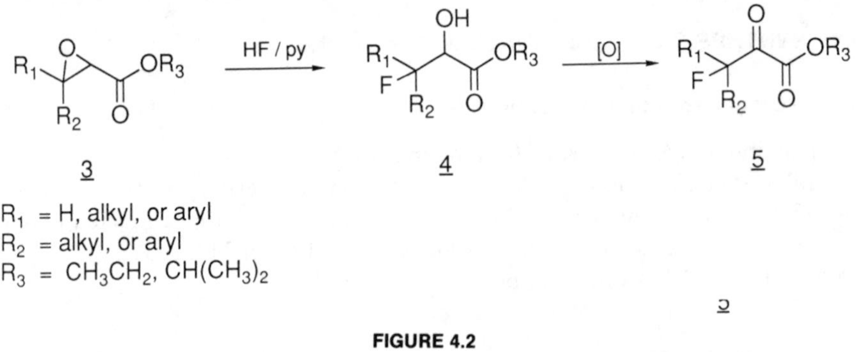

$$R_1 = H, \text{ alkyl, or aryl}$$
$$R_2 = \text{alkyl, or aryl}$$
$$R_3 = CH_3CH_2, CH(CH_3)_2$$

FIGURE 4.2

Deaminative Fluorination[6–10]

Deaminative fluorination, in which the amino group is displaced by fluoride, offers a convenient, simple preparation of α-fluorocarboxylic acids from easily available α-amino acids.[6–8] Reaction of α-amino acids (**6**) in hydrogen fluoride–pyridine with excess sodium nitrite led via in situ diazotization followed by nucleophilic dediazoniation to the formation of α-fluorocarboxylic acids (**7**)[6,7] In the case of some α-amino acids, total or partial rearrangement to β-fluorocarboxylic acids (**8**) is observed due to anchimeric assistance during diazotization reaction.[6c,7,8]

The concentration of hydrogen fluoride–pyridine reagent is critical in determining the ratio of α to β acids. Glycine, alanine, and α-amino butanoic acid give α-fluoro acids exclusively with 70% as well as 48% hydrogen fluoride–pyridine. However, valine and isoleucine undergo partial rearrangement and phenyalanine, tyrosine, and threonine, total rearrangement to β-fluoro acids with the 70% reagent. The rearrangement can be fully suppressed with 48:52 w/w hydrogen fluoride–pyridine for valine, isoleucine, and phenylalanine. For tyrosine and threonine, 15–20% rearrangement to β product is observed even with the reagent of lower hydrogen fluoride concentration.

Both the substitution and rearrangement reactions are found to be stereo-specific by NMR studies and ORD measurements.

Alkyl carbamates (**9**) react with excess $NaNO_2$ in hydrogen fluoride–pyridine at room temperature to form the corresponding fluoroformates (**10**)[6b] The reaction, as in the case of amino acids, proceeds via in situ diazotization followed by dediazoniation.

Aliphatic α-amino acid esters can be converted to α-diazo esters by reaction with isoamyl nitrite and acetic acid.[9] Treatment of these diazo esters (**11**) with hydrogen fluoride–pyridine affords α-fluoro esters (**12**).[6b] In the presence of added halide ions (N-chlorosuccinimide, N-bromosuccinimide, or N-iodosuccinimide), α-fluoro-α-halo esters (**13**) are obtained.

Fluorodeamination of α-amino esters (**14**) with sodium nitrite in hydrogen fluoride–pyridine gives mainly the β-fluoro esters (**15**). The molar ratio of hydrogen fluoride–pyridine is the main parameter that determines the α to β regioselectivity.[10] The higher concentration of hydrogen fluoride favors the β-fluoro ester (**15**), whereas the lower concentration of hydrogen fluoride favors the α- fluoro ester (**16**).

Loss of nitrogen from the diazonium ion (**17**) forms the α-carbocation (**18**), which then rearranges to the more stable tertiary position.

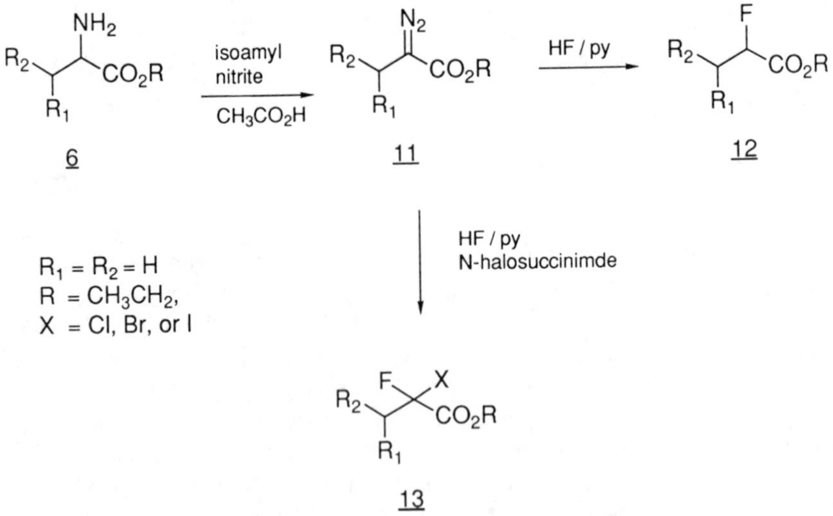

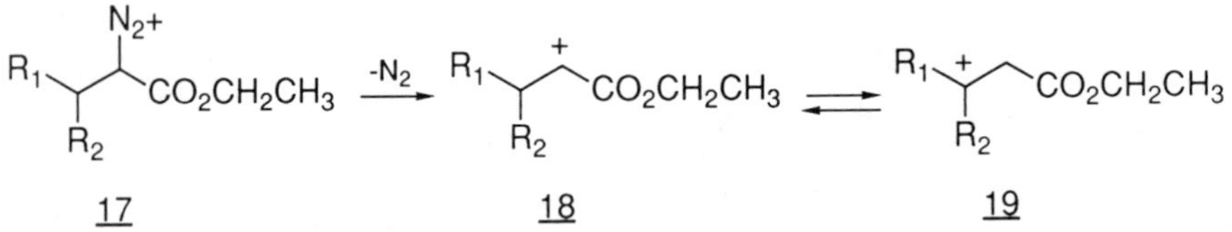

$R_1 = R_2 = H$
$R = CH_3CH_2,$
$X = Cl, Br, or I$

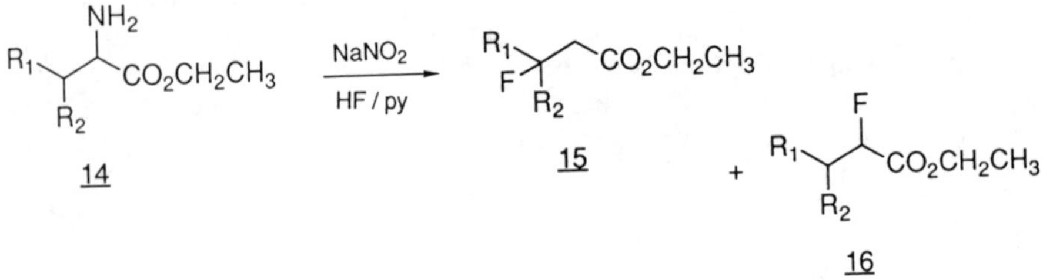

4.1.4   Potassium Fluoride

N-substituted α-halo amides (**20**) undergo halogen exchange with potassium fluoride in bis(2-hydroxyethyl) ether to form the α-fluoro amides (**21**).[11]

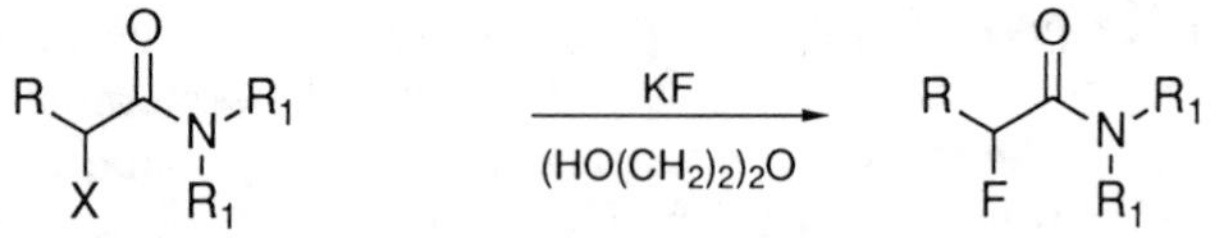

20 X = Cl, Br, OTs
   R_1 = H, CH_3CH_2, anilide,
         or benzoyl anilide

4.2 SYNTHESIS WITH FLUORINATED C$_1$ UNITS

4.2.1 Dichlorofluoromethane

Aminomalonates (**22**) react smoothly with chlorofluorocarbene to form the chlorofluoromethylated adduct (**23**), which can be easily converted to fluoropyruvic acid (**24**) under acidic conditions through hydrolysis and decarboxylation.[12] The reaction probably occurs through the intermediates **25** and **26**, which exist in the enamine and imine equilibrium.

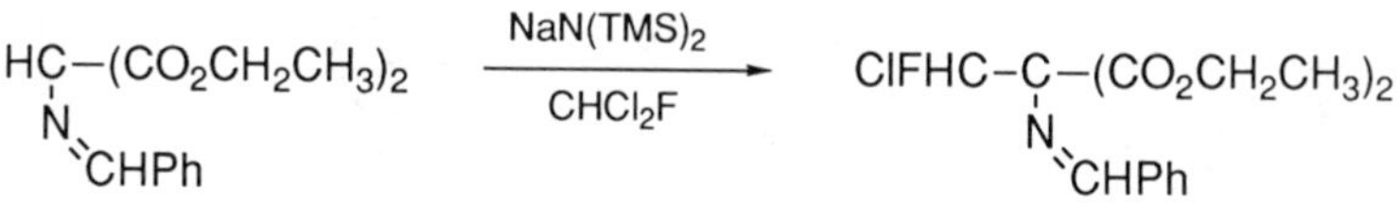

$$HC-(CO_2CH_2CH_3)_2 \xrightarrow[CHCl_2F]{NaN(TMS)_2} ClFHC-C-(CO_2CH_2CH_3)_2$$

         N   N
          `CHPh                                                `CHPh

         22   **23**

$$\xrightarrow[80\ °C]{3N\ HCl}$$

O
‖
F $\diagdown$ CO$_2$H

24

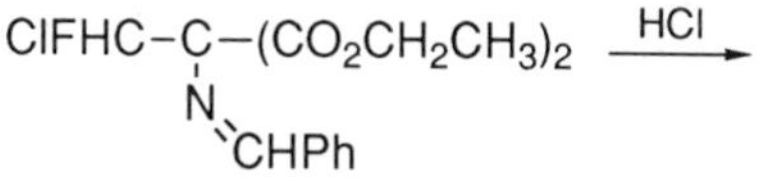

$$ClFHC-C-(CO_2CH_2CH_3)_2 \xrightarrow{HCl}$$

         N
          `CHPh

         23

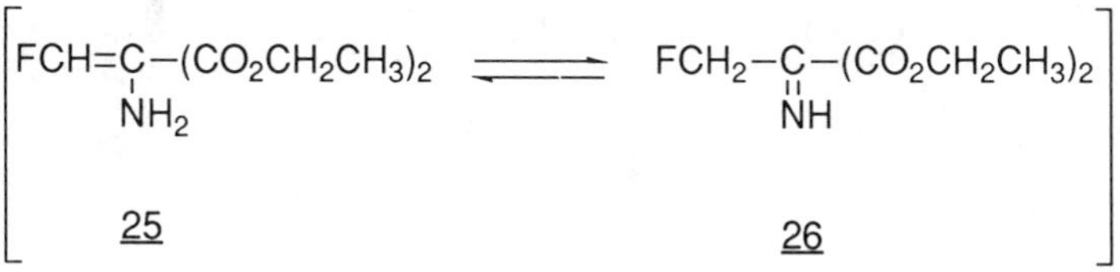

$$\left[\begin{array}{ccc} FCH=C-(CO_2CH_2CH_3)_2 & \rightleftharpoons & FCH_2-C-(CO_2CH_2CH_3)_2 \\ NH_2 & & NH \end{array} \right.$$

         25   **26**

$$\xrightarrow[CO_2]{hydrolysis}$$

O
‖
F $\diagdown$ CO$_2$H

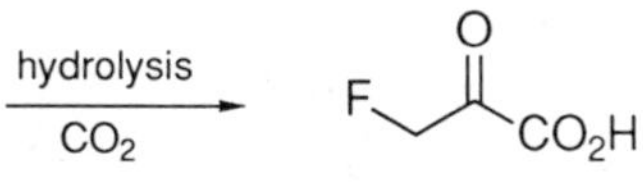

24

4.2.2 Chlorodifluoromethane

Alkylation with chlorodifluoromethane of the sodio derivative of monoalkyl-malonic ester (**27**) affords the difluoromethylated derivative (**28**)[13]. The anion of diphenylacetonitrile (**29**) can be difluoromethylated with CHF_2Cl. Subsequent hydrolysis of **30** gives α-difluoromethyldiphenylacetic acid (**31**). It is suggested that difluoromethylene is the active intermediate in these reactions since CHF_2Cl is shown to be unreactive in SN_2-type displacements.

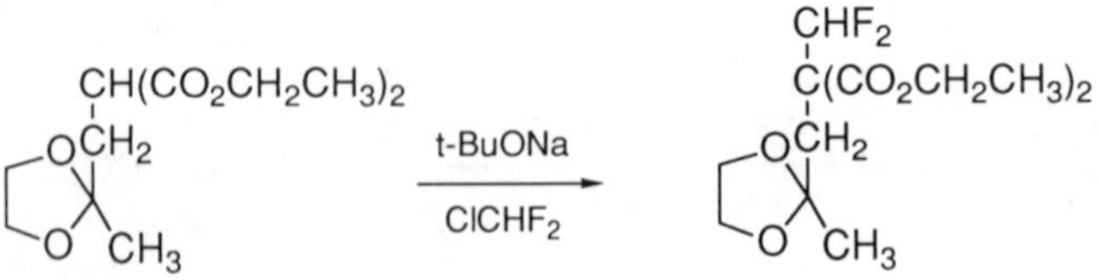

$$PhCH(CO_2CH_2CH_3)_2 \xrightarrow[\text{CICHF}_2]{\text{t-BuONa}} PhC(CHF_2)(CO_2CH_2CH_3)_2$$

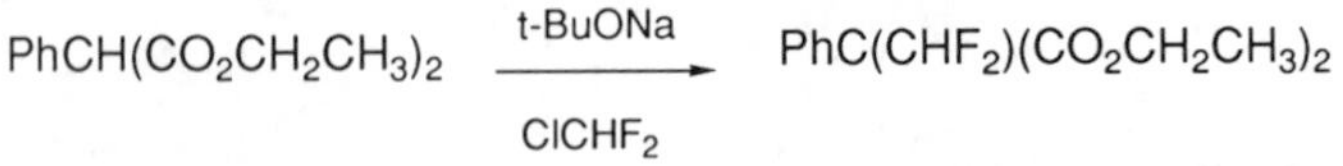

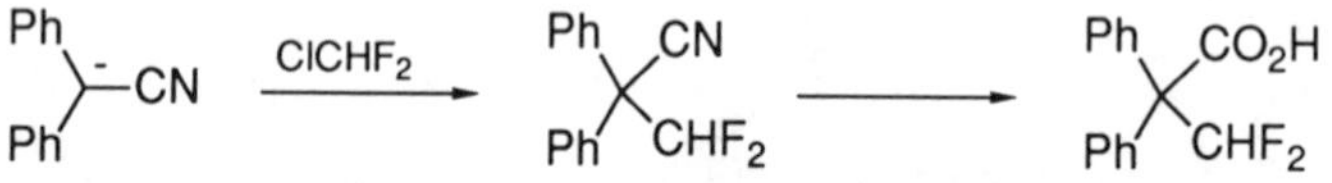

4.3 SYNTHESIS WITH FLUORINATED C₂ UNITS

4.3.1 Ethyl Fluoroacetate

Synthesis of Fluoropyruvic Acid (34)

Fluoropyruvic acid (**34**) can be synthesized by hydrolysis and "ketonic fission" of the diethyl ester of fluorooxaloacetic acid (**33**), which in turn is obtained by Claisen condensation of ethyl fluoroacetate (**32**) and diethyl oxalate in the presence of alcohol-free sodium ethoxide.[14] The enol content of the ester determines the extent of "ketonic fission"; the smaller the amount, the greater the yield of the keto acid.

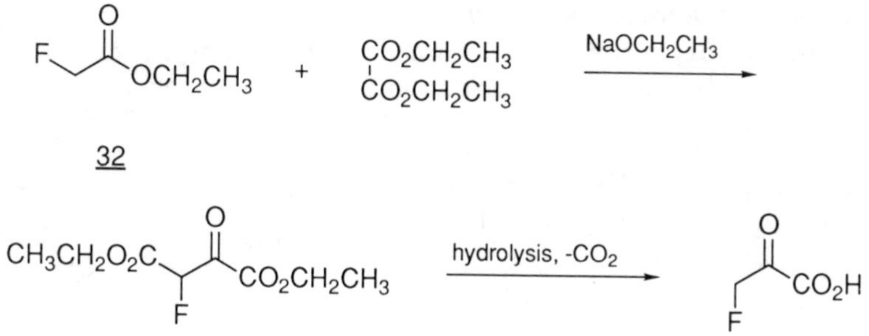

32 33 34

Synthesis of 2-Fluorocitrate Esters

Reformatsky reaction between the diethyl ester of fluorooxaloacetic acid (**33**) the ethyl bromofluoroacetate gives triethyl 2-fluorocitrate (**35**)[31]

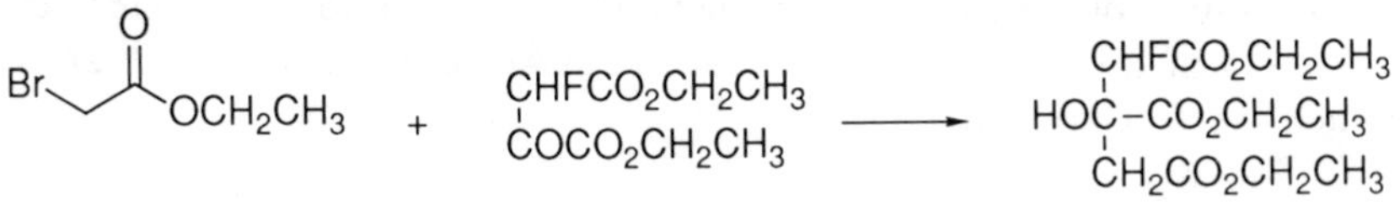

33 35

Condensation of monofluorooxaloacetate (**33**) with malonic acid (**36**) yields the diethyl ester of fluorocitric acid (**37**).[32]

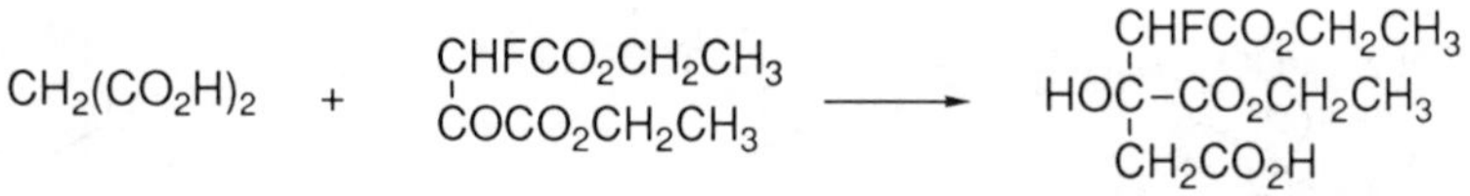

36 33 37

Synthesis of 2-Fluoro-2-alkenoic Acids (38)[15]

Reaction of diethyl sodiofluorooxaloacetate with aldehydes in the presence of aqueous alkali affords 2-fluoro-2-alkenoic acids (**38**).[15]

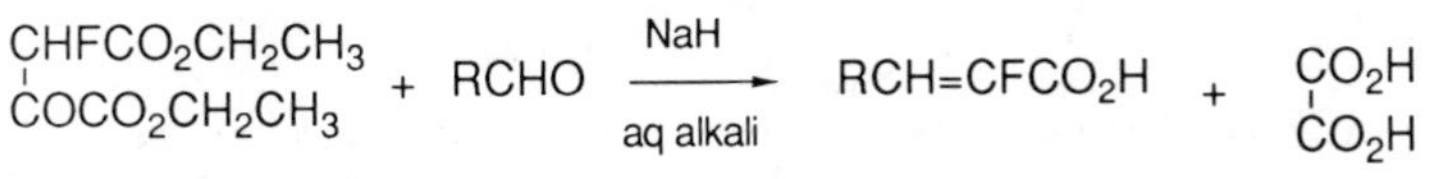

33 38

R = CH_3, $CH_3(CH_2)_2$, $CH_3(CH_2)_5$, $(CH_3)_2CH$, C_6H_5, p-$CH_3C_6H_4$, p-ClC_6H_4, o-$OCOCH_3C_6H_4$, $PhCH_2$, -CO_2Bu, 2- or 3-C_5H_4N, $CH_3CH_2COCHCH_3$

Synthesis of 3-fluorocoumarin (40)[15]

α-Fluoro-2-hydroxycinnamic acid, formed from *o*-acetoxybenzaldehyde (**39**), may be cyclized by hydrochloric acid–acetic acid mixture to 3-fluorocoumarin (**40**).[15]

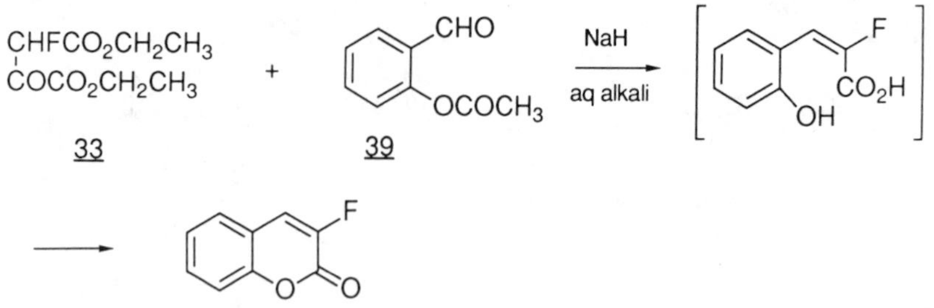
CHFCO2CH2CH3
COCO2CH2CH3
33
+
CHO
OCOCH3
39
NaH
aq alkali
F
CO2H
OH
F
O
O

40

Synthesis of 4-fluoro-2,6-dimethyl resorcinol (43)[15]

Condensation of 2-methyl-3-oxopentanal (**41**) gives the normal product, ethyl 2-fluoro-4-methyl-5-oxo-hept-2-enoate (**42**), as well as the cyclized 4-fluoro-2,6-dimethyl resorcinol (**43**).[15]

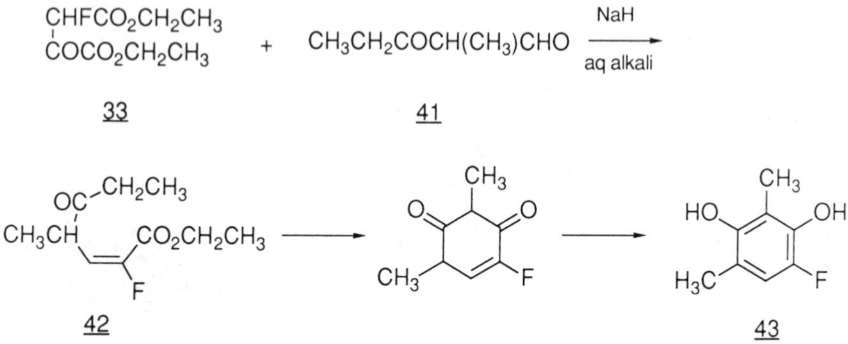
CHFCO2CH2CH3
COCO2CH2CH3
33
+
CH3CH2COCH(CH3)CHO
41
NaH
aq alkali
OC
CH2CH3
CH3CH
CO2CH2CH3
F
42
CH3
O
O
CH3
F
CH3
HO
OH
H3C
F
43

4.3.2 Ethyl Bromodifluoroacetate

Commercially available ethyl bromodifluoroacetate (**44**) undergoes Reformatsky reaction with aldehydes and ketones to form 2,2-difluoro-3-hydroxy esters (**45**).[16]

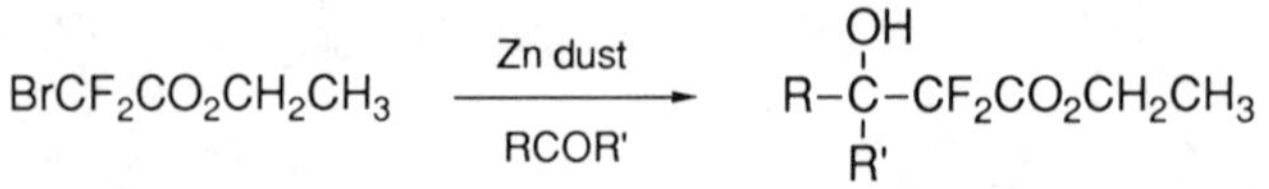
BrCF2CO2CH2CH3
Zn dust
RCOR'
OH
R–C–CF2CO2CH2CH3
R'

44

45

R = Ph, R' = H

R, R' = c-C_5H_{10}

R = C_5H_{11}, R' = H

The synthesis of difluorooxetane–oxane ring system (**47**) required for the construction of the fluoro analog of thromboxane A$_2$ makes use of the Reformatsky reaction of ethyl bromofluoroacetate with the aldehyde (**46**).[33]

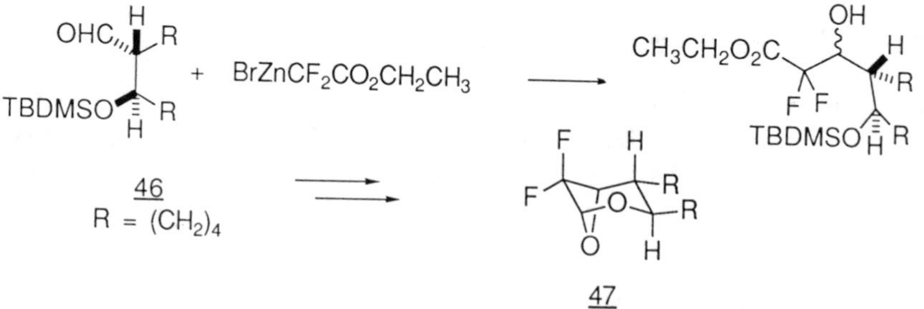

4.3.3   Methyl Iododifluoroacetate

Reactions of the methyl iododifluoroacetate (**48**)–copper system with various organic halides in aprotic solvent yield 2,2-difluoro esters (**49**).[17] With alkenyl halides, the reaction is stereospecific with retention of double bond geometry.

$$ICF_2CO_2CH_3 \; + \; Cu \; + \; RX \quad\longrightarrow\quad RCF_2CO_2CH_3$$

$$\underline{48} \hspace{7cm} \underline{49}$$

However, alkynyl iodide produces a complex mixture of products. The reaction proceeds well with alkyl and aryl halides as well as with allylic bromides.

4.4   SYNTHESIS WITH FLUORINATED C$_3$ UNITS

4.4.1   Fluoropyruvic Acid

An efficient high-yield synthesis of fluoromethyl glyoxal (**52**) from fluoropyruvic acid (**50**) is outlined in Fig. 4.3.[18, 19] The reduction of the ketal ester **51** with LAH

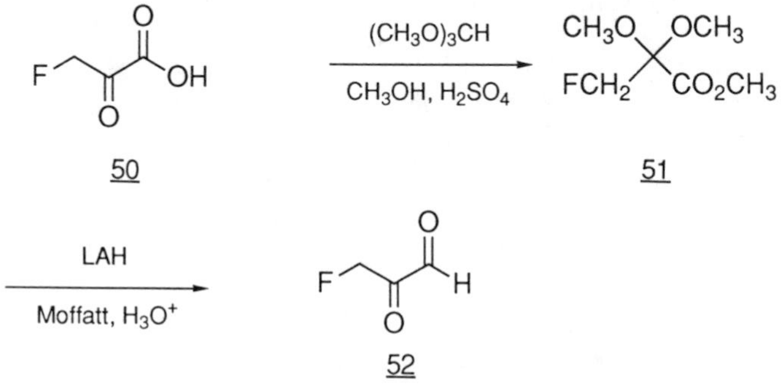

FIGURE 4.3

is particularly useful for the introduction of deuterium or tritium into the molecule via labeled reducing agent, the label ultimately resulting in the aldehydic position.

4.4.2 Trifluoropropene

Hydrocarboxylation and Hydroesterification
The hydroesterification and hydrocarboxylation of trifluoropropene with palladium complexes with phosphine ligands as catalysts at 100–125°C and 100–120 atm of carbon monoxide give the corresponding esters (**54**) and acids (**55**).[20]

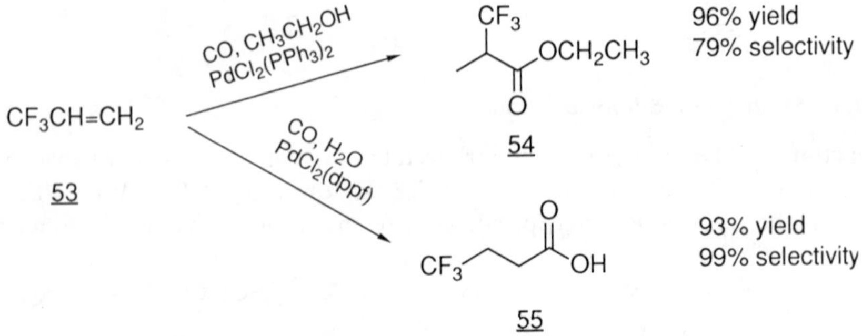

dppf 1,1'-bis (diphenylphospheno)ferrocene

Under optimum conditions, very high yields of the products with very high regioselectivities are obtained. It is suggested that the hydrocarboxylation may involve hydroxycarbonyl palladium (π) intermediates, whereas the hydroesterification may proceed via alkyl palladium (π) and acyl palladium (π) intermediates.

Bromination and Carboxylation
The bromination of trifluoropropene (**53**) on irradiation followed by dehydrobomination with potassium hydroxide gives 2-bromotrifluoropropene (**56**). On carboxylation of 2-bromotrifluoropropene (**56**) with palladium catalysts [$PdCl_2(PPh_3)_2$ or $PdCl_2(dppf)$], α-trifluoromethyl acrylic acid (**57**) is obtained.

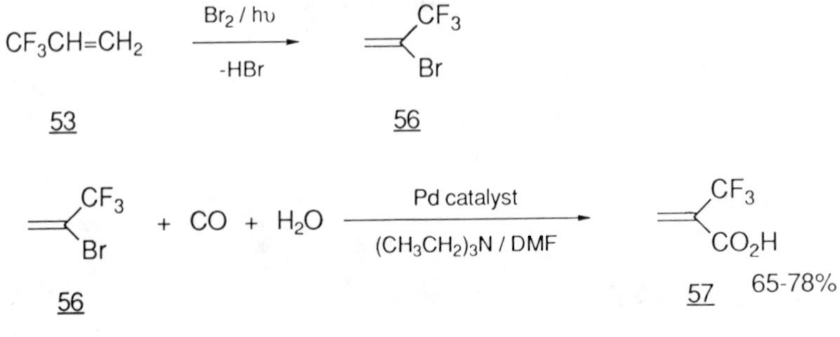

4.5   SYNTHESIS WITH FLUORINATED C₄ UNITS

4.5.1   1-Fluoro-3-Bromo and 1-Fluoro-4-Bromo-butane

Syntheses of 6-fluorocaproic acid (**59**) and 5-fluoro-3-methylvaleric acid (**60**) starting from diethyl malonate (**58**) and fluorobromobutanes are represented in Fig. 4.4.[21]

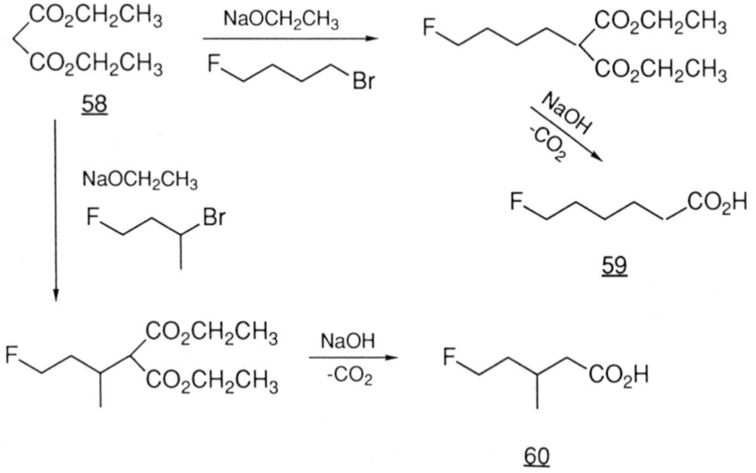

FIGURE 4.4

4.5.2   Perfluoro-2-methylpropene

Trifluoromethyl malonic ester (**62**) is a useful building block for the synthesis of various trifluoromethylated aliphatic or heterocyclic compounds. It can be synthesized from perfluoro-2-methylpropene (**61**).[22] An improved method for its synthesis is via the enolate anion (**63**), which can undergo alkylation as well as Michael addition to vinyl ketones (Fig. 4.5).

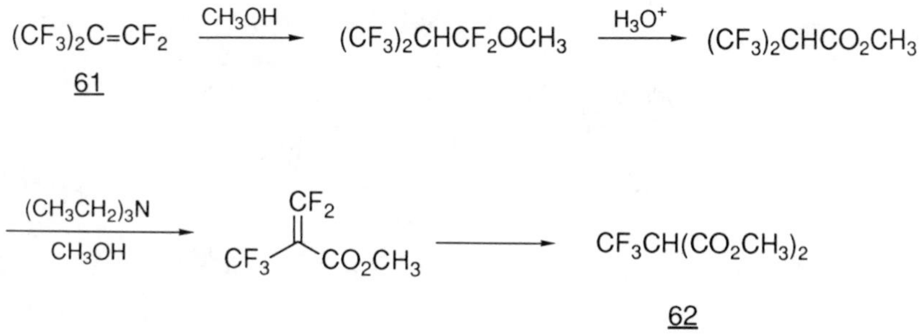

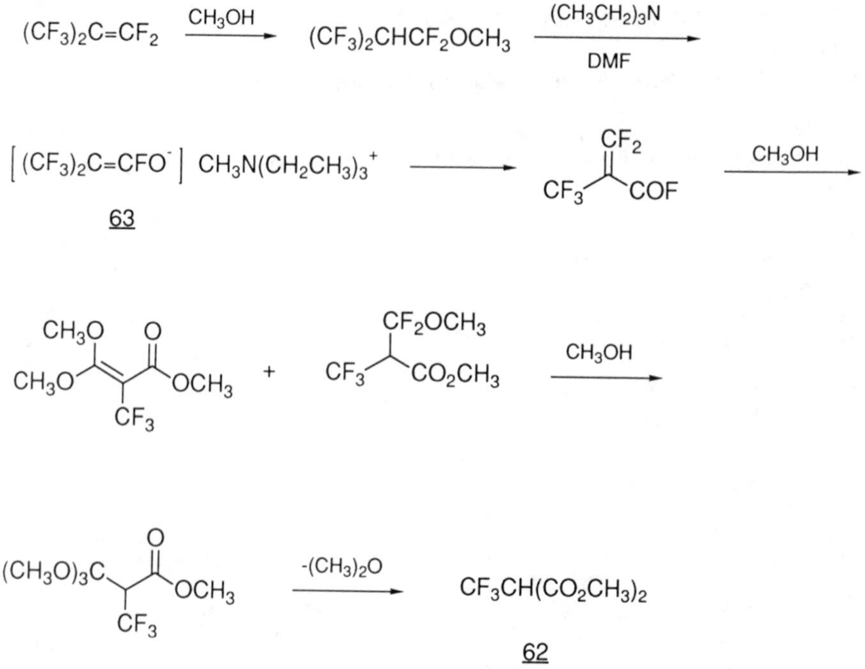

$$(CF_3)_2C=CF_2 \xrightarrow{CH_3OH} (CF_3)_2CHCF_2OCH_3 \xrightarrow[DMF]{(CH_3CH_2)_3N}$$

$$\left[(CF_3)_2C=CFO^- \right] CH_3N(CH_2CH_3)_3{}^+ \longrightarrow \underset{CF_3}{\overset{CF_2}{\diagdown}}{C}{\diagup}{COF} \xrightarrow{CH_3OH}$$

63

CH₃O–C(=)–C(CF₃)(C(=O)OCH₃) ... $+$ CF₃–C(CF₂OCH₃)(CO₂CH₃) $\xrightarrow{CH_3OH}$

$$(CH_3O)_3C-\underset{CF_3}{C}-\overset{O}{C}-OCH_3 \xrightarrow{-(CH_3)_2O} CF_3CH(CO_2CH_3)_2$$

62

FIGURE 4.5

4.5.3 4,4,4-Trifluorocrotonate

The addition of unsymmetrical reagents to the double bond of trifluorocrotonate yields β-substituted derivatives.[23] The trifluoromethyl group may be expected to direct the entering anion to the α-carbon through inductive or hyperconjugative effects as in **68**, whereas the carboxethoxy group directs the nucleophile to the β

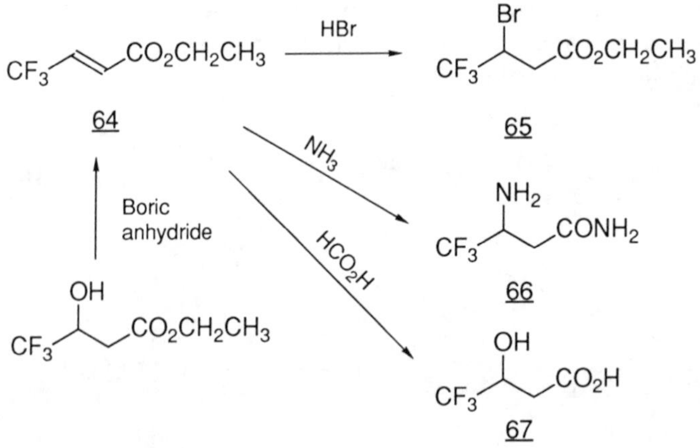

CF₃–CH=CH–CO₂CH₂CH₃ $\xrightarrow{HBr}$ CF₃–CH(Br)–CH₂–CO₂CH₂CH₃

64   **65**

CF₃–CH(OH)–CH₂–CO₂CH₂CH₃   (via Boric anhydride)

NH₃ → CF₃–CH(NH₂)–CH₂–CONH₂ **66**

HCO₂H → CF₃–CH(OH)–CH₂–CO₂H **67**

$$\bar{F} \quad CF_2=CH\overset{+}{C}HCO_2CH_2CH_3 \qquad\qquad CF_3\overset{+}{C}H\overset{..}{=}CH\overset{..}{=}\overset{\overset{O^-}{|}}{C}-OCH_2CH_3$$

68   69

position as in **69**. The products indicate that the resonance of carboalkoxy group with the double bond plays the important role in directing the addition to the double bond. The CF_3 group is unable to reverse the normal mode of addition.

4.5.4   4,4,4-Trifluoro-3-oxobutanoate

Reduction of 4,4,4-trifluoro-3-oxobutanoate (**70**) by fermenting baker's yeast (Saccharomyces cervisiae) affords the hydroxy ester (**71**).[34]

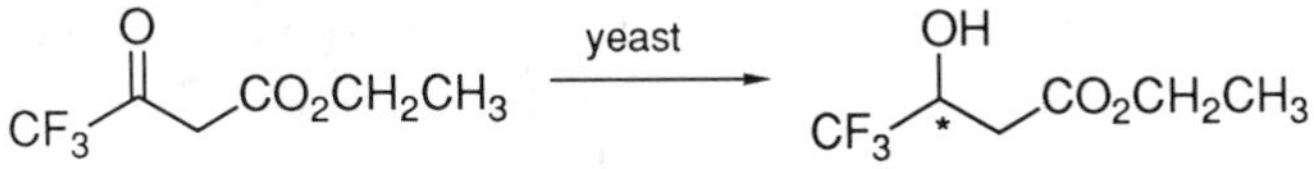

4.6   SYNTHESIS WITH LARGER FLUORINATED SYNTHONS

4.6.1   3-Methyl-4,4,4-trifluorobutyrate

α-Hydroxy-β-trifluoromethyl esters (**73, 74**) are synthesized with excellent stereoselectivity by reaction of the enolate of 3-methyl-4,4,4-trifluorobutyrate (**72**) with MoO_5-pyridine–hexamethylphosphoramide complex (MoOPH).[24] The

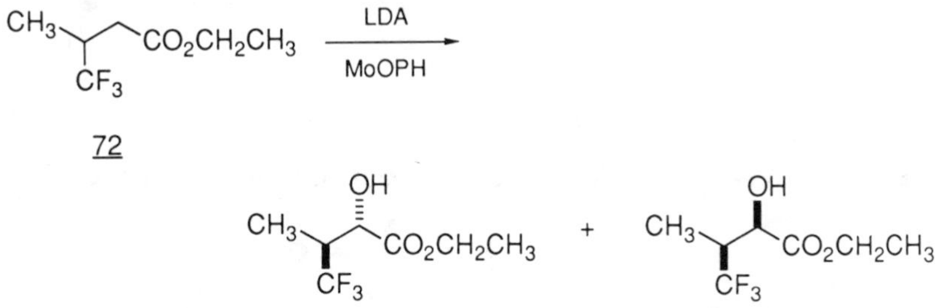

high stereoselectivity is achieved possibly due to the electronic effects of the trifluoromethyl group. The lithium atom may be chelated to trifluoromethyl group. The bulky electrophile, MoOPH, attacks on the less hindered side.

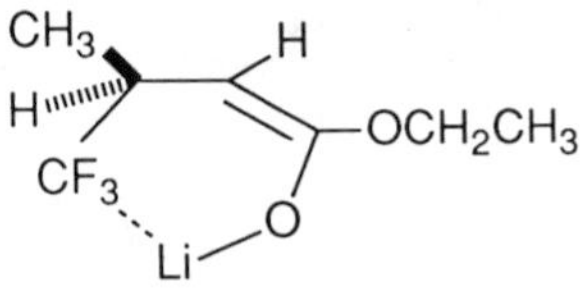

The trifluoromethyl group also effects selective reduction of keto ester (**75**).

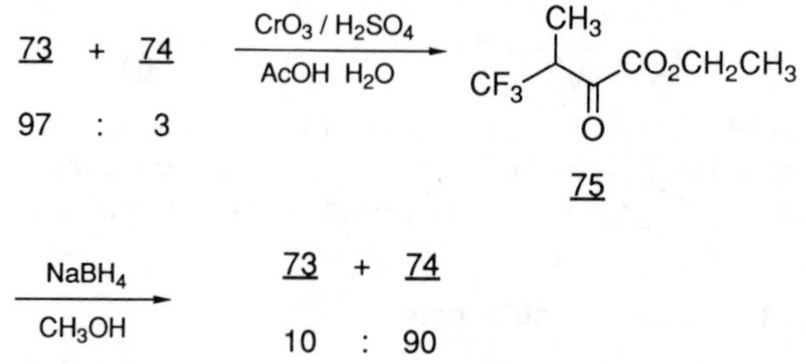

$$73 + 74 \quad \xrightarrow[\text{AcOH } H_2O]{\text{CrO}_3 / H_2SO_4}$$

97 : 3

75

$$\xrightarrow[\text{CH}_3\text{OH}]{\text{NaBH}_4} \quad \begin{array}{ccc} 73 & + & 74 \\ 10 & : & 90 \end{array}$$

4.6.2   6-Benzyloxy-3-Trifluoromethyl-Hex-4-Enoic Acid

Iodolactonization of carboxylic acids (**76**) with iodine–potassium iodide followed by treatment with base gives the hydroxylactones **77** and **78** (X = OH).[30] The iodolactones **77** and **78** (X = I) react with potassium carbonate in methanol to

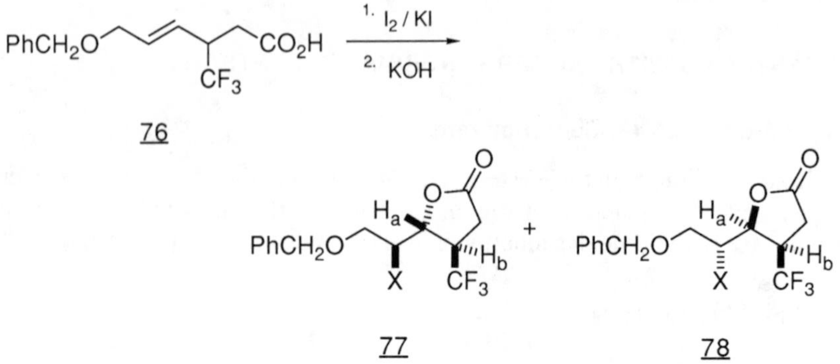

76

77   **78**

X = I or OH

afford the epoxy esters, which could then be debenzylated and hydrolyzed to give **79** and **80**.

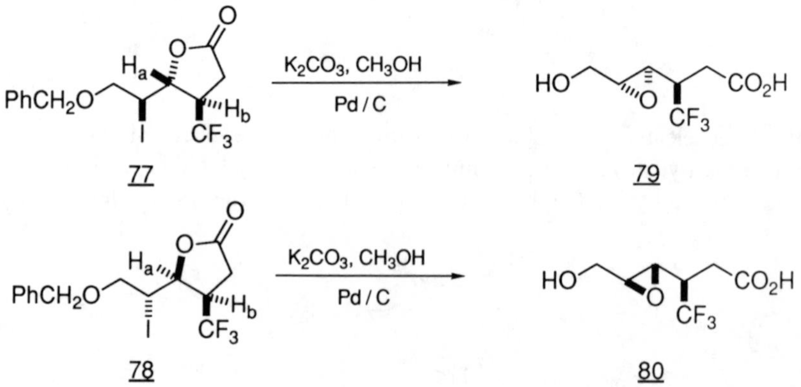

77   K$_2$CO$_3$, CH$_3$OH   Pd / C   **79**

78   K$_2$CO$_3$, CH$_3$OH   Pd / C   **80**

4.6.3 Pentafluorostyrene

Hydrocarboxylation of pentafluorostyrene (**81**) catalyzed by $PdCl_2(dppf)$ gives 3-pentafluorophenylpropionic acid (**82**) in 93% yield and 99% regioselectivity. Hydroesterification of pentafluorostyrene (**81**) catalyzed by $PdCl_2(PPh_3)_2$ affords 2-pentafluorophenylpropionate (**83**) in 89% yield and 95% selectivity.[20]

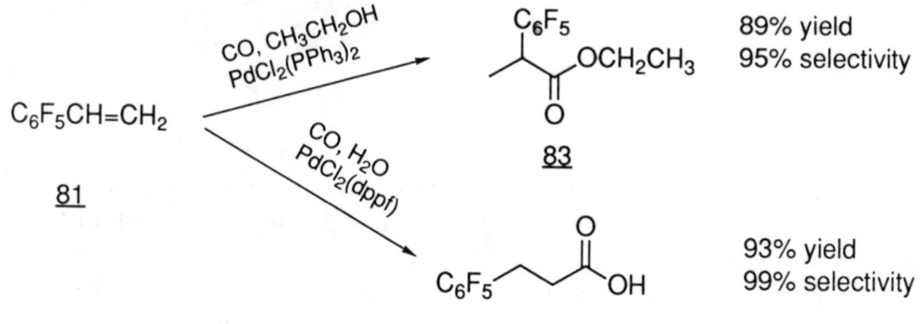

4.6.4 ω-Fluoro-Alcohols, Acids & Nitriles

Synthesis of a number of ω-fluorocarboxylic acids has been reported.[25] Some of the lower members of these aliphatic acids are obtained by oxidation of alcohols (**84**) or hydrolysis of nitriles (**86**). Methyl 18-fluorostearate (**90**) is synthesized by

$$F(CH_2)_nOH \xrightarrow[\text{or } CrO_3]{K_2Cr_2O_7} F(CH_2)_{n-1}CO_2H$$

84   **85**

n = 3, 4, 10

$$F(CH_2)_nCN \xrightarrow[CH_3OH]{HCl} F(CH_2)_nCO_2CH_3 \xrightarrow{5\% \ H_2SO_4} F(CH_2)_nCO_2H$$

86   **87**

n = 4, 5, 6

unsymmetrical coupling of 10-fluorodecanoic acid (**88**) and methyl hydrogen sebacate (**89**). The isomeric 9(10)-fluorosteric acid (**92**) is constructed by addition

$$F(CH_2)_9CO_2H \ + \ HO_2C(CH_2)_8CO_2CH_3 \xrightarrow{\text{anodic synthesis}}$$

88   **89**

$$F(CH_2)_{17}CO_2CH_3 \ + \ 2 \ CO_2 \ + \ H_2 \xrightarrow{\text{hydrolysis}} F(CH_2)_{17}CO_2H$$

90

of anhydrous hydrogen fluoride to ethyl oleate (**91**) followed by hydrolysis.

$$CH_3(CH_2)_7CH=CH(CH_2)_7CO_2H \quad \xrightarrow[\text{hydrolysis}]{\text{HF}} \quad CH_3(CH_2)_nCHF(CH_2)_mCO_2H$$

$$n = 8 \text{ or } 7$$
$$m = 7 \text{ or } 8$$

91   **92**

4.7 ENZYMATIC METHODS

Microbial hydrolysis of 2-fluoro-2-substituted malonic acid diesters (**93**) with several kinds of esterases and cellulases gives the optically active ($+$)- or ($-$)-2-fluoro-2-substituted malonic acid monoesters (**94**), which are very useful as practical chiral synthons.[26] Asymmetric hydrolysis based on the utility of the

$$RCF(CO_2R')_2 \quad \xrightarrow[\text{or cellulases}]{\text{esterases}} \quad \begin{array}{c} CO_2H \\ RCF \\ CO_2R' \end{array}$$

              93   **94**

$$R = CH_3, \ H, \ CH_3CH_2CH_2,$$
$$CH_3CH_2, \ CH_3CH_2CH_2CH_2$$
$$R' = CH_3, \ CH_3CH_2$$

"mimic effect" of the fluorine atom is the most convenient one-step process for the synthesis of monofluorinated chiral units. Use of the fluorine atom is advantageous as no racemization is observed under the reaction conditions.

4.8 BIOLOGICAL ACTIVITY

Much research has recently been devoted to the biological effects of fluorinated acids and esters. It has been shown that fluoroacetic acid is metabolized in the organism to fluorocitric acid, which eventually blocks the tricarboxylic acid cycle.[14b] The toxicity of diethyl sodiofluorooxaloacetate and fluoropyruvic acid has been tested on rats and mice. The former is practically nontoxic (LD_{50} 750 mg/kg), whereas the latter as well as its methyl ester are lethal at doses of around 80 mg/kg.

A pronounced alteration in toxicity of members of the homologous series of ω-fluorocarboxylic esters has been reported.[25] Those esters whose acid moieties contain an even number of carbon atoms are toxic, whereas those that are odd-numbered are innocuous. The toxic properties have been correlated with known metabolic mechanisms. Both methyl 18-fluorostearate and the free acid are found to be toxic to mice, the ester LD_{50} 18 mg/kg and the acid, 5.7 mg/kg. The isomeric acid, 9(10)-fluorostearic acid, has a very low toxicity (LD_{50}

$> 400\,mg/kg$). The marked difference between the two isomeric fluorostearic acids emphasizes the high specificity of the ω-fluorine atom for pharmacological activity. The enzyme butyryl-CoA dehydrogenase catalyzes the elimination of hydrogen fluoride by α-proton abstraction from 3-fluoropropionyl-CoA and (3,3-difluorobutyryl)pantetheine.[29]

Fluoromethyl glyoxal, FCH_2COCHO (**52**), is accepted as a substrate for the enzyme glyoxalase I.[18,19] The deuterated compound (**95**) has been of considerable importance as a mechanistic probe for glyoxalase I. Fluoromethyl glyoxal undergoes rapid conversion to fluorolactate (**96**) under basic conditions. This occurs via an internal Cannizzaro reaction (1,2-hydride shift).[18] This is in marked

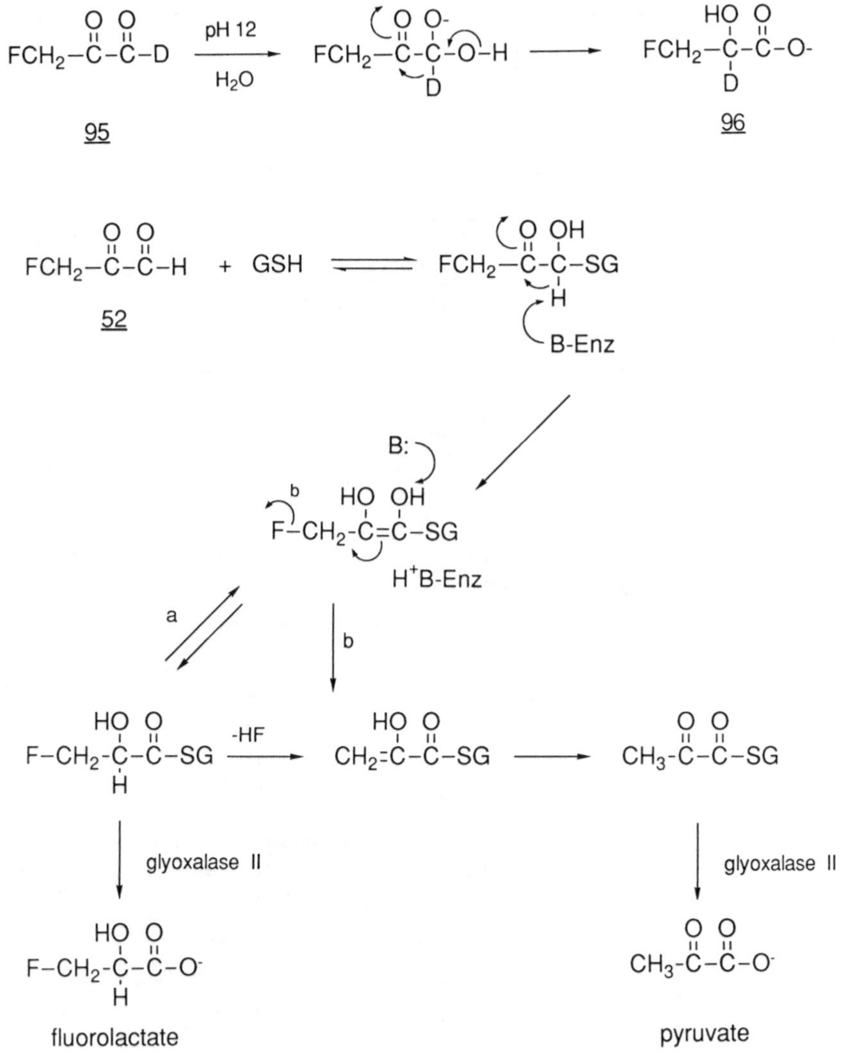

FIGURE 4.6

contrast with the enzymatic conversion of fluoromethyl glyoxal (52) by glyoxalases I and II and glutathione (GSH), which proceeds via a rapid proton
transfer–enediol mechanism with retention of the aldehydic proton (or deuterium) and partial retention of the fluoride (Fig. 4.6).[19] Fluoromethyl glyoxal
constitutes a unique case where the typical fluoride elimination path is partially
suppressed.

Citrate ($-O_2CCH_2C(OH)(CO_2-)CH_2CO_2-$) is a central component in the
primary metabolism of prokaryotes and eukaryotes.[27] The fluorinated citrate
analogs 97 and 98 have received much attention, with the better known being 2-
fluorocitrate (97). The. The (−)-*erythro*- and (+)-*erythro*-2-fluorocitrates are

F–CHCO$_2$-   CH$_2$CO$_2$-
HO–C–CO$_2$-   F–C–CO$_2$-
CH$_2$CO$_2$-   CH$_2$CO$_2$-

<u>97</u>   <u>98</u>

substrates for the cytoplasmic ATP citrate lyase.[35] The (−)-*erythro*-2-
fluorocitrate is the toxic species generated by "lethal synthesis" and is responsible
for the toxicity of fluoroacetate.[31] It is observed that malate synthase, in contrast
to citrate synthase, processes one of is two substrates achirally.[35]

3-Fluoro-3-deoxycitrate (98) interacts with the four citrate-processing enzymes; it is a competitive inhibitor of citrate synthase, a substrate for ATP
splitting only with ATP citrate lyase, an efficient inactivator of klebsiella
aerogenes citrate lyase, and a substrate for hydrogen fluoride elimination with
aconitase. The fate of 3-fluorocitrate with each enzyme is uniquely related to the
mechanism of action.[27]

3-Fluoropyruvate, $FCH_2COCO_2^-$, has been very successfully employed in
determining the stereochemistry of the transcarboxylase-catalyzed carboxylation of 3-fluoropyruvate to form 3-fluorooxaloacetate (Fig 4.7).[28] The

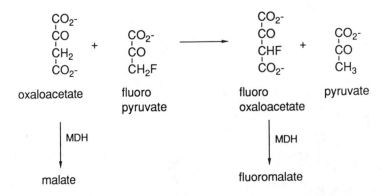

FIGURE 4.7

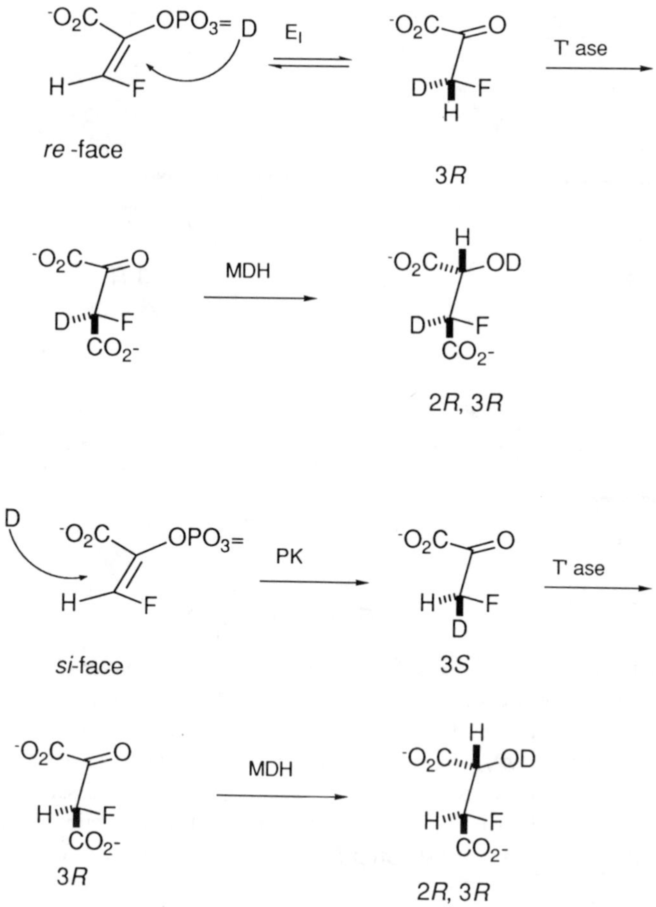

FIGURE 4.8

stereochemistry of the carboxylation, as determined using ^{19}FNMR, is found to require retention of absolute configuration. Transcarboxylase coupled to malate dehydrogenase has been used to analyze samples of chiral fluoropyruvate obtained by dephosphorylation of (Z)-fluorophosphoenolpyruvate in D_2O with either pyruvate kinase or enzyme I from the *E. coli* sugar transport systems. The fluoroenolpyruvate is reductively deuterated from opposite faces by these two enzymes (Fig. 4.8).[28] The transcarboxylase is specific for one of the two prochiral hydrogens in fluoropyruvate.[36] The use of the fluoromethyl group greatly simplifies the stereochemical analysis relative to the classic $^1H^2H^3H$ approach.

Stereochemical and product analysis studies on the hydration pathway, such as those of monofluorofumarate and 2,3-difluorofumarate by the enzyme fumarase, indicate that both substrates are processed chirally by the enzyme (Fig. 4.9).[37] A common active-site enzyme substrate geometry is observed.

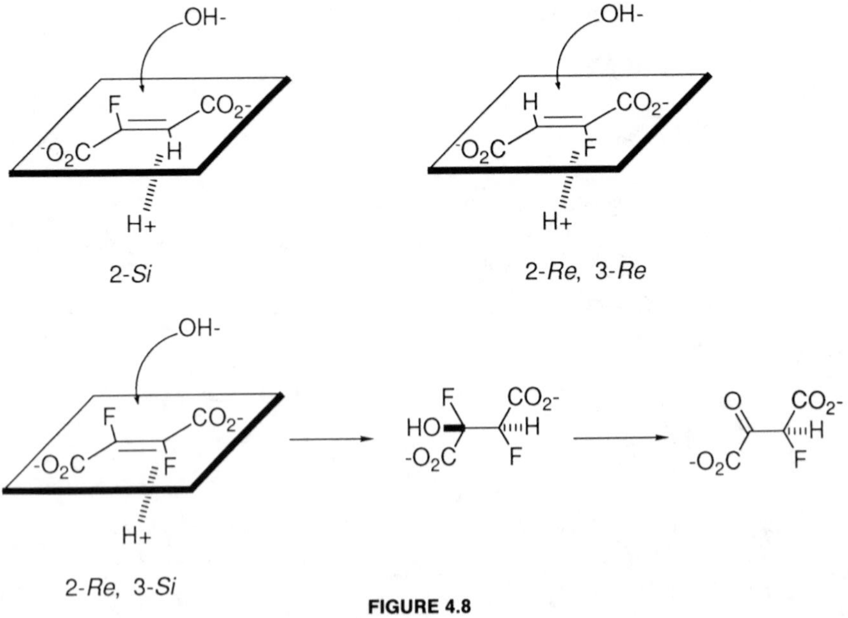

2-Si

2-Re, 3-Re

2-Re, 3-Si

FIGURE 4.8

The hydroxylactones **77** and **78** (X = OH) and the epoxyacids **79** and **80** are considered to be useful fragments in the preparation of trifluomethylated analogs of polyhydroxyl compounds processing important biological activities.[30]

Thromboxane A_2 (**99**) is a powerful vasoconstricting and platelet-aggregating principle derived from arachidonic acid. It has not been accessible by total or partial synthesis due to the unstable nature of the bicyclic oxetane–oxene structure. If fluorines are substituted for hydrogen in the oxetane ring α to the

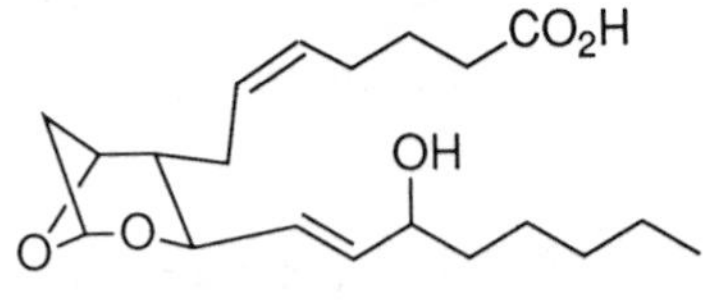

99

acetatic linkage, as in **47**, such structure may exhibit greater resistance to acid hydrolysis.[33] Also, the characteristic biological properties of thromboxane A_2 may be preserved.

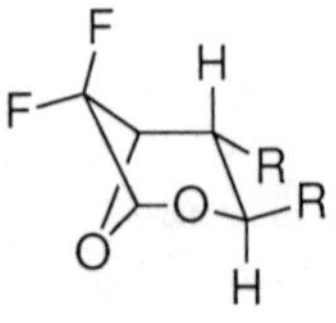

47

REFERENCES

1. S. E. Rokita, P. A. Srere, and C. T. Walsh, *Biochemistry* **21**, 3765–3774, 1982.
2. W. J. Middleton, *J. Org. Chem.* **40**, 574–578, 1975.
3. H. Gershon and M. W. McNeil, *J. Med. Chem.* **16**, 1407–1409, 1973.
4. A. Ourari, R. Condom, and R. Guedj, *Can. J. Chem.* **60**, 2707–2710, 1982.
5. R. Guedj, A. I. Ayi, and M. Remli, *Ann. Chim. Fr.* **9**, 691–694, 1984.
6. (a) G. A. Olah and J. T. Welch, *Synthesis* **1974**, 652; (b) G. A. Olah, J. T. Welch, Y. D. Vankar, M. Nojima, I. Kerekes, and J. A. Olah, *J. Org. Chem.* **44**, 3872–3881, 1979; (c) G. A. Olah, G. K. Surya Prakash, and Y. L. Chao, *Helv. Chim. Acta.* **64**, 2528–2530, 1981.
7. V. Mitteilung, R. Keck, and J. Retey, *Helv. Chim. Acta.* **63**, 769–772, 1980.
8. (a) J. Barber, R. Keck, and J. Retey, *Tetrahedron Lett.* **23**, 1549–1552, 1982; (b) F. Faustini, S. De Munari, A. Panzeri, V. Villa, and C. A. Gandolfi, *Tetrahedron Lett.* **22**, 4533–4536, 1981.
9. N. Takamura and T. Mizoguchi, *Tetrahedron Lett.* **1971**, 4495–4498.
10. S. Hamman and C. G. Beguin, *Tetrahedron Lett.* **24**, 57–60, 1983.
11. I. Shahak and E. D. Bergmann, *J. Chem. Soc. (C)* **1967**, 319–320.
12. T. Tsushima and K. Kawada, *Tetrahedron Lett.* **26**, 2445–2448, 1985.
13. T. Y. Shen, S. Lucas, and L. H. Sarett, *Tetrahedron Lett.* **1961**, 43–47.
14. (a) P. V. Nair and H. Busch, *J. Org. Chem.* **23**, 137–139, 1958; (b) I. Blank, J. Mager, and E. D. Bergmann, *J. Chem. Soc.* **1955**, 2190–2193.
15. E. D. Bergmann and I. Shahak, *J. Chem. Soc.* **1961**, 4033–4038.
16. E. A. Hallinan and J. Fried, *Tetrahedron Lett.* **25**, 2301–2302, 1984.
17. T. Taguchi, O. Kitagawa, T. Morikawa, T. Nishiwaki, H. Uehara, H. Endo, and Y. Kobayashi, *Tetrahedron Lett.* **27**, 6103–6106, 1986.
18. R. V. J. Chari and J. W. Kozarich, *J. Org. Chem.* **47**, 2355–2358, 1982.
19. J.W. Kozarich, R. V. J. Chari, J. C. Wu, and T. L. Lawrence, *J. Am. Chem. Soc.* **103**, 4593–4595, 1981.
20. I. Ojima, *Third Regular Meeting of Soviet-Japanese Fluorine Chemists*, Tokyo, 1983, pp. 81–100.
21. M. Hudlicky, E. Kraus, J. Korbl, and M. Cech, *Collect. Czech. Chem. Commun.* **34**, 833–842, 1969.
22. N. Ishikawa and T. Yokozawa, *Bull. Chem. Soc. Jpn.* **56**, 724–726, 1983.
23. H. M. Walborsky and M. Schwarz, *J. Am. Chem. Soc.* **75**, 3241–3243, 1953.
24. Y. Morizawa, A. Yasuda, and K. Uchida, *Tetrahedron Lett.* **27**, 1833–1836, 1986.
25. F. L. M. Pattison, J. B. Stothers, and R. G. Woolford, *J. Am. Chem. Soc.* **78**, 2255–2259, 1956.
26. T. Kitazume, T. Sato, and N. Ishikawa, *Chem. Lett.* **1984**, 1811–1814.
27. S. E. Rokita, P. A. Srere, and C. T. Walsh, *Biochemistry* **21**, 3765–3774, 1982.
28. H. Hoving, B. Crysell, and P. F. Leadlay, *Biochemistry* **24**, 6163–6169, 1985.
29. G. Fendrich and R. H. Abeles, *Biochemistry* **21**, 6685–6695, 1982.
30. Y. Hanzawa, K. Kawagoe, K. Kawada, and Y. Kobayashi, *Chem. Pharm. Bull.* **33**, 2579–2581, 1985.

31. D. E. A. Rivett, *J. Chem. Soc.* **1953**, 3710–3711.

32. R. J. Dummel and E. Kun, *J. Biol. Chem.* **244**, 2966–2969, 1969.

33. J. Fried, E. A. Hallinan, and M. J. Szwedo, Jr., *J. Am. Chem. Soc.* **106**, 3871–3872, 1984.

34. D. Seebach, P. Renaud, W. E. Schweizer, M. F. Züger, and M. Brienne, *Helv. Chim Acta* **67**, 1843–1853, 1984.

35. M. A. Marletta, P. A. Srere, and C. Walsh, *Biochemistry* **20**, 3719–3723, 1981.

36. J. A. Goldstein, Y. Cheung, M. A. Marletta, and C. Walsh, *Biochemistry* **17**, 5567–5575, 1978.

37. M. Marletta, Y. Cheung, and C. Walsh, *Biochemistry* **21**, 2637–2644, 1982.

FLUORINATED ALCOHOLS

Fluorinated alcohols have been used as probes of enzymatic reactions by virtue of their ^{19}F nuclear magnetic resonance spectra.

5.1 CHEMICAL SYNTHESIS

5.1.1 1-Phenylsulfonyl-3,3,3-trifluoropropene

1-Phenylsulfonyl-3,3,3-trifluoropropene (**1**) reacts with active methylene substances to give Michael-type addition compounds (**2**).[1] Desulfurization affords the carbinol (**3**).

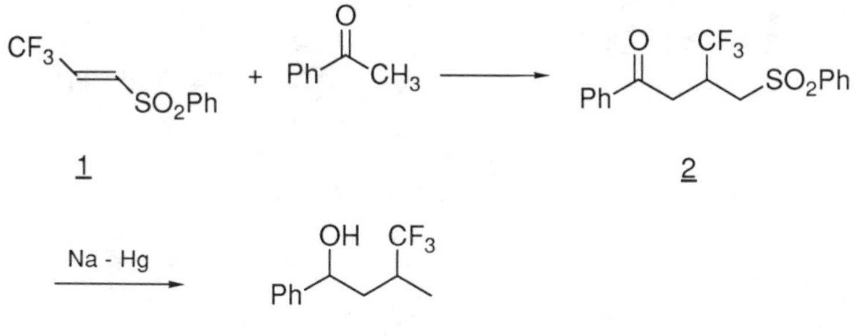

5.1.2 3-Fluoro-1,2-propanediol

3-Fluoro-1-hydroxypropane-2-one (fluorohydroxyacetone, **5**) is synthesized from 3-fluoro-1,2-propanediol (**4**). This unstable compound has been used in saline solution for pharmacological studies and was not isolated.[2]

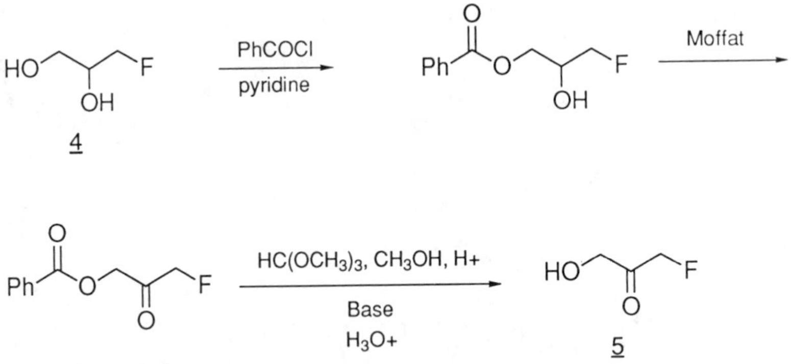

5.2 ENZYMATIC REDUCTIONS OF FLUORINATED KETONES AND KETOESTERS

Chiral fluorinated alcohols **(7, 9)** may be prepared by enzymatic reduction of fluorinated ketones **(6)** and ketoesters **(8)**.[3] Reduction of α,β-unsaturated ketones

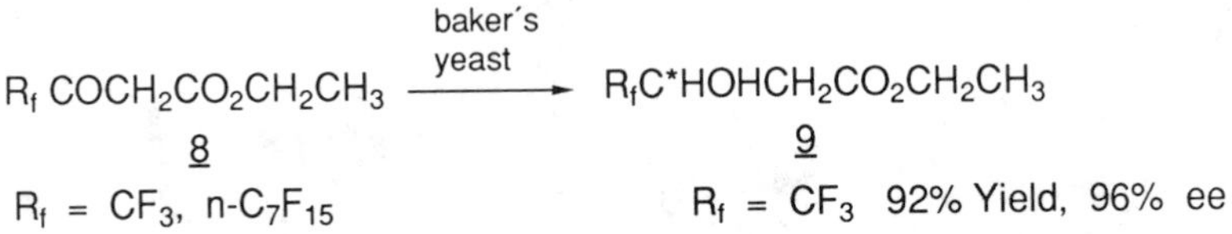

(10), in which the perfluoroalkyl group attached on carbon–carbon double bond, with baker's yeast affords the optically active carbinol **11** as well as the saturated

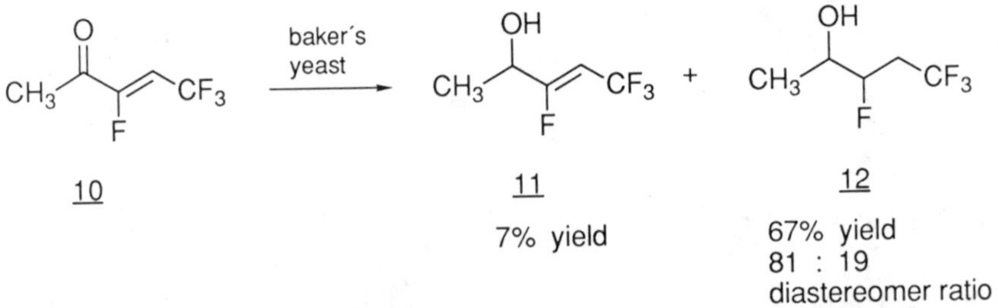

diastereomer **12**. The carbonyl group is found to be selectively reduced prior to the reduction of the double bond.

5.3   CATALYTIC REDUCTION OF FLUORINATED OLEFINS

Catalytic asymmetric hydrogenation of fluorinated olefins (**13**) can afford chiral fluorinated alcohols (**14**) in good yields and high enantiomeric excesses.[4]

100% yield
77% ee

13   **14**

cat = (Rh (1,5-c-Oct)((R,R) -di-PAMP))$^+$ BF$_4^-$

1,5 -c- Oct = 1,5- cyclooctadiene

(R,R) di - PAMP =

5.4   BIOLOGICAL ACTIVITY

The substrate analog of 2,2,2-trifluoroethanol (TFE) has been used as a probe of the active site of alcohol dehydrogenase from horse liver (LADH).[5] The sensitivity of ^{19}F chemical shifts to ionization makes it possible to monitor the ionization of the alcohol as binding occurs. This NMR probe is sensitive enough to allow the study of the effect of added NAD during the formation of a ternary complex with the enzyme.

Fluorodeoxyglycerol phosphates have been used as specific inhibitors of *sn*-glycerol phosphate dehydrogenase isolated from the locust flight muscle.[6] The hydroxy groups make little contribution to enzyme–substrate interactions. No evidence is observed for enhanced binding of any isomer, suggesting that neither the polarity of the C–F bond nor its ability to accept a hydrogen bond influences the interaction of fluorinated analogs.

Differences in glycerolipid metabolism of neoplastic and normal cells may be utilized in cancer chemotherapy.[2] 1-Fluoro analogs of glycerol-3-phosphate are studied to exploit differences in levels of cytosolic NAD-linked glycerol-3-phosphate dehydrogenase.

REFERENCES

1. T. Taguchi, G. Tomizawa, M. Nakajima, and Y. Kobayashi, *Chem. Pharm. Bull.* **33**, 4077–4080, 1985.

2. R. W. Pero, P. Babiarz-Tracy, and T. P. Fondy, *J. Med. Chem.* **20**, 644–647, 1977.

3. (a) T. Kitazume and N. Ishikawa, *Chem. Lett.* **1983**, 237–238; (b) T. Kitazume and N. Ishikawa, *Chem. Lett.* **1984**, 587–590.

4. K. E. Koenig, G. L. Bachman, and B. D. Vienyard, *J. Org. Chem.* **45**, 2362–2365, 1980.

5. D. C. Anderson and F. W. Dahlquist, *Biochemistry* **21**, 3569–3578, 1982.

6. W. J. Lloyd and R. Harrison, *Arch. Biochem. Biophys.* **163**, 185, 1974.

FLUORINATED PHOSPHONIC ACIDS AND DERIVATIVES

Fluorinated analogs of biologically important phosphonic acids have received much attention in recent years. Mono- and difluorinated phosphonates have great potential for biological activity as isosteric and isopolar analogs of phosphate esters.[1] Mono- and difluoromethylene biphosphonic acids are of interest as selective chelating agents and as analogs of pyrophosphoric acid of significant biological potential.[2]

6.1 SYNTHESIS OF DIFLUOROPHOSPHONOACETIC ACID AND DERIVATIVES

6.1.1 Dibromodifluoromethane

Dibromodifluoromethane reacts with triethylphosphite to give diethyl (bromodifluoromethyl) phosphonate (**1**), which with zinc dust affords the stable bromide (**2**). Acylation of **2** with ethyl chloroformate under cuprous bromide catalysis followed by selective silylation and hydrolysis affords compound (**3**).[3,4]

6.1.2 Fluoromethane Phosphonates

Fluoromethane phosphonates (**4**) are deprotonated with LDA at low temperature to yield lithiated α-fluorophosphonate carbanions which with carbon dioxide give carboxylic acids (**5**) and with carbonyl sulfide give thiocarboxylic acids (**6**).[1] Further transformations to the amide (**7**) or acid (**8**) are also possible.

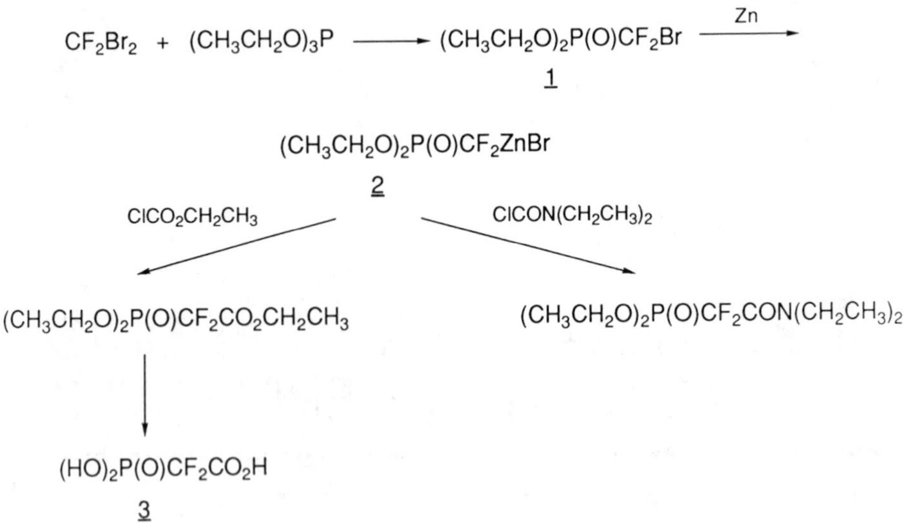

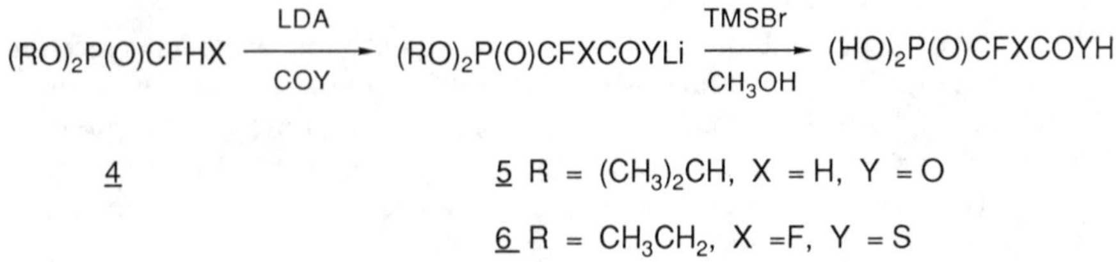

$$CF_2Br_2 \ + \ (CH_3CH_2O)_3P \ \longrightarrow \ (CH_3CH_2O)_2P(O)CF_2Br \ \xrightarrow{\ Zn\ }$$

$$\underline{1}$$

$$(CH_3CH_2O)_2P(O)CF_2ZnBr$$

$$\underline{2}$$

$ClCO_2CH_2CH_3$ ↙   ↘ $ClCON(CH_2CH_3)_2$

$(CH_3CH_2O)_2P(O)CF_2CO_2CH_2CH_3$ $(CH_3CH_2O)_2P(O)CF_2CON(CH_2CH_3)_2$

↓

$$(HO)_2P(O)CF_2CO_2H$$

$$\underline{3}$$

$$(RO)_2P(O)CFHX \ \xrightarrow[\text{COY}]{\text{LDA}} \ (RO)_2P(O)CFXCOYLi \ \xrightarrow[\text{CH}_3\text{OH}]{\text{TMSBr}} \ (HO)_2P(O)CFXCOYH$$

$$\underline{4} \qquad\qquad\qquad \underline{5}\ R = (CH_3)_2CH,\ X = H,\ Y = O$$

$$\underline{6}\ R = CH_3CH_2,\ X = F,\ Y = S$$

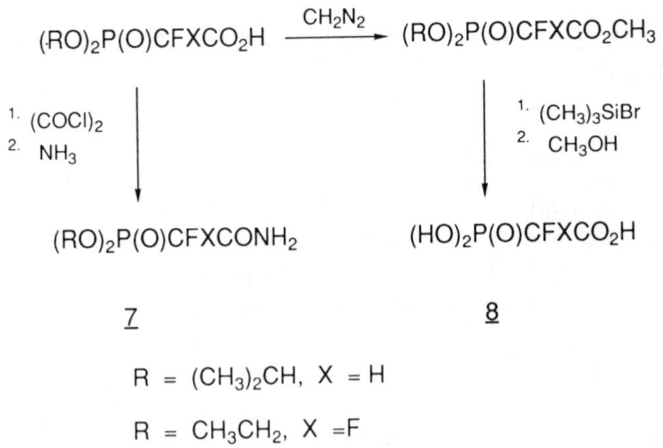

$$(RO)_2P(O)CFXCO_2H \ \xrightarrow{\ CH_2N_2\ } \ (RO)_2P(O)CFXCO_2CH_3$$

$\begin{array}{l} 1.\ (COCl)_2 \\ 2.\ \ NH_3 \end{array}$ ↓   ↓ $\begin{array}{l} 1.\ (CH_3)_3SiBr \\ 2.\ \ CH_3OH \end{array}$

$$(RO)_2P(O)CFXCONH_2 \qquad\qquad (HO)_2P(O)CFXCO_2H$$

$$\underline{7} \qquad\qquad\qquad\qquad \underline{8}$$

$$R = (CH_3)_2CH,\ X = H$$

$$R = CH_3CH_2,\ X = F$$

6.1.3 Acylation

Acylation of the anion (**9**) of diisopropyl fluoromethyl phosphonate with ethyl chloroformate gives **10**.[5]

$$((CH_3)_2CHO)_2P(O)CFHLi \xrightarrow{\quad ClCO_2CH_2CH_3 \quad} ((CH_3)_2CHO)_2P(O)CFHCO_2CH_2CH_3$$

$$\underset{\underline{9}}{} \qquad\qquad\qquad\qquad\qquad\qquad \underset{\underline{10}}{}$$

Acylation of **11** using ethyl isocyanate gives diethyl *N*-ethyl phosphonodifluoroacetamide (**12**).[5]

$$(CH_3CH_2O)_2P(O)CF_2Li \xrightarrow{\quad CH_3CH_2NCO \quad} (CH_3CH_2O)_2P(O)CF_2\overset{\overset{\displaystyle O}{\|}}{C}NHCH_2CH_3$$

$$\underset{\underline{11}}{} \qquad\qquad\qquad\qquad\qquad\qquad \underset{\underline{12}}{}$$

6.2 SYNTHESIS OF MONOFLUOROPHOSPHONATE

6.2.1 Perchloryl Fluoride

Direct fluorination of diethyl lithiomethanephosphonate (**13**) with perchlorylfluoride gives a good yield of monofluorophosphonate (**14**).[5]

$$(CH_3CH_2O)_2P(O)CH_2Li \xrightarrow[\quad KH_2PO_4 \quad]{\quad FClO_3 \quad} (CH_3CH_2O)_2P(O)CH_2F$$

$$\underset{\underline{13}}{} \qquad\qquad\qquad\qquad\qquad\qquad \underset{\underline{14}}{}$$

$$\uparrow\ \text{LDA}$$

$$(CH_3CH_2O)_2P(O)CH_3$$

6.3 SYNTHESIS OF FLUOROMETHYLENEBISPHOSPHONATES

6.3.1 Phosphorylation

Phosphorylation of the anion (**15**) with diethyl phosphorochloridate yields *P,P*-diethyl-*P',P'*-diisopropyl fluoromethylenebisphosphonate (**16**).[5]

$$((CH_3)_2CHO)_2P(O)CFHLi \xrightarrow{\quad (CH_3CH_2O)_2POCl \quad} ((CH_3)_2CHO)_2P(O)CFHP(O)(OCH_2CH_3)_2$$

$$\underset{\underline{15}}{} \qquad\qquad\qquad\qquad\qquad\qquad \underset{\underline{16}}{}$$

6.3.2 Carboxylation

Acylation of anion (**17**) of difluoromethanephosphonate ester with ethyl chloroformate leads to tetraethyl difluoromethylenebisphosphonate (**18**) as the major product, along with another novel compound (**19**).[5]

$$(CH_3CH_2O)_2P(O)CF_2Li \xrightarrow{\text{ClCO}_2\text{CH}_2\text{CH}_3} (CH_3CH_2O)_2P(O)CF_2CO_2CH_2CH_3$$

<u>17</u>

$$\xrightarrow{(CH_3CH_2O)_2P(O)CF_2Li} (CH_3CH_2O)_2P(O)CF_2P(O)(OCH_2CH_3)_2$$

<u>18</u>

$$+ \quad (CH_3CH_2O)_2OPCF_2\text{-}\overset{\overset{\displaystyle OH}{|}}{\underset{\underset{\displaystyle OH}{|}}{C}}\text{-}CF_2PO(OCH_2CH_3)_2$$

<u>19</u>

6.3.3 Electrophilic Displacement

The anion (**21**) of chlorofluoromethanephosphonate ester (**20**) reacts with ethyl chloroformate, trimethylsilyl bromide, and other electrophiles to afford **22** as the sole product.[5] The mechanism for the formation of the bisphosphonate ester is

$$((CH_3)_2CHO)_2P(O)CFHCl \longrightarrow ((CH_3)_2CHO)_2P(O)CFClLi$$

<u>20</u>   <u>21</u>

$$\xrightarrow[\text{KH}_2\text{PO}_4]{\text{electrophile}} ((CH_3)_2CHO)_2P(O)CFHP(O)(OCH(CH_3)_2)_2$$

<u>22</u>

shown in Fig. 6.1. Transfer of chlorine from **23** to anion **24** forms the anion **25** and dichlorofluoromethanephosphonate ester (**26**). The ester **26** then undergoes a displacement reaction giving **22**.

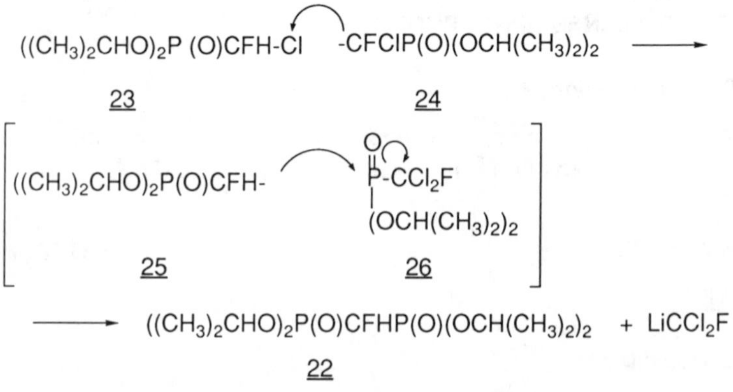

$$((CH_3)_2CHO)_2P(O)CFH\text{-}Cl \quad \overset{\frown}{} \quad \text{-}CFClP(O)(OCH(CH_3)_2)_2 \longrightarrow$$

<u>23</u>   <u>24</u>

$$\left[((CH_3)_2CHO)_2P(O)CFH\text{-} \quad \overset{\frown}{} \quad \overset{\overset{\displaystyle O}{\|}}{P}\text{-}CCl_2F \atop (OCH(CH_3)_2)_2 \right]$$

<u>25</u>   <u>26</u>

$$\longrightarrow ((CH_3)_2CHO)_2P(O)CFHP(O)(OCH(CH_3)_2)_2 \quad + \quad LiCCl_2F$$

<u>22</u>

FIGURE 6.1

6.3.4. Perchlorylfluoride

Treatment of the sodium salt of tetraisopropylmethylene bisphosphonate (**27**) with an excess of perchlorylfluoride in THF gives the tetraisopropyl mono-fluoromethylenebisphosphonate (**22**) mixed with some difluoromethylene bis-phosphonate (**28**).[2]

$$((CH_3)_2CHO)_2P(O)CH_2P(O)(OCH(CH_3)_2)_2 \xrightarrow{\text{NaH, FClO}_3}$$

<u>27</u>

$$((CH_3)_2CHO)_2P(O)CFHP(O)(OCH(CH_3)_2)_2$$

$$+ \quad ((CH_3)_2CHO)_2P(O)CF_2P(O)(OCH(CH_3)_2)_2$$

<u>22</u>   <u>28</u>

4 : 1

6.3.5 Bromodifluoromethylphosphonate

Bromodifluoromethylphosphonate ester (**29**) reacts with the sodium salt of dibutylphosphonate (**30**) to give difluoromethylenebisphosphonic acid as the tetraalkyl ester **31**.[2,6]

$$(CH_3CH_2O)_2P(O)CF_2Br \quad + \quad NaP(O)(OCH_2CH_2CH_2CH_3)_2$$

<u>29</u>   <u>30</u>

$$\xrightarrow[-40°]{\text{hexane}} \quad (CH_3CH_2O)_2P(O)CF_2P(O)(OCH_2CH_2CH_2CH_3)_2 \quad + \quad NaBr$$

<u>31</u>

6.4 BIOLOGICAL ACTIVITY

Phosphonoacetic acid has been found to be an antiviral agent and a herbicide. The trialkyl ester of phosphonodichloroacetate is reported to have useful insecticidal properties.[1] α-Fluorination may enhance the potency of phosphono-acetic acid as an inhibitor for the replication of herpes simplex virus in animals.[5] α-Fluorination of alkyl phosphonic acids and esters modifies the reactivity of the parent materials without affecting their steric bulk relative to the corresponding nonfluorinated phosphonic acid esters.[2] The diphosphonates are structurally similar to pyrophosphates but have increased stability toward chemical hydrolysis and enzymatic degradation. Difluoromethylenebisphosphonic acid may be a therapeutically useful diphosphonate for the complexation of Ca^{2+} and other metal ions.[7]

REFERENCES

1. G. M. Blackburn, D. Brown, and S. J. Martin, *J. Chem. Res(S).,* **1985**, 92–93.
2. G. M. Blackburn, D. A. England, and F. Kolkmann, *J. Chem. Soc. Chem. Commun.* **1981**, 930–932.
3. D. J. Burton, L. G. Sprague, D. J. Pietrzyk, and S. H. Edelmuth, *J. Org. Chem.* **49**, 3438–3440, 1984.
4. D. J. Burton and L. G. Sprague, *J. Org. Chem.* **53**, 1523–1527, 1988.
5. G. M. Blackburn, D. Brown, S. J. Martin, and M. J. Paratt, *J. Chem. Soc. Perkin Trans.* **1**, 181–186, 1987.
6. D. J. Burton, D. J. Pietrzyk, T. Ishihara, T. Fonong, and R. M. Flynn, *J. Fluorine Chem.* **20**, 617–626, 1982.
7. T. Fonong, D. J. Burton, and D. J. Pietrzyk, *Anal. Chem.* **55**, 1089–1094, 1983.

FLUORINATED KETONES

Selective introduction of a fluorine atom into the α position of a carbonyl moiety is of practical interest. Fluoroketones have been successfully employed as suicide enzyme inhibitors.

7.1 HYDROGEN FLUORIDE–PYRIDINE

7.1.1 Diazoketones

α-Fluoroketones (2) can be prepared successfully by the reaction of diazoketones (1) with hydrogen fluoride–pyridine at 0°C.[1,2] In the presence of added halide ions using N-halosuccinimides, the fluorohaloketones were obtained.

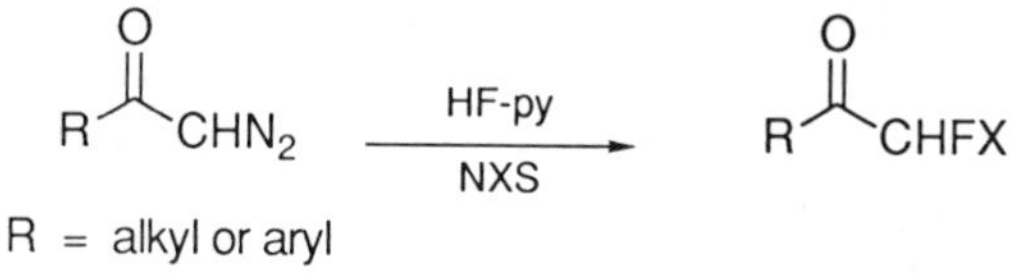

R = alkyl or aryl

7.1.2 α-Haloketones

When α-haloketones (3) were added to a suspension of yellow mercuric oxide in hydrogen fluoride–pyridine, the halides were exchanged for fluorides. Bromides were more easily exchanged than chlorides.

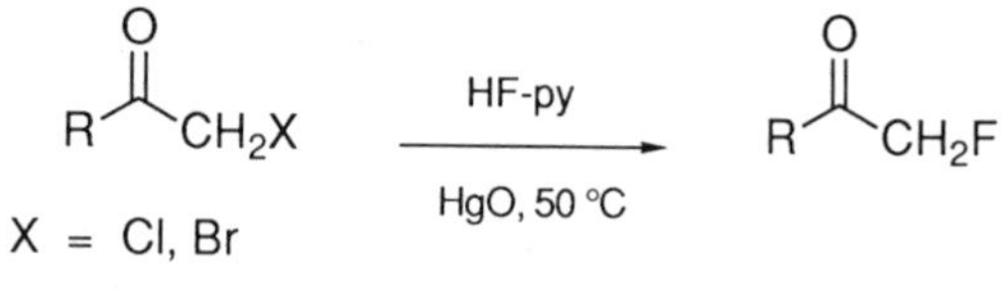

X = Cl, Br

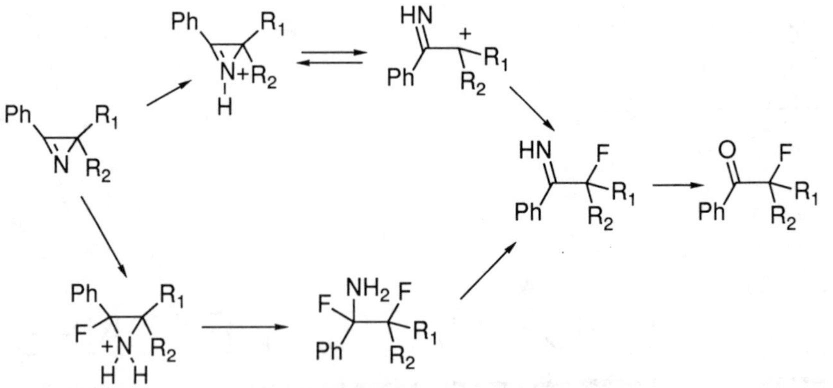

FIGURE 7.1

7.1.3 Azirine Ring Opening

Phenyl azirines (**5**) add hydrogen fluoride to give, after hydrolysis, α-fluoro-ketones (**6**).[3,4] The formation of the α-fluoroketones can be rationalized as shown in Fig. 7.1.[4]

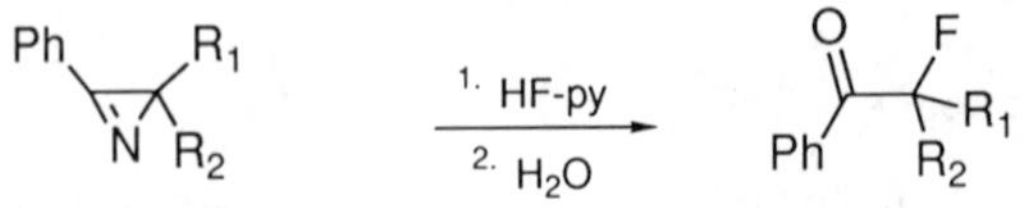

$$R_1 = R_2 = CH_3$$
$$R_1, R_2 = c\text{-}C_5H_{10}$$
$$R_1 = Ph, R_2 = H$$

7.2 MOLECULAR FLUORINE

Direct α-fluorination of ketones to form **8** via the free enol compound (**7**) is possible by means of molecular fluorine (10% fluorine in nitrogen) in an inert solvent at low temperature.[5] Those ketones existing in the keto form give complex mixtures of products under the same conditions.

$$R_2CH=C(OH)CO_2R_1 \xrightarrow{F_2} R_2CHFCOCO_2R_1 \xrightarrow{BSA}$$

$$\underline{7} \qquad\qquad\qquad \underline{8}$$

$$R_1 = CH_3CH_2, CH_3, H$$
$$R_2 = Ph, p\text{-}Ph, 3\text{-indole, Acyl}$$

$$R_2CF=C(OTMS)CO_2R_1 \xrightarrow{\text{hydrolysis}} R_2CF=C(OH)CO_2R_1$$

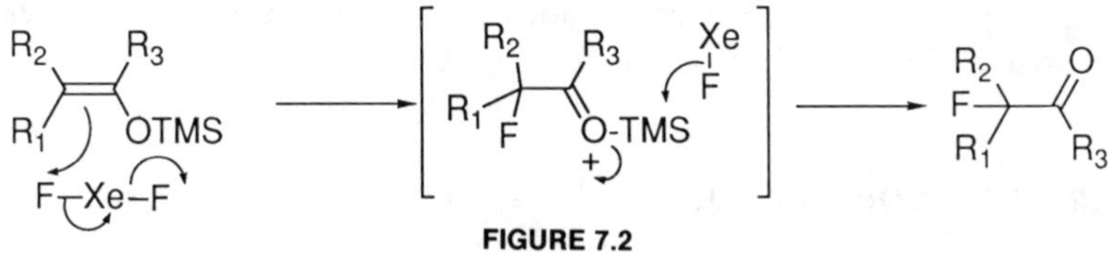

FIGURE 7.2

This suggests that direct fluorination involves a rate-determining enolization step after which the electrophilic step proceeds smoothly. However, for the keto-type compounds which have a slow enolization step, rapid side reactions occur.

7.3 XENON DIFLUORIDE

Xenon difluoride reacts smoothly with various steroidal silyl enol ethers in the absence of acid catalyst to afford stereoselectively α-oriented α-fluoroketones (**9**).[6]

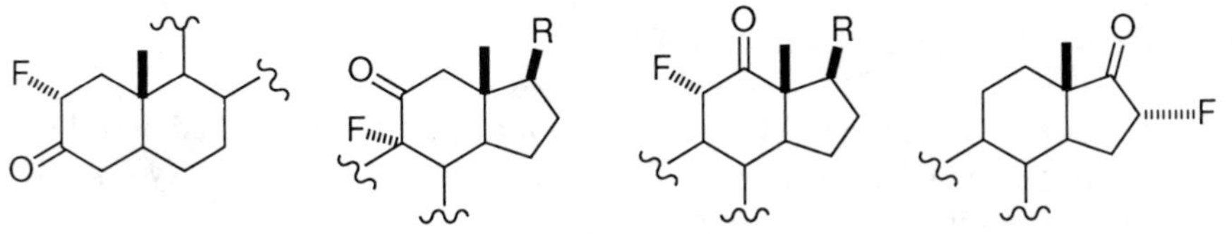

_9_

An electrophilic fluorination mechanism is suggested in Fig. 7.2. The reagent must approach the double bond from the less hindered side to explain the stereochemistry of the product.

7.4 ACETYL HYPOFLUORITE

A quick one-step conversion of a carbonyl moiety into an α-fluorocarbonyl derivative (**11**) can be effected using acetyl hypofluorite (AcOF).[7] This mild electrophilic reagent reacts with metal enolates (**10**) cleanly and efficiently. Since

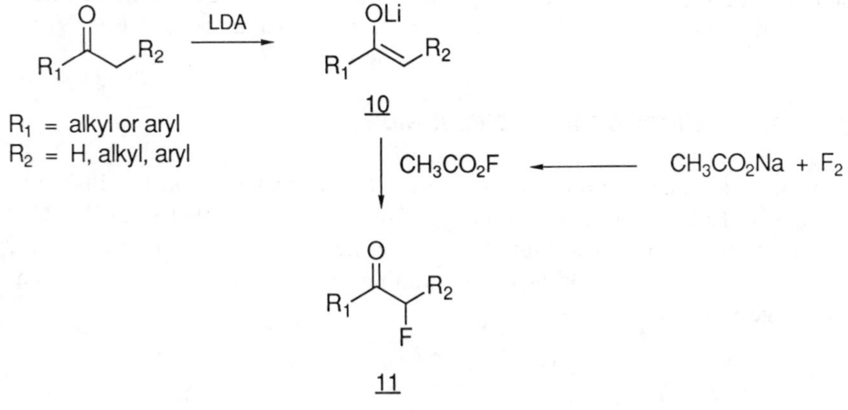

R_1 = alkyl or aryl
R_2 = H, alkyl, aryl

_10_

$CH_3CO_2F \longleftarrow CH_3CO_2Na + F_2$

_11_

the reagent is prepared from elemental fluorine, the method can be adapted easily to synthesize ^{18}F-labeled fluoroketones using $AcO^{18}F$.

7.5 2,2,2-TRIFLUORETHANOL

Chiral fluorinated ketones (**14**) have been prepared by yeast-promoted additions of 2,2,2-trifluoroethanol (**12**) to α,β-unsaturated ketones (**13**).[8]

13 12 14

38-50% yield
93% ee

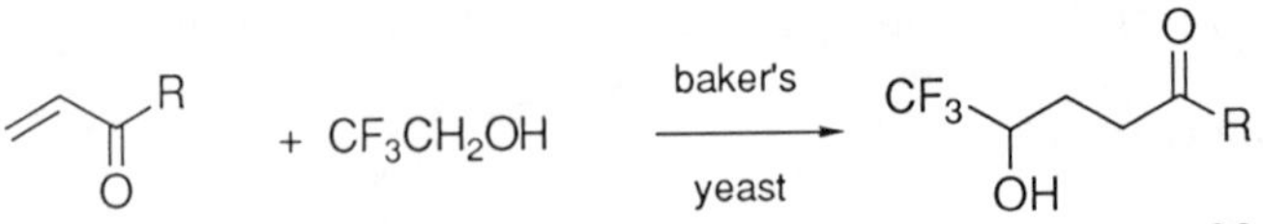

7.6 3,3,3-TRIFLUOROPROPIONIC ACID

3,3,3-Trifluoropropionic acid (**15**) may be converted to the fluorinated diazobutanone (**16**), which on treatment with phosphoric acid formed 1-hydroxy-2-keto-4,4,4-trifluorobutane phosphoric acid (HTFP) (**17**).[9]

$$\underset{\underline{15}}{CF_3CH_2CO_2H} \xrightarrow[CH_2N_2]{PCl_5} \underset{\underline{16}}{CF_3CH_2COCHN_2}$$

$$\xrightarrow{H_3PO_4} \underset{\underline{17}}{CF_3CH_2COCH_2OPO_3^{=}}$$

7.7 METHYL 2,2-DIFLUOROBUTYRATE

The difluoromethylene ketone phospholipid analog **19** has been synthesized in the racemic form using methyl-2,2-difluorobutyrate (**18**) as seen in Fig. 7.3.[10]

7.8 4-METHYL IODOBENZENEDIFLUORIDE

Substituted iodobenzenedifluoride (*p*-methyl iodobenzenedifluoride) (**20**) reacts with silyl enol ethers of steroids to produce β-oriented α-fluoroketones (**21**), (**22**).[6] mechanism involving an iodonium ion intermediate and its subsequent S_N2 nucleophilic substitution or β-proton elimination by either F^- or OH^- has been suggested (Fig. 7.4).

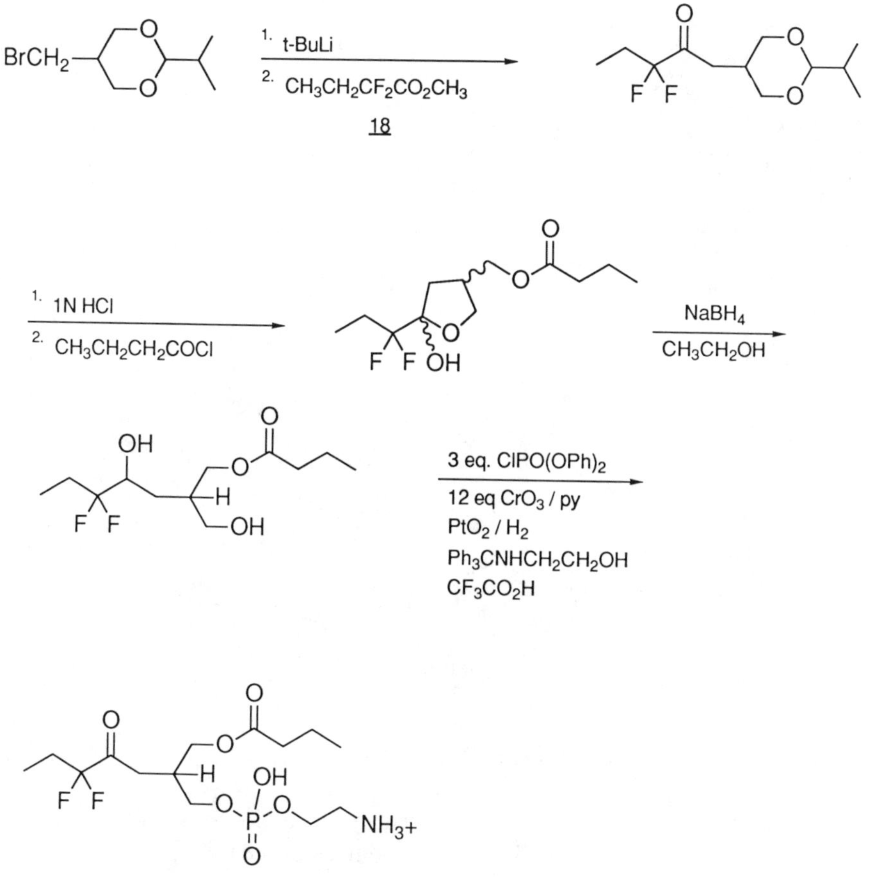

FIGURE 7.3

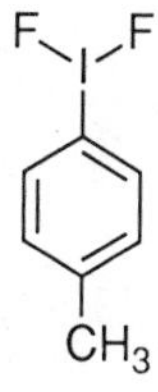

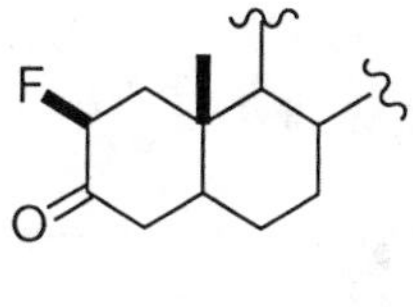

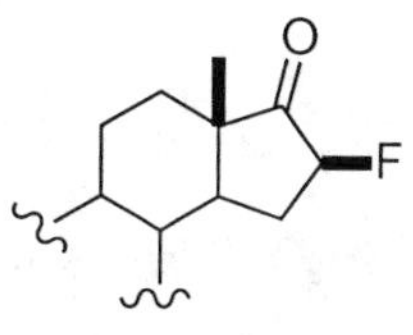

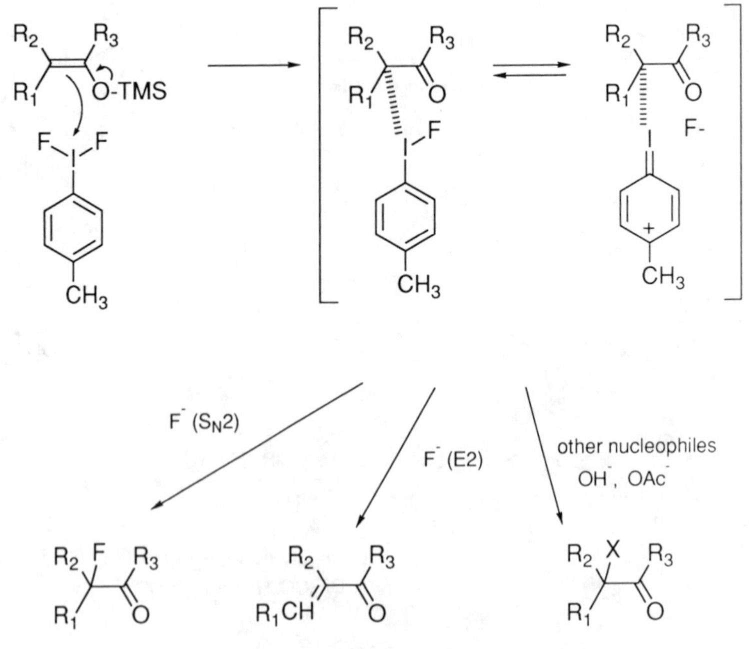

FIGURE 7.4

7.9 PERFLUOROCARBOXYLIC ACID ESTERS

Fluorinated ketones (**25**) can be prepared by the action of Grignard reagents, such as **24**, on fluorinated acids (**23**).[11] Oxidation of the methylene ketones with

$$C_3F_7CO_2H + C_6H_5CH_2MgCl \longrightarrow C_3F_7COCH_2C_6H_5 \xrightarrow{SeO_2}$$
$$\underline{23} \qquad\qquad \underline{24} \qquad\qquad\qquad \underline{25}$$

$$C_3F_7COCOC_6H_5$$
$$\underline{26}$$

Also,

$$HO_2C(CF_2)_4CO_2H + C_6H_5CH_2MgCl \longrightarrow$$

$$C_6H_5CH_2CO(CF_2)_4COCH_2C_6H_5 \xrightarrow{SeO_2}$$

$$C_6H_5COCO(CF_2)_4COCOC_6H_5$$
$$\underline{27}$$

selenium dioxide gives di- and tetraketones (**26**, **27**). Similarly, addition of organolithium reagents (**28**) also afford fluorinated ketones (**29**).[12]

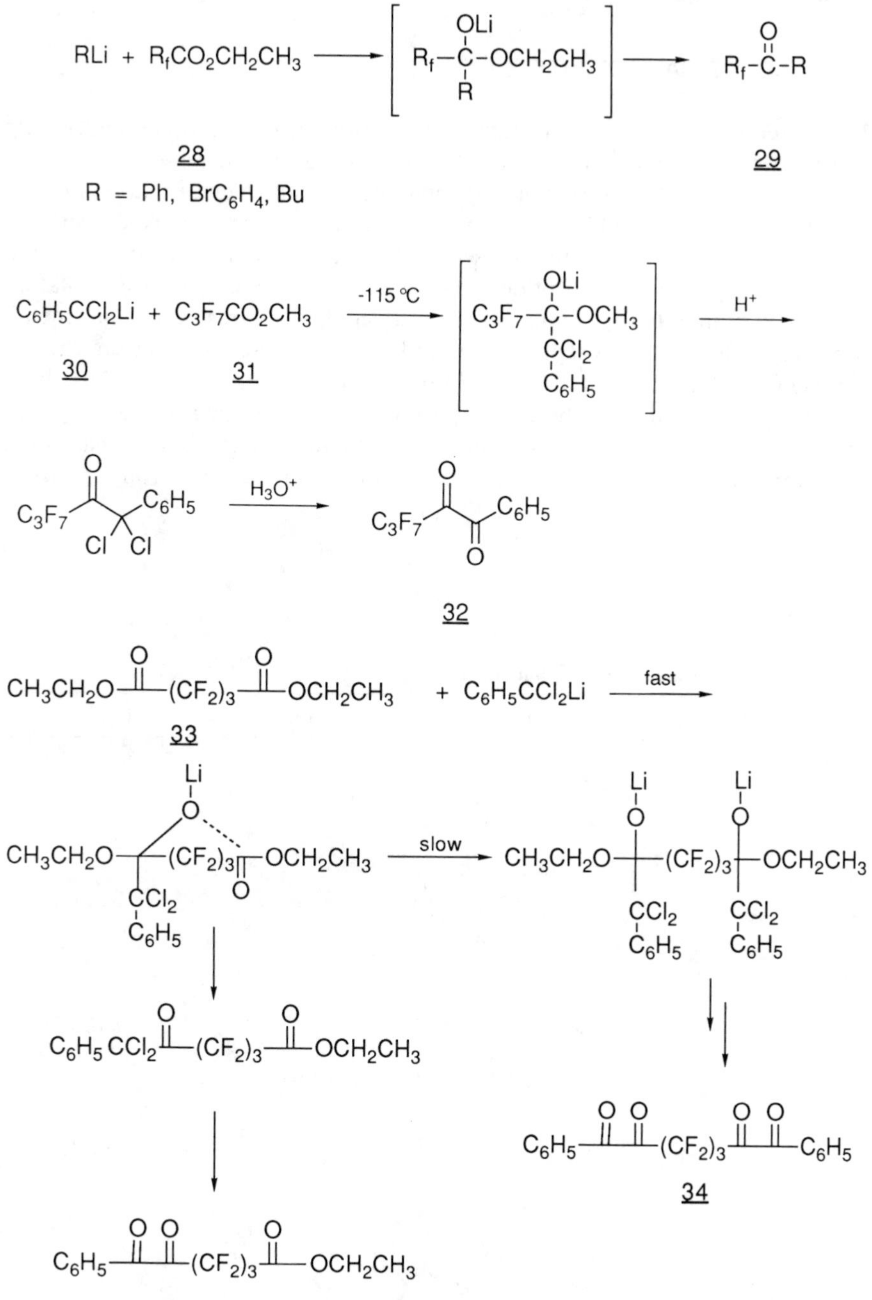

$$RLi + R_fCO_2CH_2CH_3 \longrightarrow \left[R_f-\underset{\underset{R}{|}}{\overset{\overset{OLi}{|}}{C}}-OCH_2CH_3 \right] \longrightarrow R_f-\overset{\overset{O}{\|}}{C}-R$$

$$\underline{28} \qquad\qquad\qquad\qquad\qquad\qquad\qquad\qquad \underline{29}$$

R = Ph, BrC$_6$H$_4$, Bu

$$C_6H_5CCl_2Li + C_3F_7CO_2CH_3 \xrightarrow{-115\,°C} \left[C_3F_7-\underset{\underset{\underset{C_6H_5}{|}}{\overset{|}{CCl_2}}}{\overset{\overset{OLi}{|}}{C}}-OCH_3 \right] \xrightarrow{H^+}$$

$$\underline{30} \qquad\qquad \underline{31}$$

$$\underset{\underset{Cl\quad Cl}{}}{C_3F_7}\overset{\overset{O}{\|}}{}\!\!\!\diagdown\!C_6H_5 \xrightarrow{H_3O^+} C_3F_7\overset{\overset{O}{\|}}{}\overset{}{\underset{\underset{O}{\|}}{C}}C_6H_5$$

$$\underline{32}$$

$$CH_3CH_2O\overset{\overset{O}{\|}}{C}-(CF_2)_3-\overset{\overset{O}{\|}}{C}-OCH_2CH_3 \; + \; C_6H_5CCl_2Li \xrightarrow{fast}$$

$$\underline{33}$$

$$CH_3CH_2O-\underset{\underset{\underset{C_6H_5}{|}}{\overset{|}{CCl_2}}}{\overset{\overset{\overset{Li}{|}}{\overset{O}{|}}}{C}}(CF_2)_3-\overset{\overset{O}{\|}}{C}-OCH_2CH_3 \xrightarrow{slow} CH_3CH_2O-\underset{\underset{\underset{C_6H_5}{|}}{\overset{|}{CCl_2}}}{\overset{\overset{\overset{Li}{|}}{\overset{O}{|}}}{C}}(CF_2)_3\underset{\underset{\underset{C_6H_5}{|}}{\overset{|}{CCl_2}}}{\overset{\overset{\overset{Li}{|}}{\overset{O}{|}}}{C}}OCH_2CH_3$$

$$C_6H_5 CCl_2\overset{\overset{O}{\|}}{C}-(CF_2)_3-\overset{\overset{O}{\|}}{C}-OCH_2CH_3$$

$$C_6H_5-\overset{\overset{O\;\;O}{\|\;\|}}{}-(CF_2)_3-\overset{\overset{O\;\;O}{\|\;\|}}{}-C_6H_5$$

$$\underline{34}$$

$$C_6H_5-\overset{\overset{O\;\;O}{\|\;\|}}{}-(CF_2)_3-\overset{\overset{O}{\|}}{C}-OCH_2CH_3$$

FIGURE 7.5

The reaction of phenyldichloromethyllithium (**30**) with perfluorinated mono- and diesters yields fluorochloroketones in high yield.[13] Hydrolysis of such compounds produces α,β-fluorinated di- and tetraketones (Fig. 7.5).

7.10 BIOLOGICAL ACTIVITY

Fluoroketones have been investigated as inhibitors of hydrolytic enzymes.[14] Fluorine on a carbon adjacent to the carbonyl group increases the electrophilicity of the carbonyl carbon and therefore the tendency of nucleophiles to add. Inhibition is probably due to the formation of a stable hemiketal with the active site. The hydrated ketones resemble the tetrahedral intermediates formed during the hydrolysis of peptide substrates. Fluoroketone substrate analogs have been found to be inhibitors of acetylcholine esterase and zinc metalloproteases like carboxypeptidase A and angiotensin-converting enzyme. A synthetic analog of pepstatin containing difluorostatone was found to be a transition state analog inhibitor of aspartyl proteases like pepsin (Table 7.1). In all cases, the fluoroketones are much more effective inhibitors than their nonfluorinated analogs. The fluorinated ketones show a very strong tendency

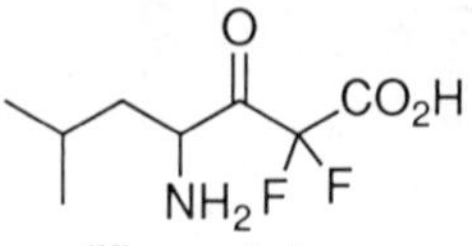

difluorostatone

TABLE 7.1 Fluoroketone Enzyme Inhibitors

Enzyme	Compound	K_i nM
A	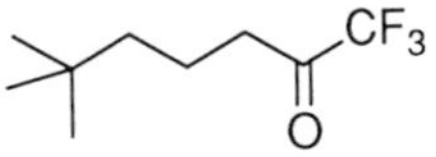	16
A	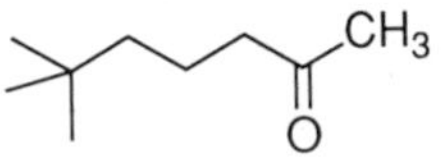	310,000
A	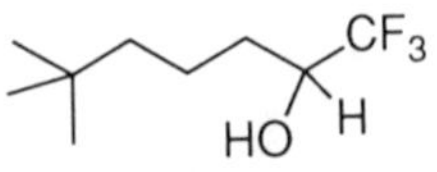	410,000
A	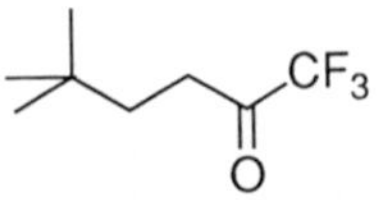	8,000
A	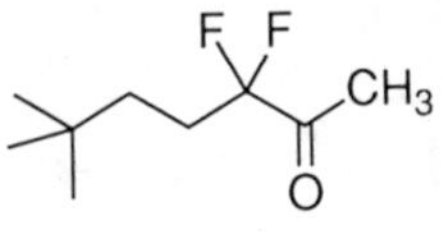	1.6

TABLE 7.1   (*continued*)

Enzyme	Compound	K_i nM
A	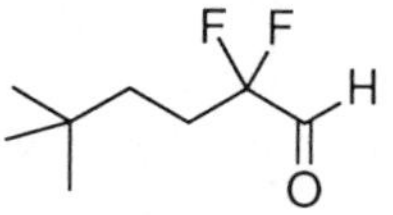	600
A	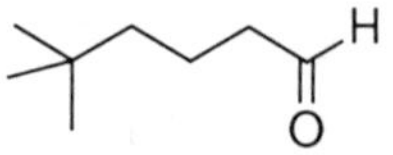	$>10^6$
B	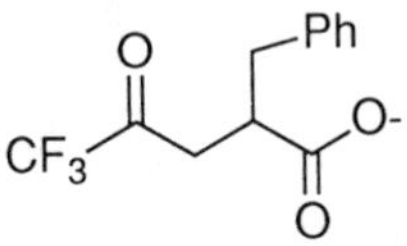	200
B	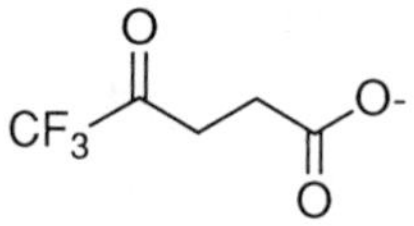	700,000
B	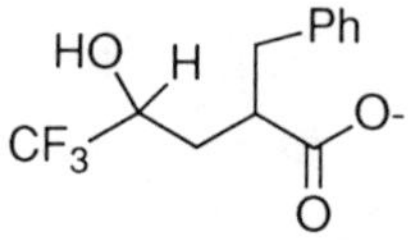	90,000
B	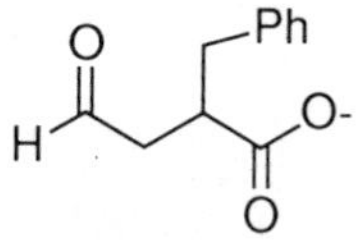	480
C	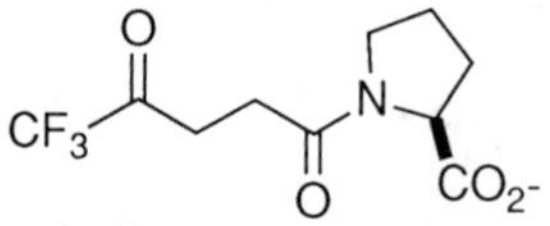	15,000
C	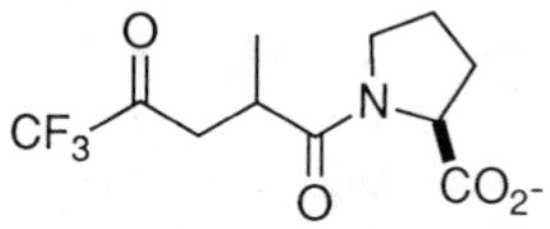	200
C	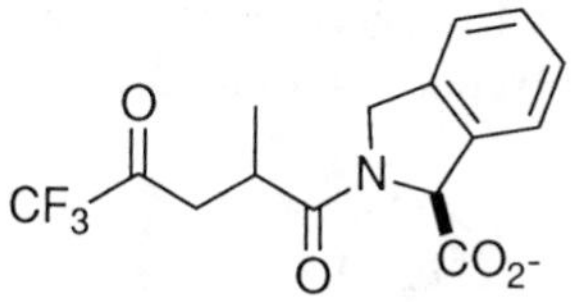	12
C	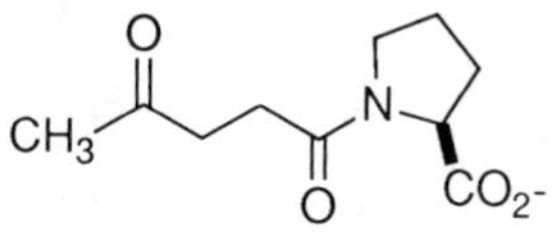	4,500,000

TABLE 7.1 (*continued*)

Enzyme	Compound	K_i nM

D — 1.1

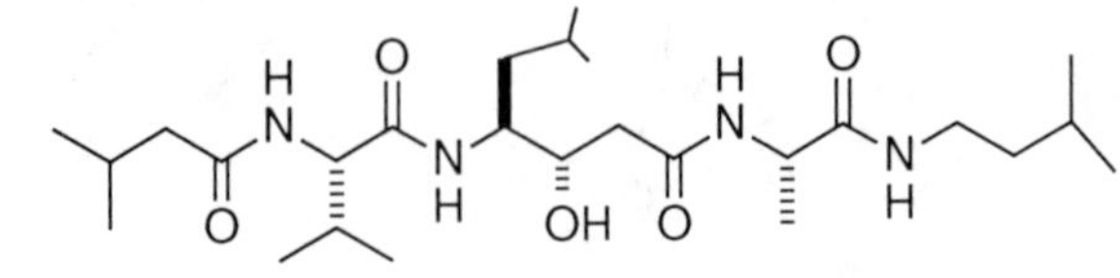

D — 56

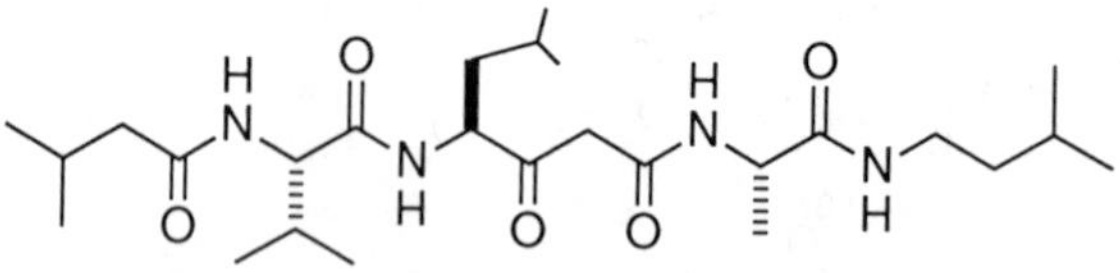

D — 0.06

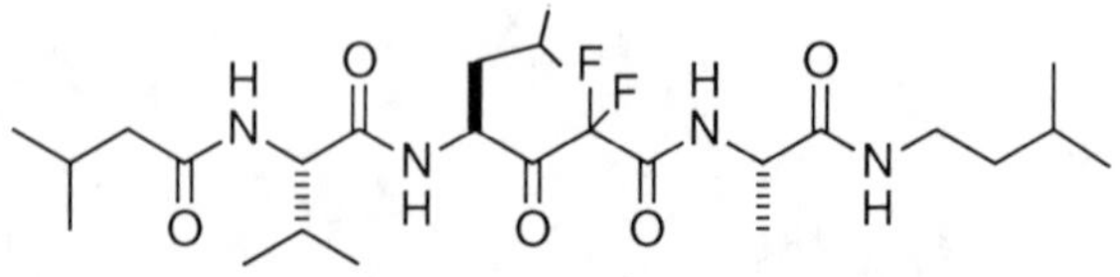

D — 0.5

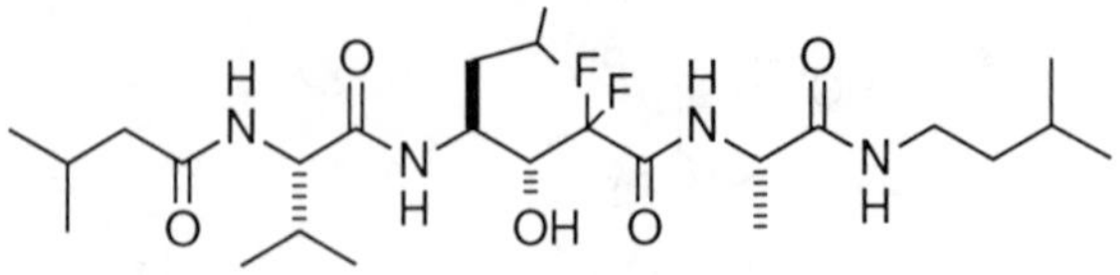

A = acetylcholinesterase
B = carboxypeptidase A
C = angiotensin converting enzyme
D = pepsin

to remain associated with the enzymes. The strong binding of fluoroketones to angiotensin-converting enzyme suggests their potential use as antihypertensives, since angiotensin is a potent vasoconstrictor.

Rabbit muscle aldolase, which promotes the reversible cleavage of fructose 1,6-diphosphate (**35**) into dihydroxyacetonephosphate (**36**) and D-glyceraldehyde-3-phosphate (**37**), is irreversibly inhibited by HTFP.[9] Presumably the

$$^=O_3POCH_2CH(OH)CH(OH)CH(OH)COCH_2OPO_3^= \rightleftharpoons HOCH_2COCH_2OPO_3^=$$

$$\underline{35} \qquad\qquad \underline{36}$$

$$+ \ OHCCH(OH)CH_2OPO_3^=$$

$$\underline{37}$$

intermediate enamine (**39**), formed via the Schiff base (**38**), loses fluoride ion to form the Michael acceptor (**40**). Conjugate addition of a nucleophilic residue on the enzyme to the activated substrate inactivates the enzyme.

The enzyme phospholipase A_2 catalyzes the liberation of arachidonic acid from the phospholipid membrane pool.[10] It cleaves sn-glycero phospholipids specifically at the 2 position, as seen in Fig. 7.6. The difluoromethylene ketone phospholipid analog (**41**) is a transition state analog inhibitor of phospholipase A_2. It binds about 300 times tighter than the nonfluorinated substrate. The ability of **41** to act as a tight-binding inhibitor may be due to its ability to form

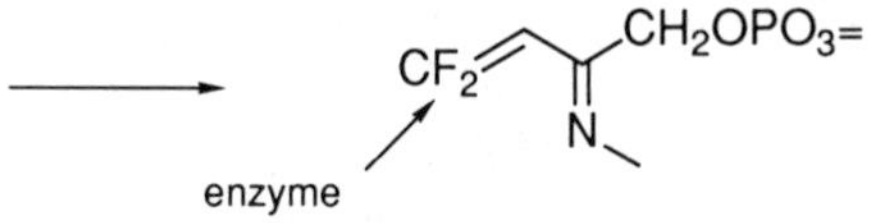

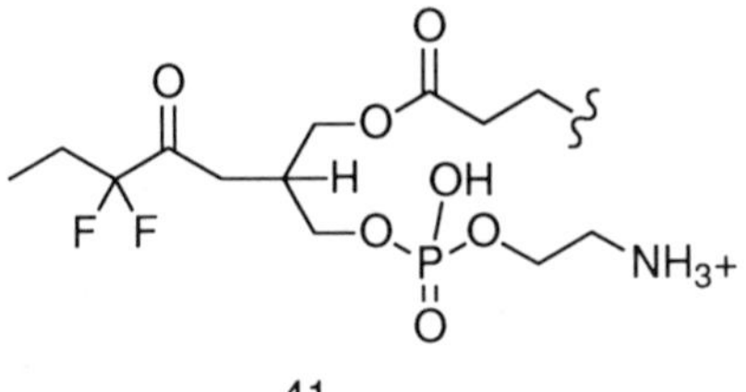

tetrahedral species, which mimics the tetrahedral intermediate formed during phospholipase-catalyzed lipolysis.

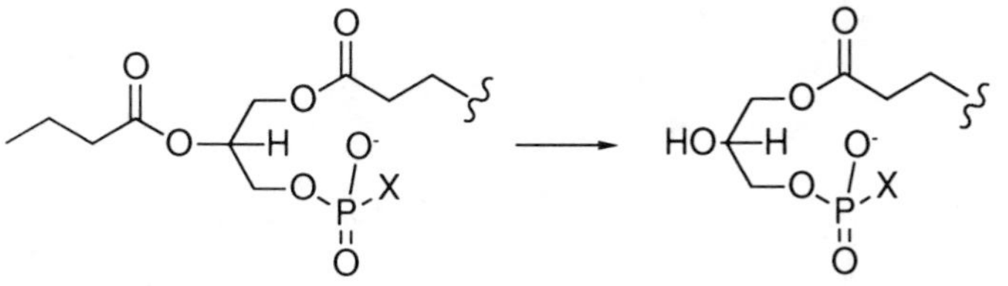

X = OH, choline,
    ethanolamine, etc

FIGURE 7.6

REFERENCES

1. G. A. Olah, J. T. Welch, Y. D. Vankar, M. Nojima, I. Kerekes, and J. A. Olah, *J. Org. Chem.* **44**, 3872–3881, 1979.

2. G. A. Olah and J. T. Welch, *Synthesis* **1974**, 896–897.

3. T. N. Wade and R. Kheribet, *J. Org. Chem.* **45**, 5333–5335, 1980.

4. G. Alvernhe, E. Kozlowska-Gramsz, S. Lacombe-Bar, and A. Laurent, *Tetrahedron Lett.* **1978**, 5203–5206.

5. T. Tsushima, K. Kawada, and T. Tsuji, *J. Org. Chem.* **47**, 1107–1110, 1982.

6. T. Tsushima, K. Kawada, and T. Tsuji, *Tetrahedron Lett.* **23**, 1165–1168, 1982.

7. S. Rozen and M. Brand, *Synthesis* **1985**, 665–667.

8. T. Kitazume and N. Ishikawa, *Chem. Lett.* **1984**, 1815–1818.

9. X. Le Clef, A. Magnien, and J. F. Biellmann, *Ann. Chim. Fr.* **1984**, 703.

10. M. H. Gelb, *J. Am. Chem. Soc.* **108**, 3146–3147, 1986.

11. P. M. Hergenrolher and M. Hudlicky, *J. Fluorine Chem.* **12**, 439, 1978.

12. (a) L. S. Chen, G. J. Chen, and C. Tamborski, *J. Fluorine Chem.*, **18**, 117–121, 1981; (b) L. S. Chen and C. Tamborski, *J. Fluorine Chem.* **19**, 34–53, 1982.

13. L. S. Chen and C. Tamborski, *J. Fluorine Chem.* **26**, 269–279, 1984.

14. M. H. Gelb, J. P. Svaren, and R. H. Abeles, *Biochemistry* **24**, 1813–1817, 1985.

8

FLUORINATED SUGARS

8.1 INTRODUCTION

Selectively fluorinated carbohydrates have found utility in probing biochemical mechanisms or in modifying the activity of glycosides.[1] Although these materials have many applications in biochemistry, medicinal chemistry, and pharmacology, only one naturally occurring fluorinated carbohydrate is known,[1a] that is, the 4-deoxy-4-fluorosugar constituent of nucleocidin (Fig. 8.1). The biochemical rationale for incorporating fluorine in the carbohydrate residue is that replacement of a hydroxyl by fluorine would cause only a very minor steric perturbation of the structure or conformation while at the same time having a profound electronic effect on neighboring groups. The substitution is possible while retaining the capacity of the position as an acceptor in hydrogen bonding. Yet these same attributes make the synthesis of fluorinated carbohydrates difficult. The synthesis of fluorinated carbohydrates affords a particularly fruitful field for the combination of modern chemical and enzymatic synthetic techniques. Early on it was recognized that total synthesis would be difficult because of the stereochemical control required at the multiple adjacent asymmetric centers of a fluorinated carbohydrate.[2]

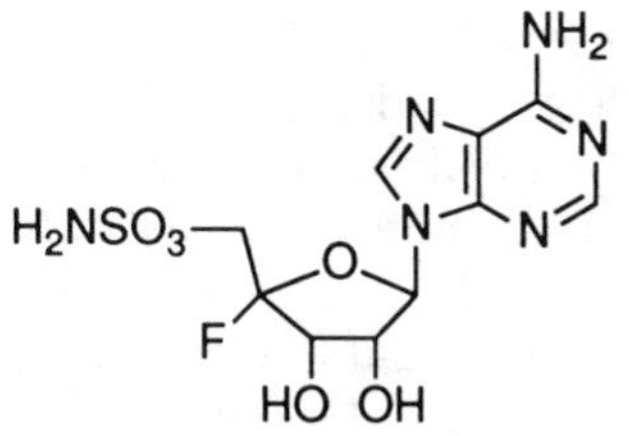

nucleocidin

FIGURE 8.1

The synthetic strategies employed in the preparation of fluorinated carbo-hydrates are presented in the following order: fluoride displacement reactions of sulfonate esters, fluorinative dehydroxylations with diethylaminosulfur tri-fluoride (DAST), epoxide and aziridine ring opening by fluoride ion, additions of fluorine, halogen fluorides, or hypofluorites to olefins. In separate sections, total syntheses based on aldol condensations and sigmatropic rearrangements, enzymatic methods, and the preparation of glycosyl fluorides are discussed.

8.2 DISPLACEMENT REACTIONS

Nucleophilic displacements by fluoride ions are highly dependent on the source of the fluoride ion and the nature of the leaving group. Fluoride ions are poor nucleophiles but good bases. Additionally, since fluoride salts are frequently poorly soluble, we also emphasize the nature of the fluoride ion source. Although various sulfonate leaving groups have been employed, we focus on the use of the most successfully and more commonly employed groups, such as trifluoro-methanesulfonates (triflates), methanesulfonates (mesylates), and arenesulfonates. Equally important and perhaps obvious from the point of view of the carbohydrate chemist are neighboring group effects. The nature of the carbo-hydrate can have a profound effect on the course of the reaction because of neighbouring group participation in the development of cationic charge. Also 1,3-diaxial or 1,2-steric effects can obstruct or modify the course of reaction.

8.2.1 Sulfonate Esters

The use of sulfonate esters and halide ions as nucleophiles for the preparation of halocarbohydrates has been reviewed.[3]

Trifluoromethanesulfonates
Of these esters the most widely used because of their intrinsically high reactivity are the trifluoromethanesulfonate (triflate) esters. Triflates are easily prepared and, in contrast to somewhat less expensive methanesulfonate (mesylate) and arenesulfonate esters, undergo displacement reactions with fluoride ions without the necessity for forcing or destructive reaction conditions.

2-Deoxy-2-fluoropyranoses 2-Deoxy-2-fluoroglucose (^{18}F) has been esta-blished as a useful radiopharmaceutical for studying glucose metabolism in both normal and diseased tissue.[4] Efficient syntheses are required because of the short half-life of ^{18}F, 110 minutes. Because of the difficulty associated with preparing the radionuclide, syntheses which utilize as much of the available ^{18}F as possible are required. Synthesis of 2-deoxy-2-fluoroglucose (1) requires displacement of a leaving group from the 2 position of a protected mannose (2). Clearly such a reaction will be very sensitive to the anomeric configuration of the pyranose. Although the 4,6-*O*-benzylidene-2-*O*-methanesulfonyl-3-*O*-methyl-α-D-*manno*-

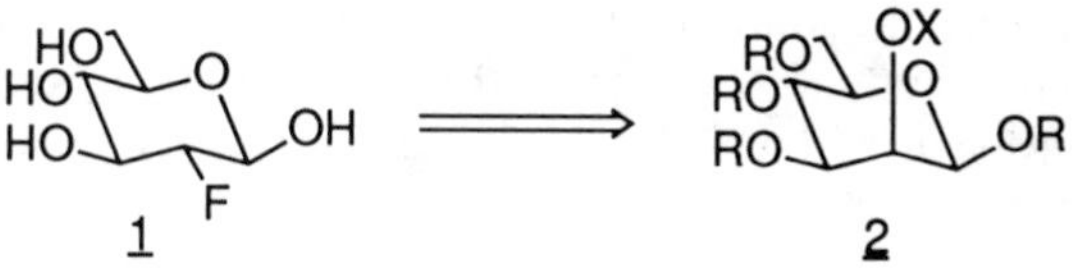

pyranoside reacts sluggishly in displacement reactions, the β-anomer is much more reactive. It was found that in 4,6-O-benzylidene-3-O-methyl-2-O-trifluoromethanesulfonyl-α-*manno*-pyranoside (**3**) the trifluoromethanesulfonate ester was easily displaced by cesium fluoride in dimethylformamide to form 2-fluoroglucopyranose (**4**) in 42% yield.[5] This reaction was also adapted to work with $Cs^{18}F$.[6]

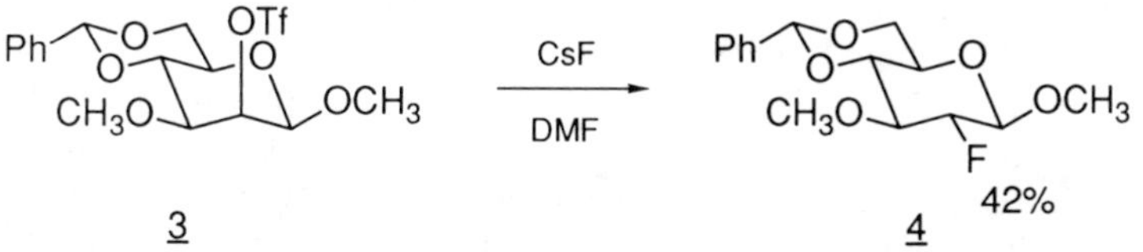

Other workers found that cleavage of the methyl ether was very difficult. When the protecting groups were varied, the most effective protection was the 3-O-benzyl-4,6-O-benzylidene combination as in **5**. The yields were consistently good and the deprotection facile.[7] By employing 1,6-anhydro-3,4-di-O-benzyl-2-O-

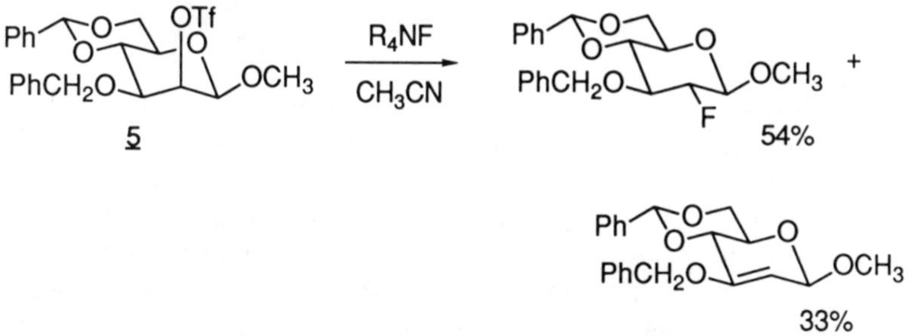

trifluoromethanesulfonyl-β-D-*manno*-pyranose (**6**) as the substrate for the displacement reaction, the tendency of the substrate to form hex-2-eno-pyranosides by elimination was suppressed.[8] The 1,6-anhydro bridge locks the carbohydrate into a conformation where the leaving group is trans diequatorial to the neighboring hydrogen. Displacement with tetramethylammonium fluoride resulted in 91% yield of the desired fluoride. Cleavage and deprotection proceeded in 70% yield for a 64% overall yield of the desired sugar (**7**).

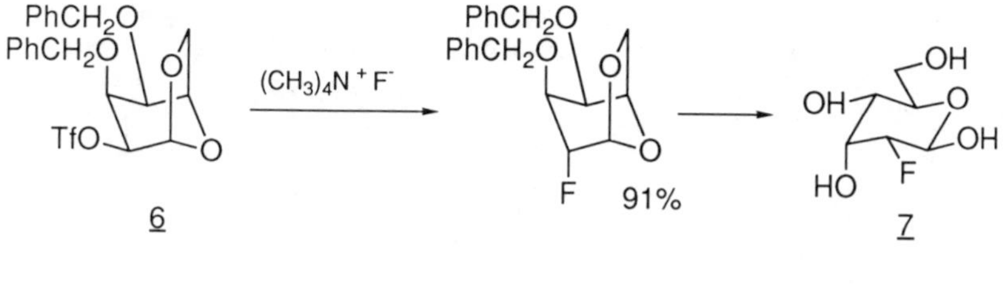

As described earlier, the source of the fluoride ion can also have an effect on the yield of the reaction. When **3** was treated with tris(dimethylamino)sulfonium difluorotrimethylsilicate[9] (TAS F⁻) the yield of the reaction did improve, although only to 61%.[10]

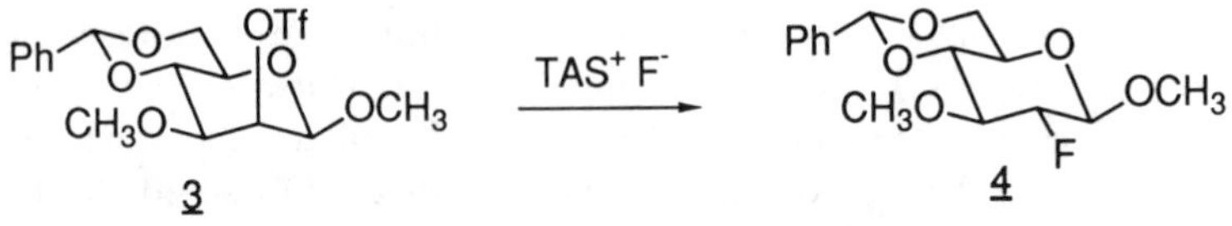

Preparation of 2-deoxy-2-fluoromannose requires reaction of the protected carbohydrate epimeric at carbon-2. In positron emission tomography using radiolabeled fluorinated carbohydrates, it was found that 2-deoxy-2-fluoromannose is also an effective agent for imaging subcutaneous AH109A tumors in rabbits.[11] Protected 2-deoxy-2-fluoro-D-mannose (**8**) was readily prepared by tetra-n-butylammonium fluoride treatment of benzyl 3,4,6-tri-O-benzyl-2-O-trifluoromethanesulfonyl-*manno*-pyranoside (**9**) in 77% yield.[12] As

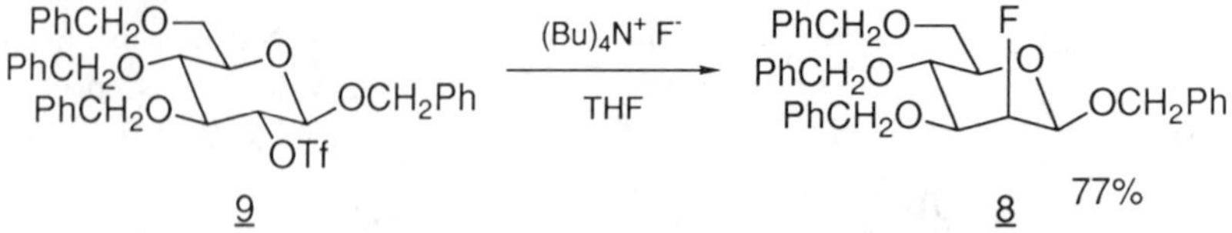

described for the preparation of 2-deoxy-2-fluoroglucoses, the choice of protecting group at carbon-3 and the anomeric conformation affect the yield and selectivity of the displacement reaction.[13] However, in the reactions of the β-D-*gluco* configuration, either the 3-O-methyl- or 3-O-benzyl- gives very comparable yields (81 and 77% respectively). Yet the attempted displacement reaction of methyl 3-O-benzoyl-4,6-benzylidene-2-O-trifluoromethanesulfonyl-α-D-*gluco*-pyranoside (**10**) with tetraethylammonium fluoride in acetonitrile led only

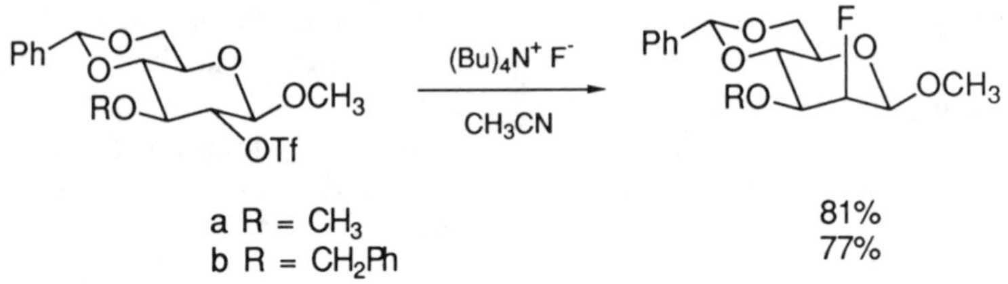

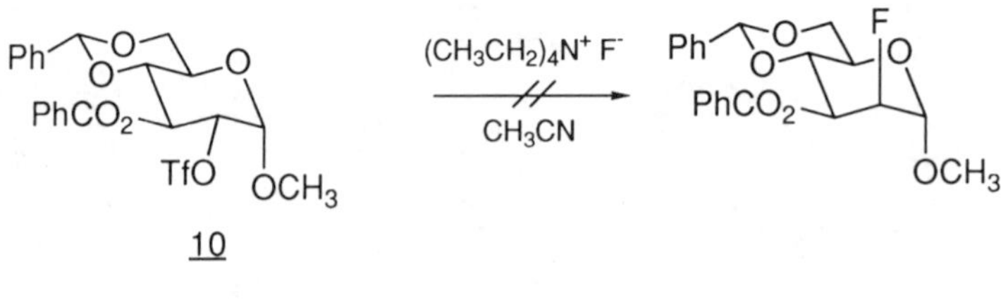

a R = CH₃
b R = CH₂Ph

81%
77%

10

to decomposition. Little difference on the displacement reactions was reported when the source of the fluoride ion was varied from tetramethylammonium fluoride, tetraethylammonium fluoride, or tetra-*n*-butylammonium fluoride. However, cesium fluoride in dimethylformamide was completely ineffective.

1-Deoxy-1-fluoro-D-fructose   The preparation of protected 1-deoxy-1-fluoro-D-fructose (**11**) was highly dependent on the source of the fluoride ion as well. Attempted direct fluorination of 2,3:4,5-di-*O*-isopropylidene-D-fructopyranose (**12**) with DAST was not successful.[14] Direct displacement of the mesylate in 2,3:4,5-di-*O*-isopropylidene-1-methanesulfonyl-D-fructopyranose (**13**) by tetra-*n*-butyl- ammonium fluoride formed the desired 1-deoxy-1-fluoro-D-fructose

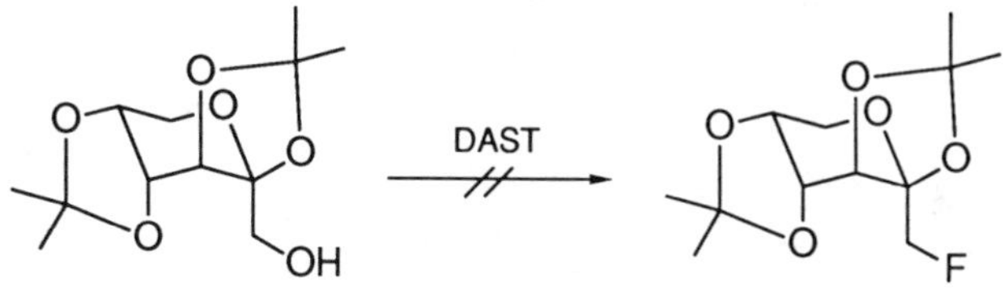

derivative (**14**) in only 9% yield. However, treatment of 2,3:4,5-di-*O*-isopropylidene-1-trifluoromethanesulfonyl-D-fructopyranose (**15**) with TAS F$^-$ in refluxing THF formed the desired fructose analog (**16**) in 80% yield.

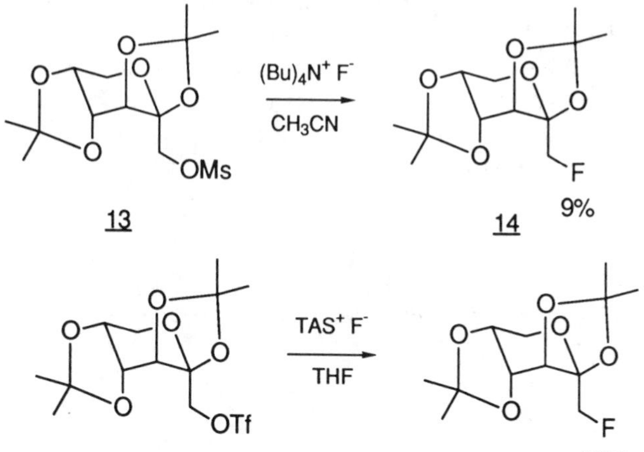

6-Deoxy-6-fluoro Carbohydrates   Displacement at primary positions is generally easier, as demonstrated by the displacement reactions to form 6-deoxy-6-fluoro carbohydrates. Fluorinated sugars have been incorporated in analogs of aminoglycoside antibiotics. Fluorination beta to an amino group of Kanamycin

A(**17**) was proposed to decrease the basicity of the amino group and thereby modify the binding of the analog. After careful manipulation of the functional groups, functionalization of Kanamycin by introduction of a trifluoromethanesulfonate at the desired 6″ position (**18**) was possible. Displacement of the leaving group by tetra-*n*-butylammonium fluoride produced the desired fluoride (**19**) in 76% yield.[15]

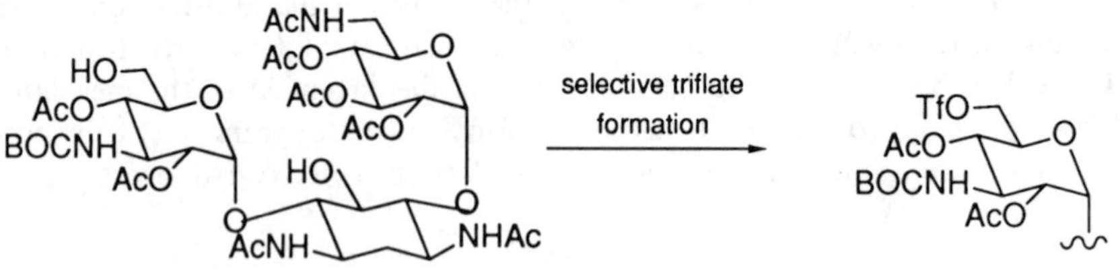

It was also possible to prepare the 6″ and 4″ ditriflate (**20**) and to successfully effect the double displacement of triflate to form the difluoride (**21**) in 79% yield.[16] When the 6″ triflate and the 4″ bromobenzenesulfonate (brosylate) (**22**) were reacted it was possible to selectively react the 6″ triflate. Differentiation of the groups was possible as the displacement of the brosylate required four days at room temperature to go to completion, whereas the triflate reacted in four hours.

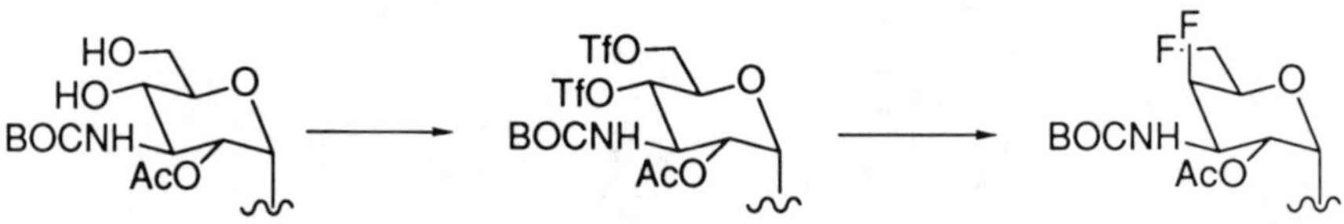

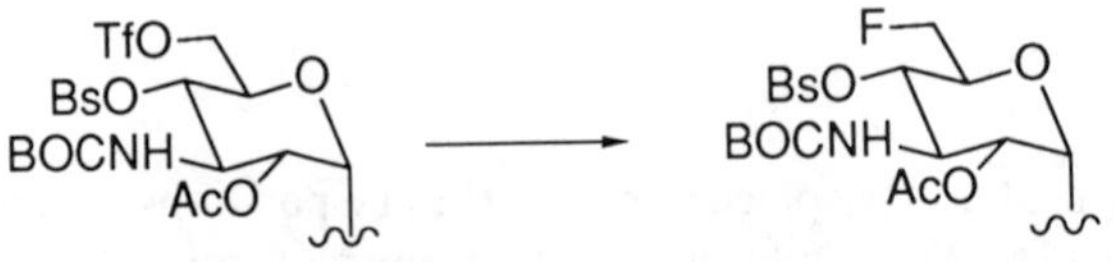

4-Deoxy-4-fluorolyxopyranoses   Synthesis of 4-deoxy-4-fluorolyxopyranoses is of particular interest in light of the importance of this configuration in the preparation of analogs of daunosamine. The preparation of these materials again illustrates the effect of neighboring groups on the displacement reaction. It was shown that epoxides will anchimerically assist the displacement by tetra-*n*-butylammonium fluoride[17] of neighboring trifluoromethanesulfonate esters. In a typical reaction **24** formed benzyl 2,3-anhydro-4-deoxy-4-fluoro-*β*-L-lyxo-pyranoside (**25**) in 60% yield and **26** formed benzyl 2,3-anhydro-4-deoxy-4-fluoro-*α*-D-lyxopyranoside (**27**) in 55% yield.

2-Deoxy-2-fluorofuranoses   Thus far the discussion of the use of triflate leaving groups in the fluorination of carbohydrates has been limited to their use either with pyranose conformations or, as in the case of 1-deoxy-1-fluorofructose, at a primary position of a furanose. Displacements at the 2 position of a furanose have been much less frequently attempted. The preparation of 2-deoxy-2-fluorofuranoses is of interest with the discovery that 2-deoxy-2-fluoroarabinose may be incorporated in a number of antiviral and anticancer nucleosides, such as 1-(2-deoxy-2-fluoro-*β*-D-*arabino*-furanosyl)-5-iodocytosine (**28**) and 1-(2-deoxy-2-fluoro-*β*-D-*arabino*-furanosyl)-5-methyluracil (**29**).[18] The 2-fluoroarabinose-containing nucleosides 1-(2-deoxy-2-fluoro-*β*-D-*arabino*-furanosyl)-5-alkenyl-cytosine (**30**)[19] and 1-(2-deoxy-2-fluoro-*β*-D-*arabino*-furanosyl)-5-alkyluracil (**31**)[20] also have good antiviral activity as well as much improved therapeutic indices (LD_{50}/ED_{90}).

The attempted displacement of 1-*O*-acetyl-3,5-*O*-dibenzoyl-2-*O*-methane-sulfonyl-*α*-*ribo*-furanoside (**32**) or 1-*O*-acetyl-3,5-*O*-dibenzoyl-2-*O*-trifluoro-methanesulfonyl-*α*-*ribo*-furanoside (**33**) with either tetra-*n*-butylammonium fluoride or potassium bifluoride in a variety of solvents led to elimination or

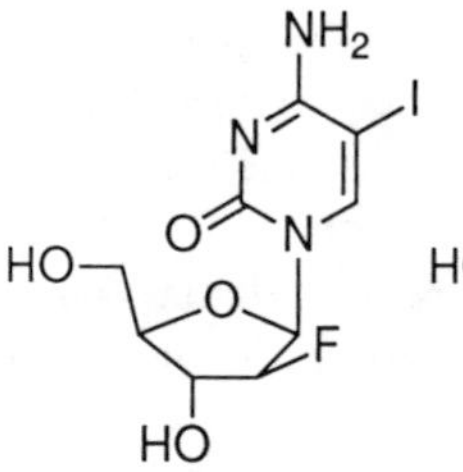

28 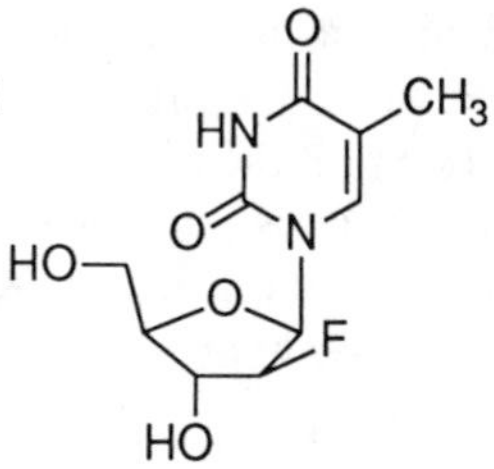   29 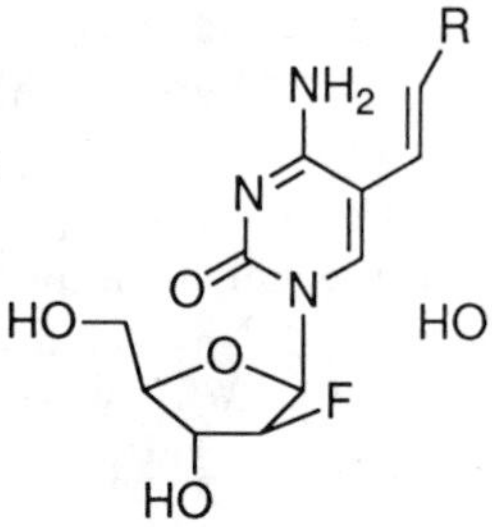   30 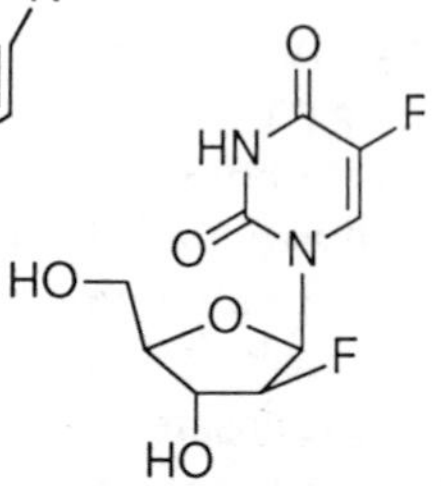   31

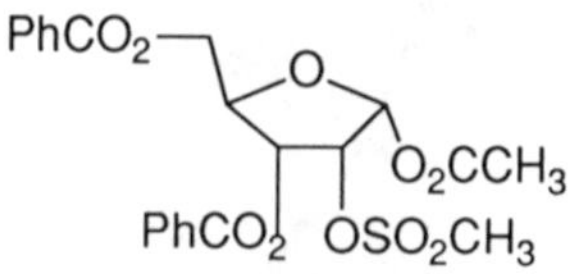

32

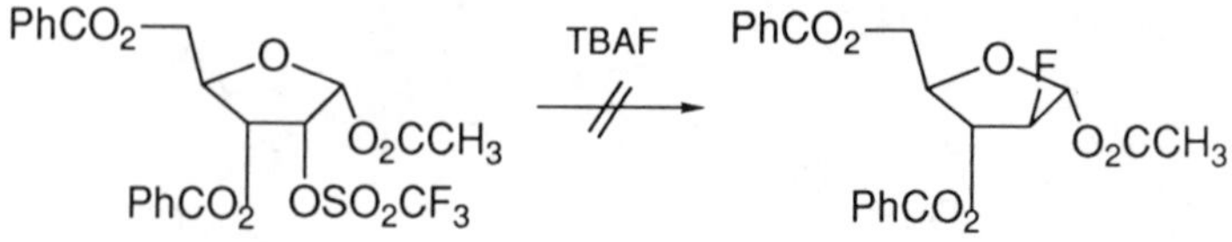

33

no reaction. As will be seen later, this obstacle can be overcome simply by a change of the leaving group in the reaction.

Synthesis of Trifluoromethanesulfonate Esters   Although the utility of tri-fluoromethanesulfonate esters in the preparation of fluorinated carbohydrates is clear, it has not been pointed out that the synthesis of these very useful esters can be difficult. The synthesis and displacement reactions of 1,2,3,4-tetra-*O*-acetyl-6-*O*-trifluoromethanesulfonyl-*gluco*-pyranose (**34**) have been studied in depth.[21] The use of a hindered base, such as 2,6-di-*t*-butyl-4-methyl-pyridine, was reported to give the highest yield of the desired ester, suppressing displacement of

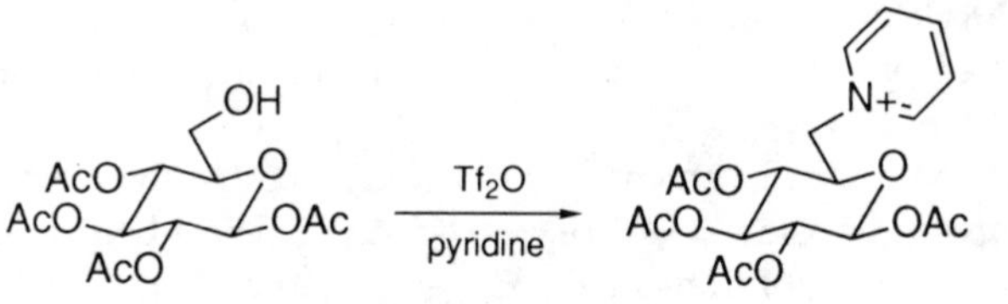

34

the ester by the base. Conversion of the 6-trifluoromethanesulfonate (**35**) prepared in this way to the fluoride (**36**) with tetra-*n*-butylammonium fluoride proceeded in 27% yield.

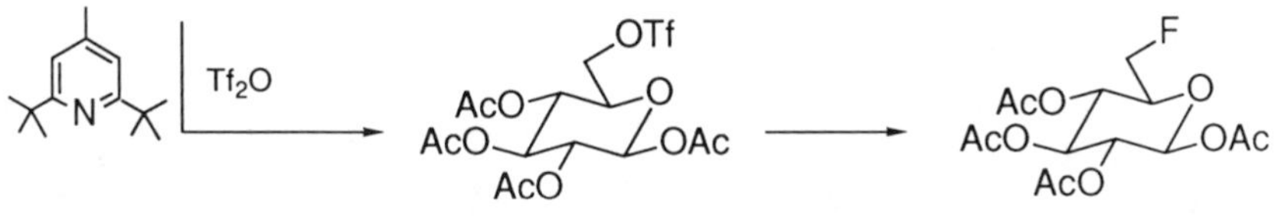

Imidazolesulfonates

As mentioned previously, the attempted displacement of 1-*O*-acetyl-3,5-*O*-dibenzoyl-2-*O*-methanesulfonyl-α-*ribo*-furanoside (**32**) or 1-*O*-acetyl-3,5-*O*-dibenzoyl-2-*O*-trifluoromethanesulfonyl-α-*ribo*-furanoside (**33**) with fluoride ion failed. However, substitution of the trifluoromethanesulfonyl group by the imidazolesulfonate group was very effective at improving the yield of the fluorination step.[22] Treatment of 1-*O*-acetyl-3,5-*O*-dibenzoyl-2-*O*-imidazole-sulfonyl-α-*ribo* furanoside (**37**) with potassium bifluoride and four equivalents of 50% aqueous hydrogen fluoride did yield 1-*O*-acetyl-2-deoxy-3,5-*O*-dibenzoyl-2-fluoro-α-*ribo*-furanoside (**39**).[23] However, treatment of **37** with tetra-*n*-butyl-

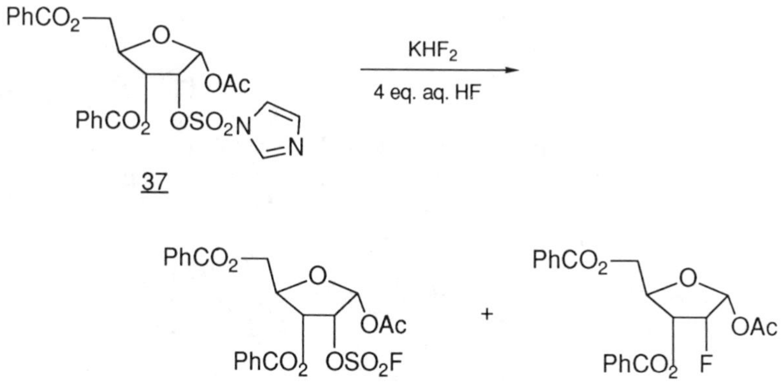

ammonium fluoride failed to yield the fluorinated product. It was suggested that the reactive intermediate in the displacement reaction is the fluorosulfate (**38**).

Cyclic Sulfite or Sulfate Esters

Both the 2,3-cyclic sulfates of methyl 4,6-*O*-benzylidene-α- and β-D-*manno*-pyranosides (**40**) and sulfites (**41**) were prepared by reaction of the 2,3-diols with sulfuryl chloride and thionyl chloride respectively.[24] The sulfates and sulfites were treated with anhydrous tetramethylammonium fluoride, prepared by

40 a = α
    b = β

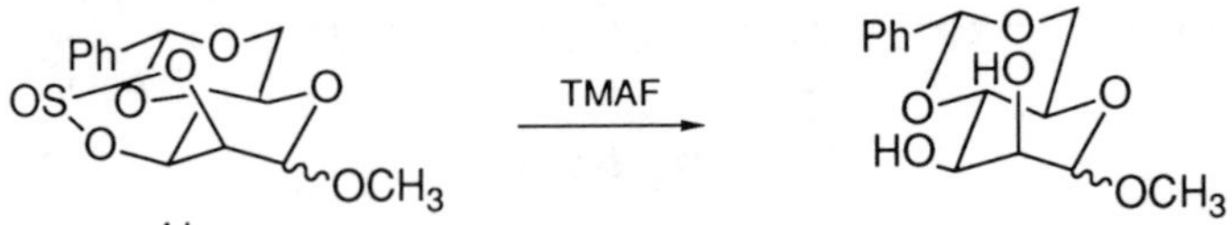

41 a = α
    b = β

azeotropic drying with acetonitrile. The reaction of the sulfites was very slow and gave only the diol products of hydrolysis. The sulfate prepared from methyl 4,6-*O*-benzylidene-α-D-*manno*-pyranoside yielded only 4,6-*O*-benzylidene-1,2-dideoxy-D-*erythro*-hex-1-enopyrano-3-ulose (**42**). However, **40b** reacted very

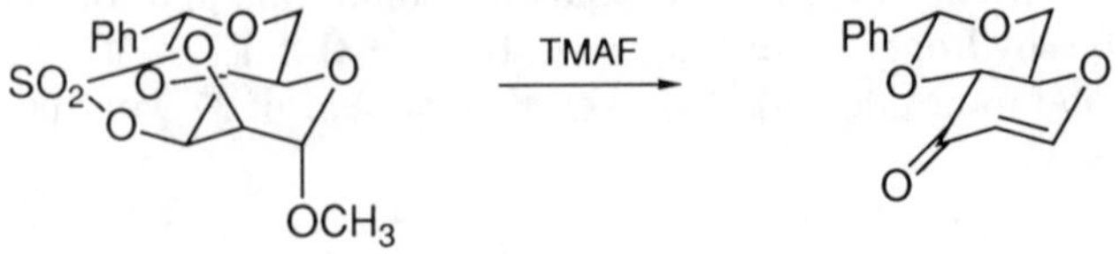

40 a   42

cleanly with tetramethylammonium fluoride (TMAF) to form methyl 2-deoxy-2-fluoro-β-D-*gluco*-pyranoside triacetate (**43**) on hydrolysis and acylation in 84% yield. The procedure also worked well with TMA^{18}F.

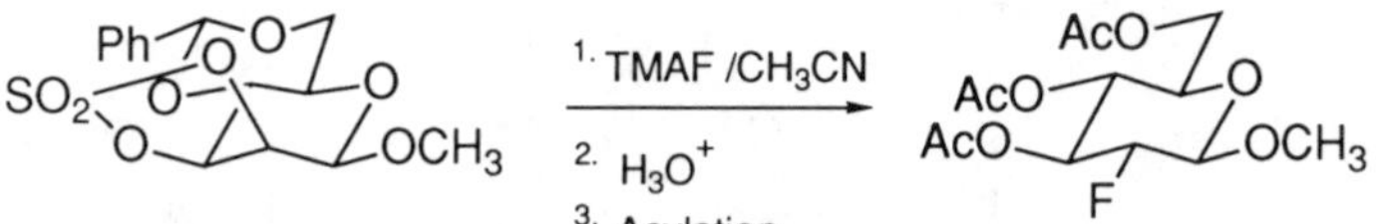

40 b   43

4-Bromobenzenesulfonates

When methyl 2,3-di-*O*-benzyl-4-(4'-bromobenzenesulfonate)-6-*O*-trityl-β-D-*gluco*-pyranoside (**44**) was treated with tetra-*n*-butylammonium fluoride it formed the corresponding 4-deoxy-4-fluoro-D-*galacto*-derivative (**45**), in only 30% yield.[26] However, treatment of **44** with fluoride from an ion exchange resin worked much better. Amberlyst A-26 (F$^-$ form) was dehydrated by azeotropic drying with benzene and was allowed to react with **44** to form **45** in 75% yield. As

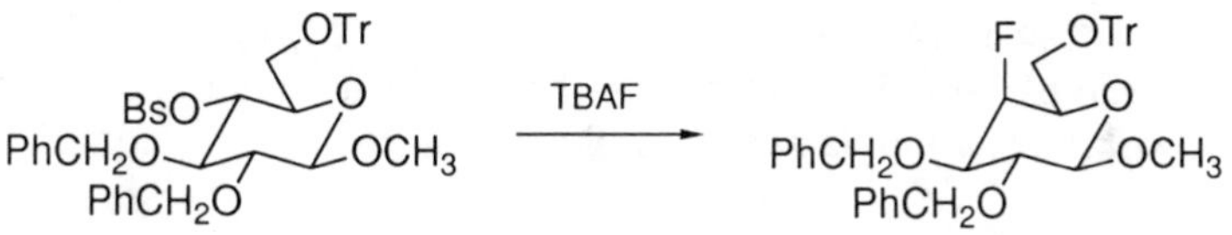

44   TBAF   45

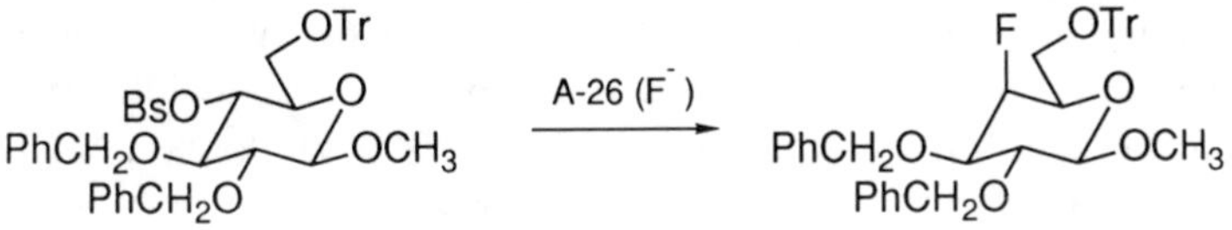

30%

44   A-26 (F⁻)   45

75%

a fluoride ion source, the resin was reported to be easier to handle and less sensitive to moisture.

The same type of displacement reaction was successfully effected with the furanose 3-O-(4'-bromobenzenesulfonyl)-1,2:5,6-di-O-isopropylidene-α-D-*gulo*-furanose (**46**)[27] where 3-deoxy-1,2:5,6-di-O-isopropylidene-3-fluoro-α-D galactofuranose (**47**) was formed in 52% yield.

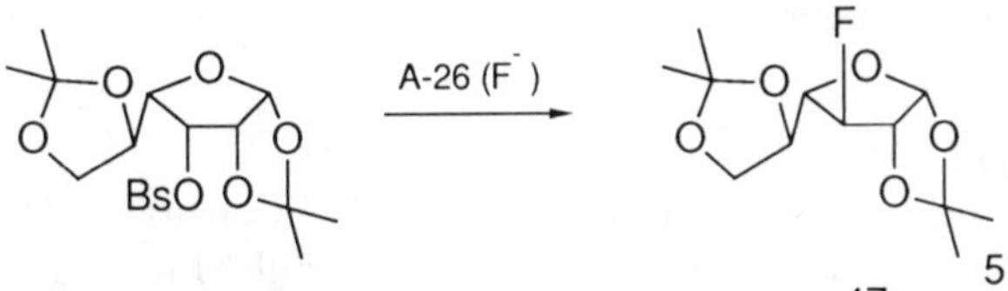

46   A-26 (F⁻)   47

51%

Methanesulfonates

Methanesulfonates (mesylates) have found less general application in the preparation of fluorinated carbohydrates, but with methanesulfonates, selective reactions are possible. The diminished reactivity of methanesulfonates relative to trifluoromethanesulfonates makes possible the selective formation of 6-deoxy-2,3-di-O-benzyl-6-fluoro-4-O-mesyl-α-D-*gluco*-pyranosyl 1,3,4,6-tetra-O-benzyl-

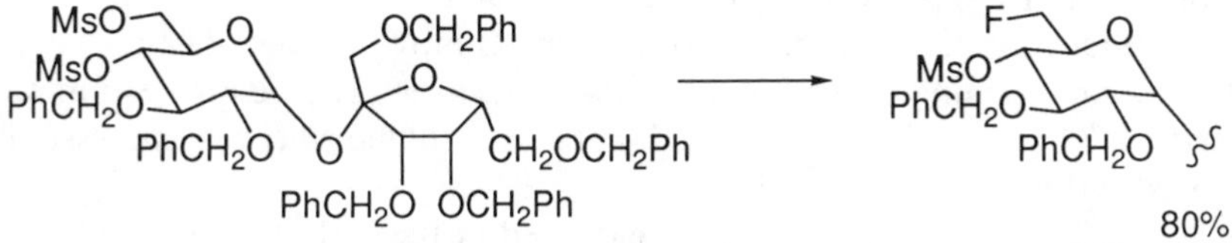

48   49

80%

β-D-fructofuranoside (**49**) in 80% yield from 2,3-di-*O*-benzyl-4,6-di-*O*-mesyl-α-D-*gluco*-pyranosyl 1,3,4,6-tetra-*O*-benzyl-β-D-fructofuranoside (**48**) on treatment with tetra-*n*-butylammonium fluoride in refluxing acetonitrile.[28]

8.3 RING-OPENING REACTIONS

The use of hydrogen fluoride to open epoxides has been an especially successful method for the formal replacement of a secondary hydroxyl group by fluorine. The tendency of epoxides to be cleaved transdiaxially is also useful in predicting product formation. Most commonly hydrogen fluoride or potassium bifluoride has been used to open the epoxide. The aziridines may be opened in a similar fashion when introduction of an amino functional group is also required.

An early application of an epoxide ring-opening reaction for the introduction of fluorine into a carbohydrate was the formation methyl 3-deoxy-3-fluoro-β-L-*xylo*-pyranoside (**50**) on treatment of methyl 2,3-anhydro-5-*O*-benzyl-β-L-*ribo*-pyranoside with hydrogen fluoride in 1,4-dioxane.[29] Another early example is

the preparation of 3-deoxy-3-fluoro-D-xylose (**52**) by treatment of benzyl 2,3-anhydro-β-D-*ribo*-pyranoside (**51**) with potassium bifluoride.[30] In many cases, however, the regiochemical control of the ring-opening reaction is poor.

Another way to apply the ring-opening reaction of epoxides is demonstrated in the preparation of 2-deoxy-2-fluoro-D-glucose. 1,2-Anhydro-3,4:5,6-di-*O*-isopropylidene-1-*C*-nitro-D-mannitol (**54**) was prepared by epoxidation of 1-deoxy-1,2-dehydro-3,4:5,6-di-*O*-isopropylidene-*C*-nitro-D-*manno*-hexitol (**53**). Following ring opening with potassium bifluoride in 79% yield, deprotection was effected in 85% yield.[31] Radiolabeled potassium bifluoride was successfully employed in the synthesis as well.[32]

When neighboring amino groups are desired in the carbohydrate synthesis, as mentioned earlier, an aziridine ring can be employed in the ring-opening

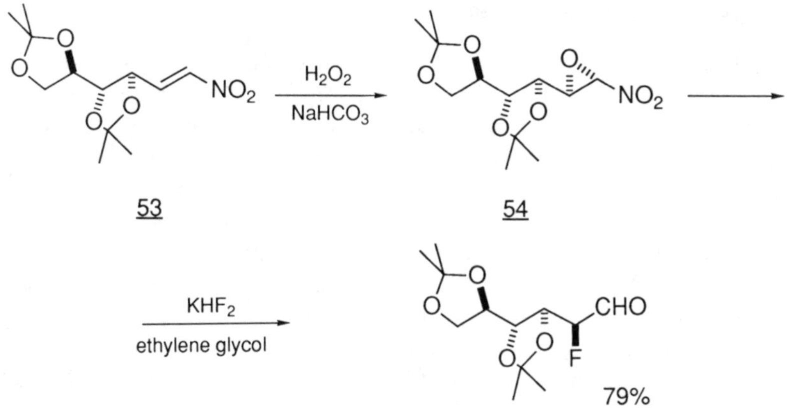

53   54

79%

process. A very clever application of this concept was employed for the preparation of methyl 4,6-*O*-benzylidene-2,3-dideoxy-3-amino-2-fluoro-altro-pyranoside (**57**).[33] When methyl 3-deoxy-4,6-*O*-benzylidene-3-diallylamino-2-*O*-methanesulfonate-altropyranoside (**55**) was treated with tetraethylammonium fluoride or triethylamine tris (hydrogen fluoride) loss of methanesulfonate was accompanied by neighboring group participation by the diallylamino function. The intermediate aziridinium ion (**56**) was regioselectively opened by the fluoride ion.

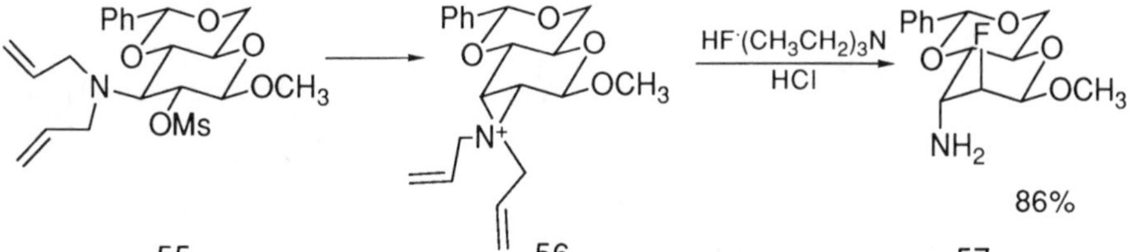

55   56   57

86%

The propensity of the diallylamino group to participate in the displacement reaction was general. The regiochemistry of the reaction was controlled by the steric effects of the remaining carbohydrate substitutents. The methyl 3-deoxy-4,6-*O*-benzylidene-3-dimethylamino-2-*O*-methanesulfonate-altro-pyranoside (**58**) will also undergo the same reaction with tetraethylammonium fluoride in acetonitrile or neat with triethylamine tris (hydrogen fluoride).

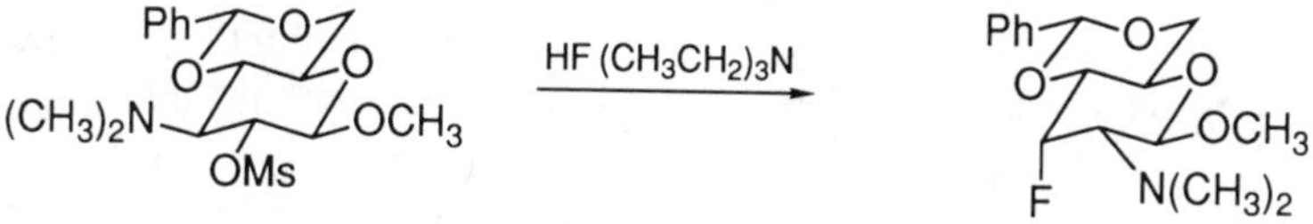

58

8.3.1   2-Fluoro-L-daunosamine

Although the anthracycline antibiotics are effective, clinically useful antitumor agents, the significant side-effects that accompany their administration have stimulated research into more specific analogs. The approach just described was employed in a very high, 70% overall, yield synthesis of 3-amino-2-fluoro-2,3,6-trideoxy-L-galactose, protected 2-fluoro-L-daunosamine (63) starting from methyl 2,3-anhydro-4,6-O-benzylidene-α-D-*manno*-pyranoside (59).[34] The *altro* compound (62) has been prepared from 59. The alcohol formed on opening of the epoxide with diallyl amine, was converted to a methanesulfonate ester (60). Treatment with triethylamine–hydrogen fluoride yielded the 2-fluoro compound (61) via the intermediacy of an aziridinium ion. Bromination, dehydrohalogenation, saponification, and reduction of 62 formed the 2-fluoro-*galacto* analog (63).

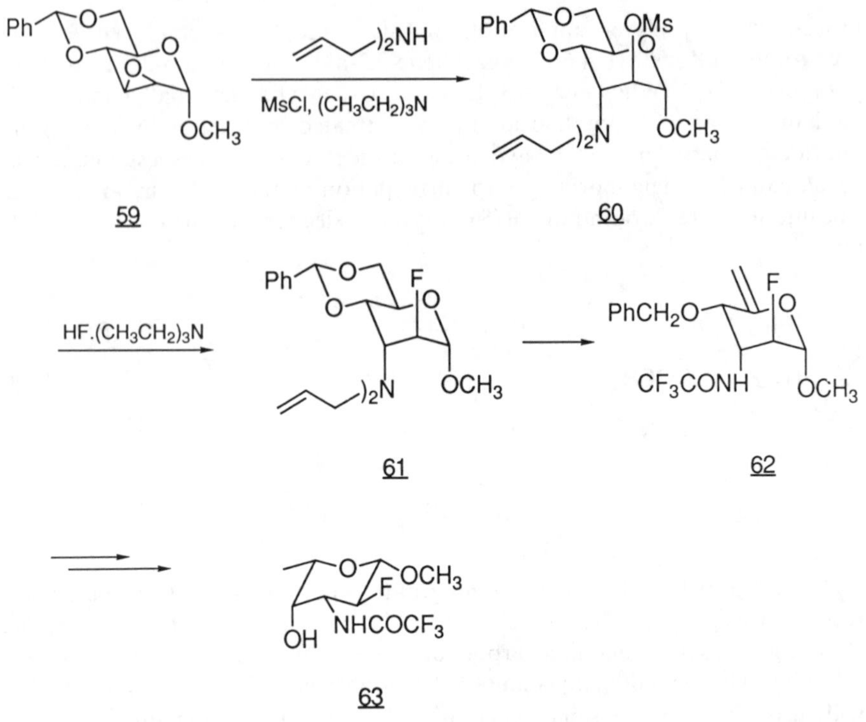

An alternative approach to the synthesis of 2-fluorodaunosamine was effected independently by two different groups, both relied on the ring-opening reaction of an aziridine[35] in the key step of the synthesis. Methyl 2-benzamido-4,6-O-benzylidene-2-deoxy-3-O-toluenesulfonyl-α-*gluco*-pyranoside (64) was treated with tetra-*n*-butylammonium fluoride to form methyl 3-benzamido-4,6-O-benzylidene-2,3-dideoxy-2-fluoro-α-D-altropyranoside in 61% yield[36] via an unisolated aziridine intermediate. Conventional manipulations including bromination, dehydrobromination, and reduction formed the desired L sugar (**63**). For efficient scaleup the excess of tetra-*n*-butylammonium fluoride had to be decreased, which lowered the yield of fluorinated product. On the larger scale, these workers treated the isolated aziridine, benzyl 4,6-*O*-benzylidene-2,3-benzoylepimino-2,3-dideoxy-α-D-*allo*-pyranoside (**65**), with tetra-*n*-butyl-ammonium fluoride in hexamethyl phosphoramide to form **66** in 38% yield while recovering 40% of the unreacted epimine.[37]

64

63 (protected)

64

65

38%

66

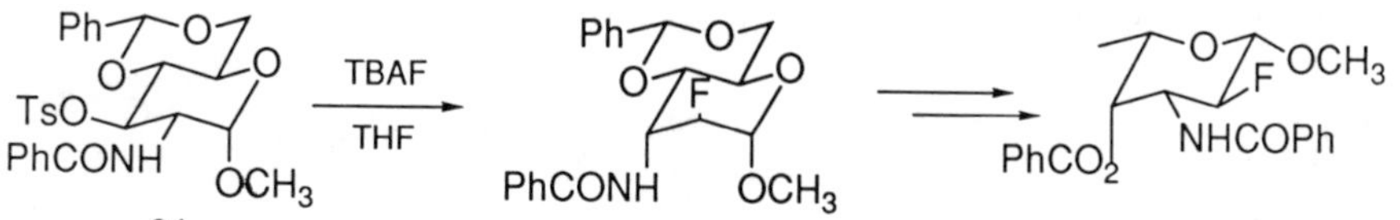

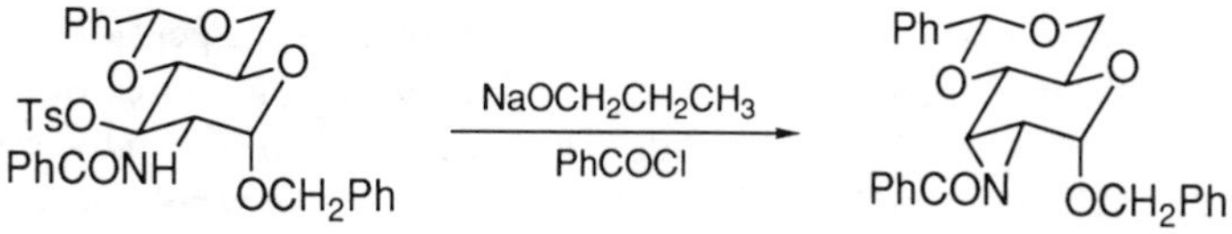

8.4 ADDITIONS TO OLEFINS

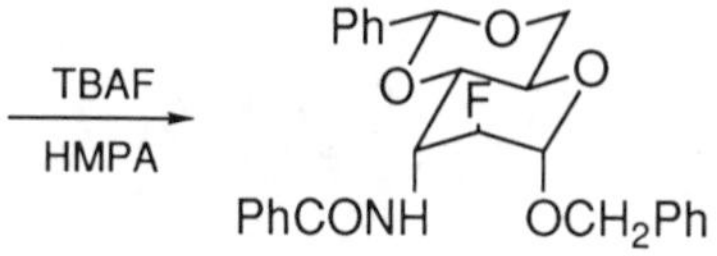

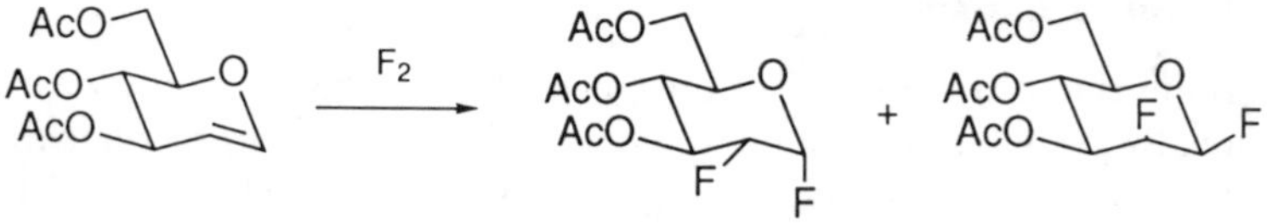

Electrophilic additions of fluorine to unsaturated carbohydrate centers has principally been applied to halogenation of glucals. Use of trifluoromethyl-hypofluorite was first reported in the reaction of glucals in 1969.[38] Later the direct addition of fluorine to triacetyl glucal (**67**) in trichlorofluoromethane soution was reported to yield 3,4,6-tri-*O*-acetyl-2-deoxy-2-fluoro-α-D-*gluco*-pyranosyl

35%

26%

67

68

69

fluoride **(68)** and 3,4,6-tri-O-acetyl-2-deoxy-2-fluoro-β-D-*manno*-pyranosyl fluoride **(69)**.[39]

8.4.1 Molecular Fluorine

Direct fluorination in water was used to prepare ^{18}F-labeled materials after the work just described.[40] Fluorination of the parent glucal in water gave the same product stereochemistry.[41] Reaction of the fluorine with water to form a reactive species is certainly likely, but products that can be definitely determined as being derived from such a species have not been isolated.

$$(67) \longrightarrow {}^{18}F^-(68) + {}^{18}F^-(69)$$

8.4.2 Acetyl Hypofluorite

The stereoselectivity of the addition is significantly improved when acetyl hypofluorite, prepared by bubbling fluorine through a suspension of sodium acetate in trichlorofluoromethane at $-78°C$, is used as a source of fluorine[42] in reactions. Addition of acetyl hypofluoride to a solution of tri-O-acetyl glucal **(67)** yields 1,3,4,6-tetra-O-acetyl-deoxy-2-fluoro-α-D-*gluco*-pyranose **(70)** in 78% yield. Treatment of the corresponding galactal **(71)** 1,3,4,6-tetra-O-acetyl-2-deoxy-2-fluoro-α-D-*galacto*-pyranose **(72)** in 84% yield.

8.4.3 Xenon Difluoride

Xenon difluoride is an especially convenient fluorinating reagent as it is both commercially available and a solid. It reacts with **67** or **71** to form, on hydrolysis of the α- and β-pyranosyl fluorides gives 3,4,6-tri-O-acetyl-2-deoxy-2-fluoro-glucose **(73)** or 2-deoxy-2-fluorogalactose **(74)** in 73% and 80% yields respectively.[43]

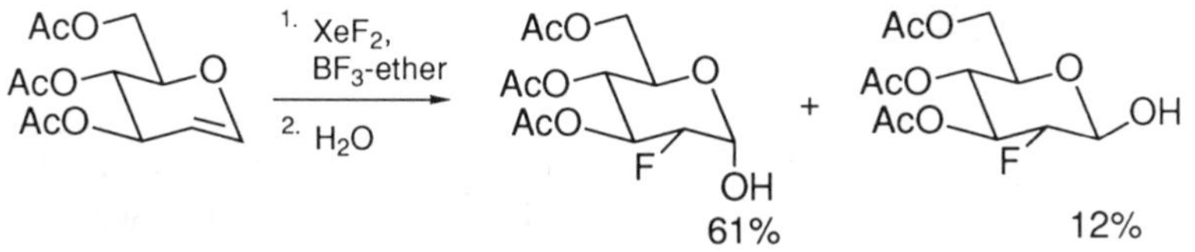

67   73a   73b

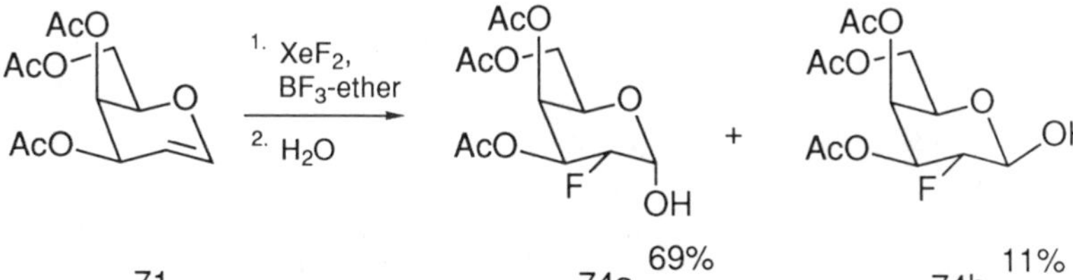

61%   12%

71   74a   74b

69%   11%

8.4.4 Trifluoromethylhypofluorite

One of the first reagents to be employed in the classic fluorination reactions of glucals, trifluoromethylhypofluorite has also been used for the preparation of branched fluorinated carbohydrates. When (E)-3-deoxy-3-C-ethoxycarbonyl-(formylamino) methylene-1,2:5,6-di-O-isopropylidene-α-D-*gluco*-furanose (**75**) was treated with trifluoromethyl hypofluorite 3-deoxy-3-C-ethoxyoxalyl-3-fluoro-1,2:5,6-di-O-isopropylidene-α-D-*gluco*-furanose (**76**) was formed in 65% yield.[44]

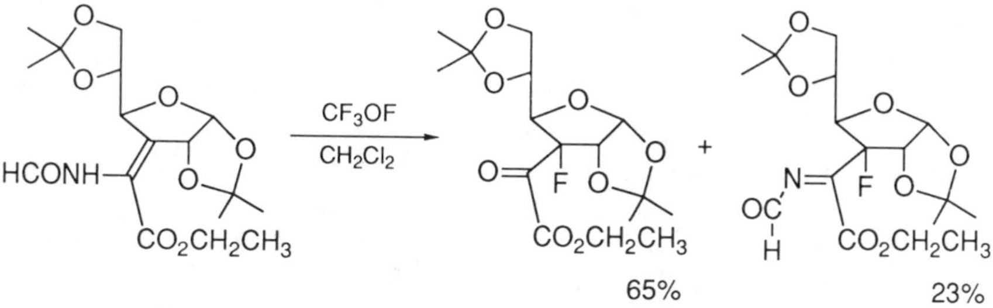

75   76

65%   23%

8.4.5 Fluorinative Dehydroxylations

An obvious extension of the use of displacement reactions of sulfonate esters by a fluoride ion is the in situ formation of the reactive ester by displacement with fluoride ion. Sulfur tetrafluoride is potentially an ideal reagent, but its low boiling point and capricious reactivity seriously limit its use. The discovery of diethylaminosulfur trifluoride (DAST), an easily handled, relatively stable liquid, has made this reaction convenient.[45]

Early experiments with DAST demonstrated that even acetates could serve as effective protecting groups for the fluorination reaction. Treatment of 1,2,3,4-tetra-O-acetyl-glucose (**77**) with DAST yielded 6-deoxy-6-fluoro-glucose (**78**).[46]

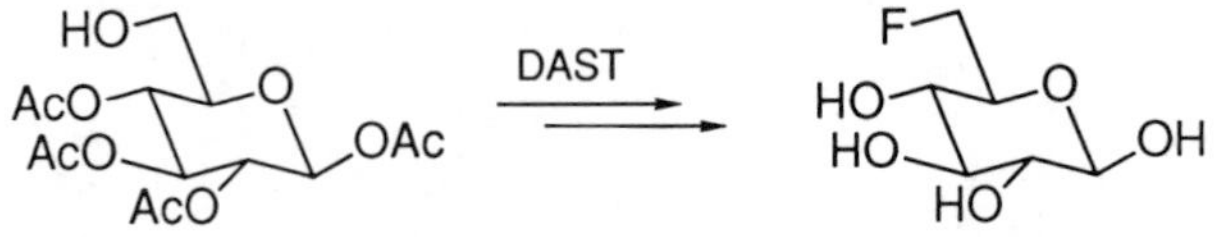

DAST was found to be particularly useful for the fluorination of unprotected or partially protected carbohydrates. Methyl α-D-*gluco*-pyranoside (**79**) with neat DAST will afford 4,6-dideoxy-4,6-difluoro-α-*galacto*-pyranoside (**80**).[47] However, treatment of **79** with DAST in methylene chloride yielded only methyl 6-deoxy-6-fluoro-α-D-*gluco*-pyranoside (**81**) in 88% yield. It was suggested that the

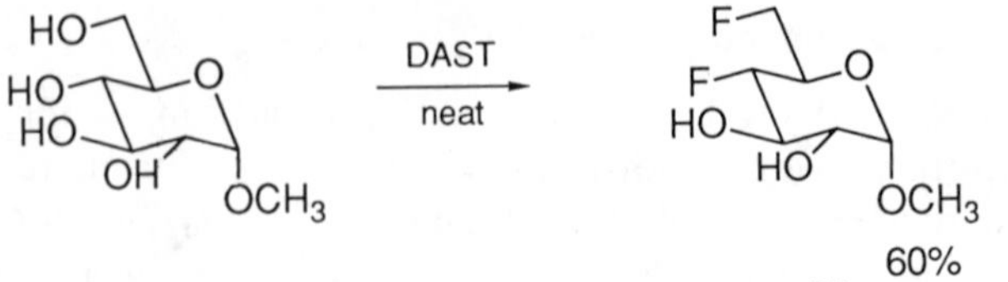

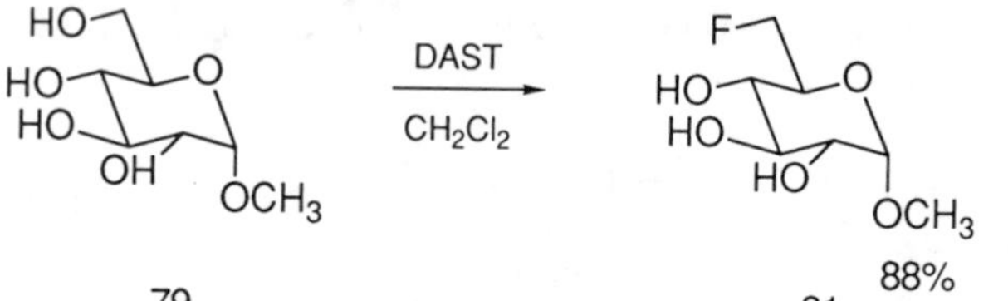

reaction required an intramolecular transfer of fluoride ion, since displacement reactions at C-4 of mannose derivatives normally are obstructed by the axial hydroxyl at C-2 (Fig. 8.2). Methyl 2,3,6-tri-O-benzoyl-4-deoxy-4-fluoro-α-D-*gluco*-pyranoside (**83**) was formed in 41% yield from methyl 2,3,6-tri-O-benzoyl-

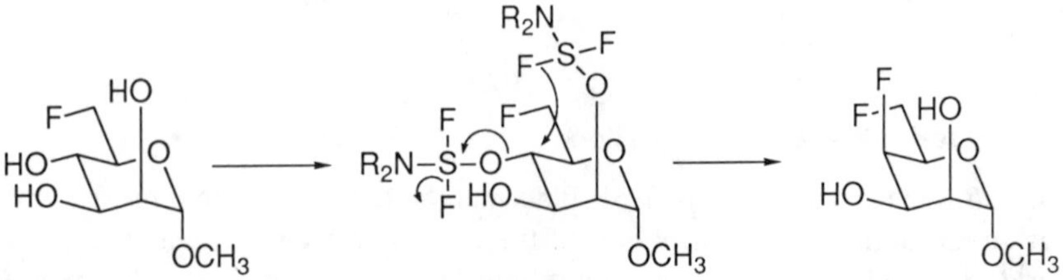

FIGURE 8.2

α-D-*galacto*-pyranoside (**82**).[48] The supposed hindrance to reaction by axial substituents was amply illustrated by the failure of methyl β-D-*galacto*-pyranoside (**84**) to react with DAST to form methyl 6-deoxy-6-fluoro-β-D-*galacto*-pyranoside (**85**).[26]

A final illustration of the effect of configuration on selectivity may be found in the reactions of methyl β-D-*xylo*-pyranoside (**86**) to form methyl 4-deoxy-4-fluoro-β-L-*arabino*-pyranoside) (**87**).[49] It has been suggested that the selectivity

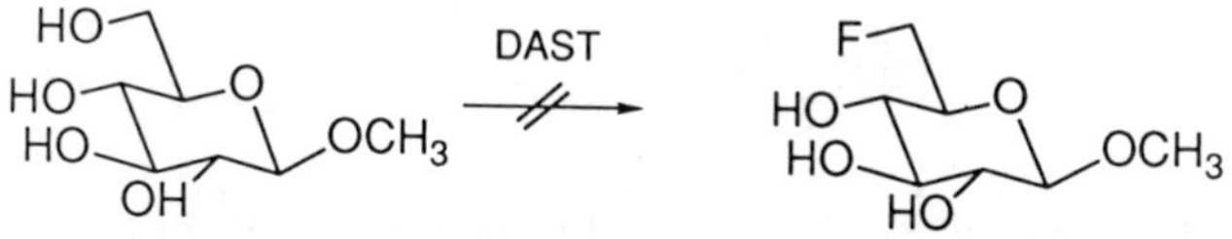

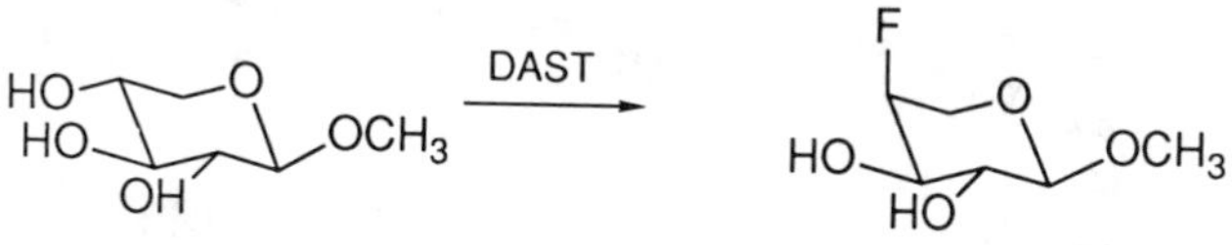

for product formation resulted from reaction occurring in the $^1C_4$ conformation where only C-4 is sterically unhindered toward S_N2 displacement (Fig. 8.3).

Fluorination of protected carbohydrates has also proven extremely useful for the synthesis of selectively fluorinated carbohydrates. 1-(5-*O*-trityl-2-deoxy-β-D-

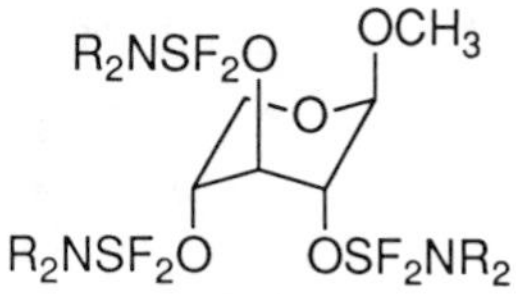

FIGURE 8.3

threo-pentofuranosyl)thymine (**88**) was fluorinated with DAST in benzene to form 3'-fluoro-3'-deoxythymidine (**89**) in 62% yield.[50] 1,6-Anhydro-2-azido-2-deoxy-4-*O*-benzyl-β-D-glycopyranose (**90**) when treated with DAST formed 1,6-anhydro-2-azido-2-deoxy-4-*O*-benzyl-3-fluoro-β-D-glycopyranose (**91**).[51]

88 89

90 91

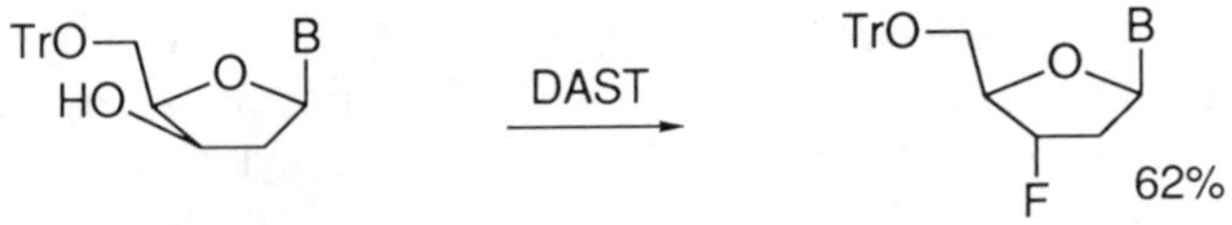

The direct fluorination of 1,2:5,6-di-*O*-isopropylidene-α-D-*gulo*-furanose (**92**) with neat DAST or DAST in methylene chloride leads to only partial reaction. However, the addition of 4-(dimethylamino)pyridine improved the isolated yield of 3-deoxy-3-fluoro-1,2:5,6-di-*O*-isopropylidene-α-D-*galacto*-furanose (**90**) in 90% yield.[27]

92 93 90%

94

59%

95

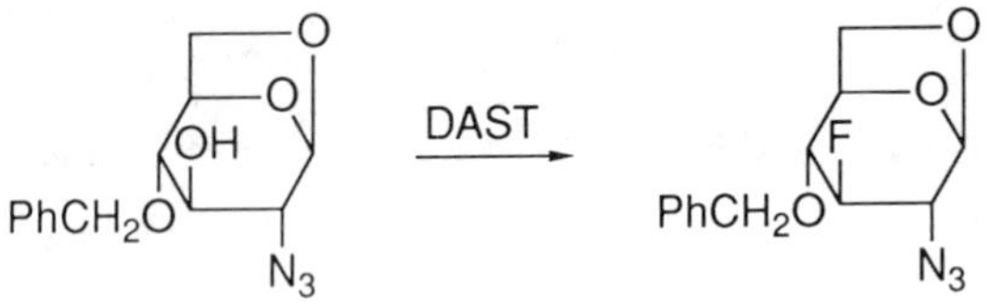

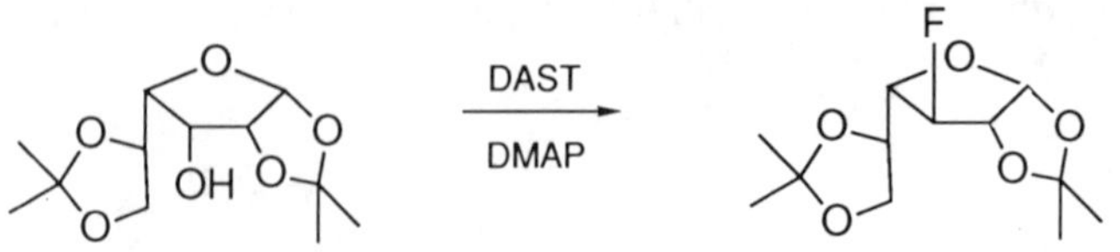

Sporaracin B (**94**) is fluorinated by treatment with DAST after protection to form the 3-epi-3-fluoro sporaracin (**95**) in 59% yield.[52]

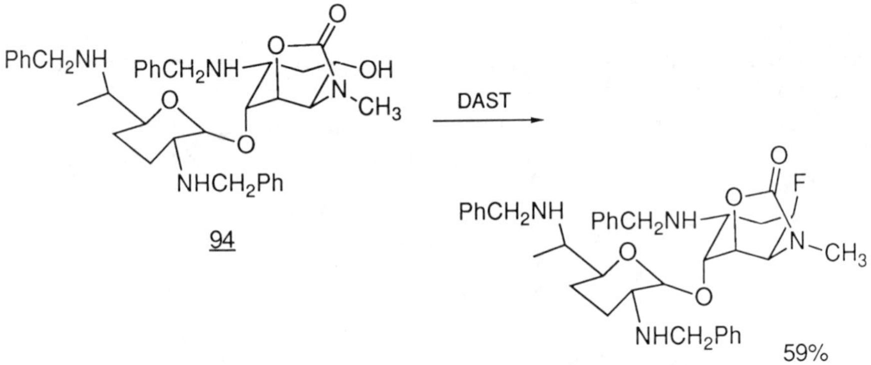

Finally, it was possible to take advantage of the ability of DAST to convert carbonyl compounds to geminal difluorides in 72% yield (Fig. 8.4).[53] However,

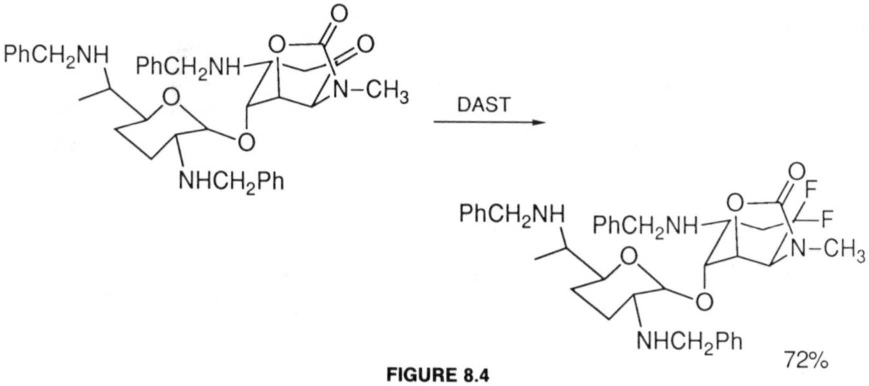

FIGURE 8.4

rearrangements do occur when neighboring group participation is possible.[54] When benzyl 3-azido-4,6-O-benzylidene-3-deoxy-α-D-*altro*-pyranoside (**96**) is treated with DAST it forms benzyl 3-fluoro-α-D-*gluco*-pyranoside (**97**) in 40% yield and 3-azido-4,6-O-benzylidene-2,3-dideoxy-2-fluoro-α-*altro*-pyranoside (**98**) in 58% yield.

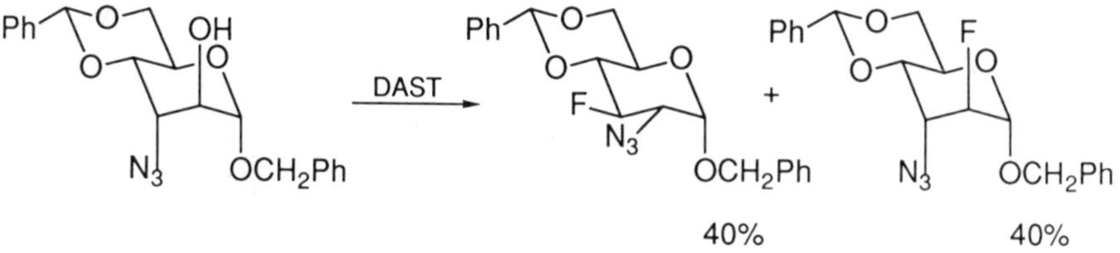

Geminal difluorination at the 2 position of carbohydrates was shown to be sensitive to the steric environment exactly as was found earlier in the dehydroxylative fluorination of hydroxyl groups. Treatment of methyl 3-azido-4,6-O benzylidene-3-deoxy-α-D-*ribo*-hexopyranosid-2-ulose (**99**) with DAST failed to form methyl 3-azido-4,6-O-benzylidene-2,3-dideoxy-2,2-difluoro-α-D-*gluco*-pyranoside (**100**).[55]

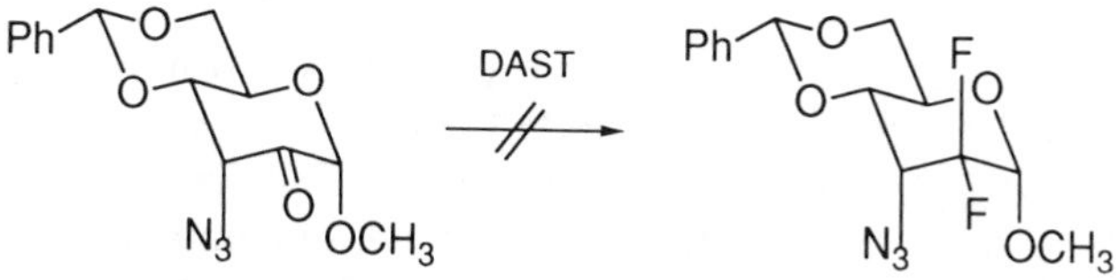

8.5 TOTAL SYNTHESIS

The total synthesis of fluorinated carbohydrates has traditionally been considered an approach of limited utility. The earliest synthetic strategies were based on Claisen or aldol condensation reactions.

8.5.1 Claisen Condensations

The Claisen self-condensation of ethyl fluoroacetate yielded ethyl 2,4-difluoro-3-oxo-butanoate (101), which on reduction with potassium borohydride formed (±)2,4-difluoro-1,3-butanediol (102).[56]

$$2 \text{ FCH}_2\text{CO}_2\text{CH}_2\text{CH}_3 \xrightarrow{\text{NaOCH}_2\text{CH}_3} \text{FCH}_2\text{COCHFCO}_2\text{CH}_2\text{CH}_3$$

101

$$\xrightarrow{\text{KBH}_4} \text{FCH}_2\text{CH(OH)CFHCH}_2\text{OH}$$

102

Claisen condensation of ethyl fluoroacetate with methyl 2,3-O-isopropylidene-D, L-glycerate (103) yielded on reduction and deprotection 2-deoxy-2-fluoro-D, L-ribitol (104).[57]

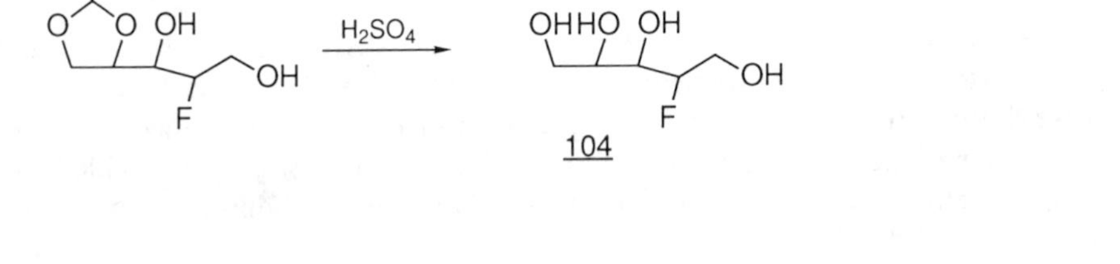

104

8.5.2 Aldol Condensations

Recently aldol condensation was utilized with preformed fluorinated enolates. The lithium enolate of ethyl fluoroacetate was prepared by treatment with lithium hexamethyldisilazide. Addition of (R)-2,3-O-isopropylidene-glyceraldehyde (105) resulted in formation of a mixture of compounds 106 and 107 with good control of stereochemistry, as predicted by Felkin–Anh. Reduction with diisobutylaluminum hydride followed by deprotection yielded directly a mixture of 2-deoxy-2-fluoro-ribo- and arabino-D-pyranoses (108, 109).[58]

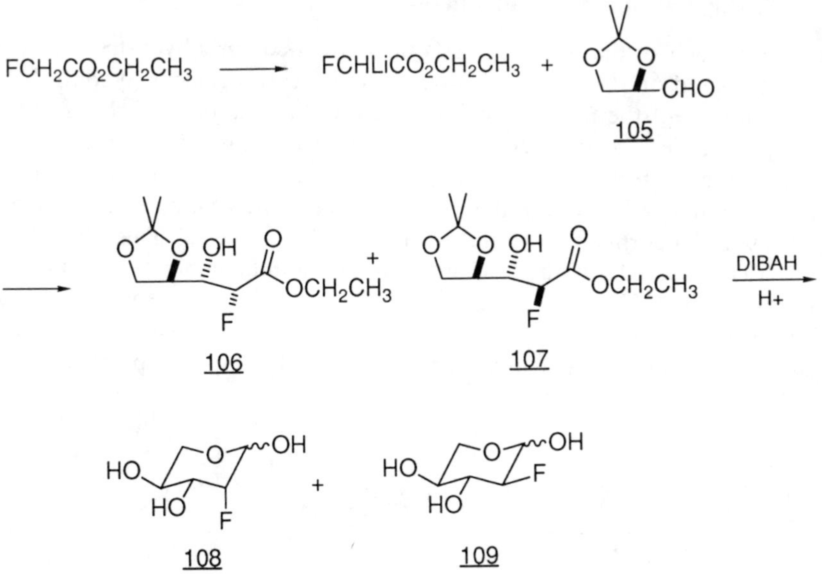

$$FCH_2CO_2CH_2CH_3 \longrightarrow FCHLiCO_2CH_2CH_3 \; + \; \underset{\underline{105}}{\text{(structure) } CHO}$$

Compounds **106** and **107**, with DIBAH / H+ arrow

Compounds **108** and **109**

Ethyl bromodifluoroacetate (**110**) was coupled with (*R*)-2,3-*O*-isopropylidene-glyceraldehyde under Reformatsky conditions to yield a 3:1 mixture of isomers **111** and **112**, again following the Felkin–Anh model for diasterefacial selectivity in the aldol process.[59]

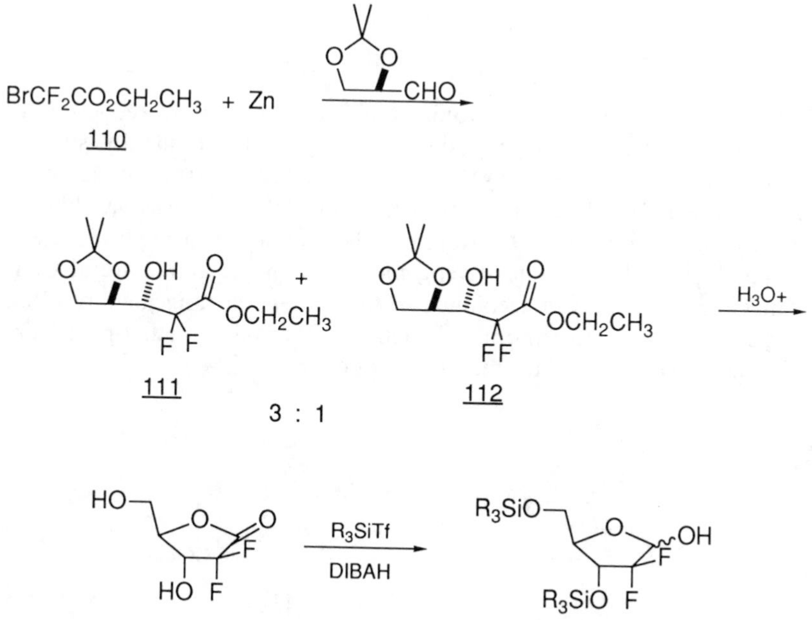

$$\underset{\underline{110}}{BrCF_2CO_2CH_2CH_3} \; + \; Zn \; \xrightarrow{\text{(structure) } CHO}$$

Compounds **111** and **112**

3 : 1

R$_3$SiTf / DIBAH

8.5.3   3,3-Sigmatropic Rearrangements

A new approach to the total synthesis of fluorinated carbohydrates has been developed based on the use of a 3,3-sigmatropic rearrangement to control the stereochemistry of the fluorinated carbon as well as the other carbons of the carbohydrate. The Ireland ester enolate Claisen rearrangement suggested an attractive approach to the stereoselective construction of the required carbon framework. Stereoselectivity in the Ireland ester enolate Claisen rearrangement requires selective formation of a fluorinated silyl ketene acetal. This silyl ketene acetal was prepared by the trapping of the lithium enolate of *E*-butenyl α-fluoro-acetate (**113**) with chlorotrimethylsilane. The *E*-butenyl-α-fluoro-α-trimethyl acetate (**114**), on warming to − 20°C, selectively formed in situ *Z*-silyl ketene acetal (**115**), which rearranged rapidly to the required pentenoic acid (**116**).[60]

Highly stereoselective iodolactonization was followed by reduction with diisobutylaluminum hydride and displacement of the iodide by silver trifluoroacetate to form 2,3-dideoxy-2-fluoro-3-*C*-methyl-5-*O*-trifluoroacetyl-*ribo*-pentofuranose (**117**).[61] The principal shortcoming of this synthesis was that it was racemic. It has been possible to overcome this limitation through the use of a chiral auxiliary with a remote stereogenic center in the amide acetal Claisen rearrangement.[62] The fluoroacetamide acetal Claisen rearrangement of **118** was used to prepare the required optically active 2-fluoro-3-*C*-methyl-pent-4-enoic acid derivative (**119**) in greater than 80% optical purity.[63]

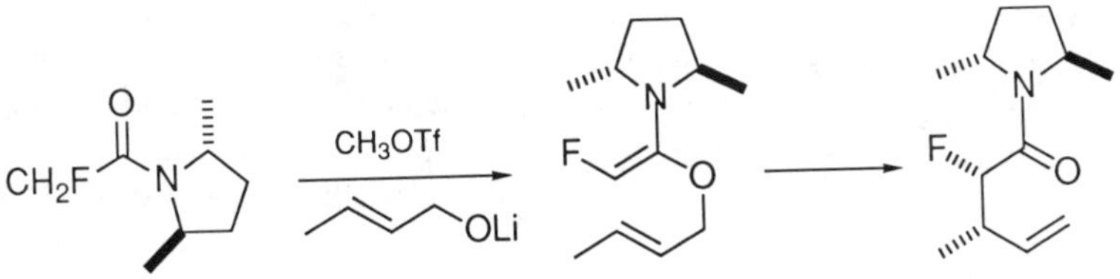

118   119

8.6   ENZYMATIC METHODS

8.6.1   Enzymatic Transformations of Fluorinated Carbohydrates

Enzymes have been used for the transformation of chemically prepared, fluorinated carbohydrates to novel sugars as well as for the conversion of one carbohydrate to another. Enzymatic methods have also been employed for the synthesis of disaccharides containing fluorinated constituents.

The combination of these reactions can be illustrated in the synthesis of a variety of fluorosucrose analogs. Protected 1-deoxy-1-fluorofructose (**120**), prepared as described in Section 8.2.1, was incubated with uridine-5'-diphospho-glucose (UDP-glucose) and sucrose synthetase to form 1-deoxy-1-fluorosucrose (**121**) in yields as high as 83%.[14]

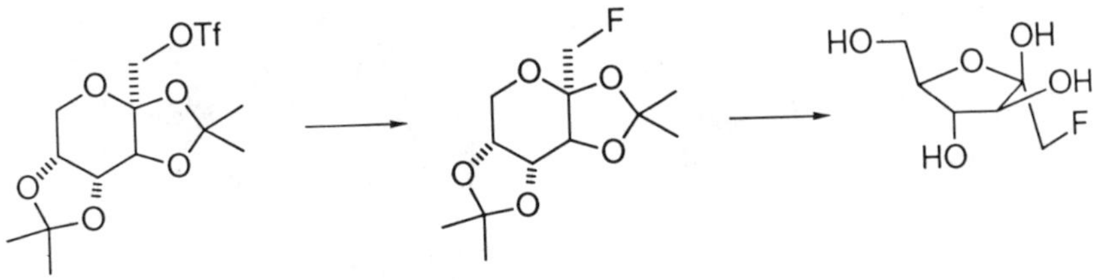

120

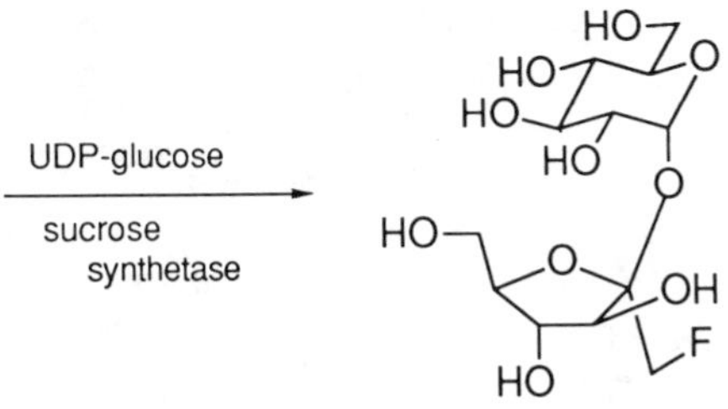

121

Sucrose synthetase was also used to combine 6-deoxy-6-fluoroglucose (**78**) with UDP-glucose in the presence of glucose isomerase.[64] The isomerase converted **78** to 6-deoxy-6-fluorofructose (**122**), which was then condensed by the synthetase with UDP-glucose to form 6-deoxy-6-fluorosucrose (**123**) in 73% isolated yield. This approach failed with 4-deoxy-4-fluoroglucose (**124**), which

78 122 123

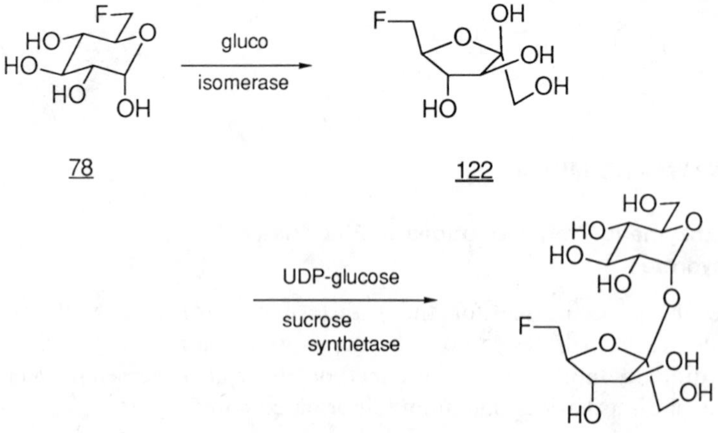

was not a substrate for glucose isomerase. However, in a series of enzymatically linked steps—treatment of **124** with hexokinase to introduce the 6-phosphate and reaction of **125** with phosphoglucoisomerase to form the furanose **126** followed by epimerization of the 3 position with fructose 6-phosphate kinase—4-deoxy-4-fluoro-fructose-1, 6-diphosphate (**127**), a substrate for sucrose synthetase, was formed.

124 125 126 127

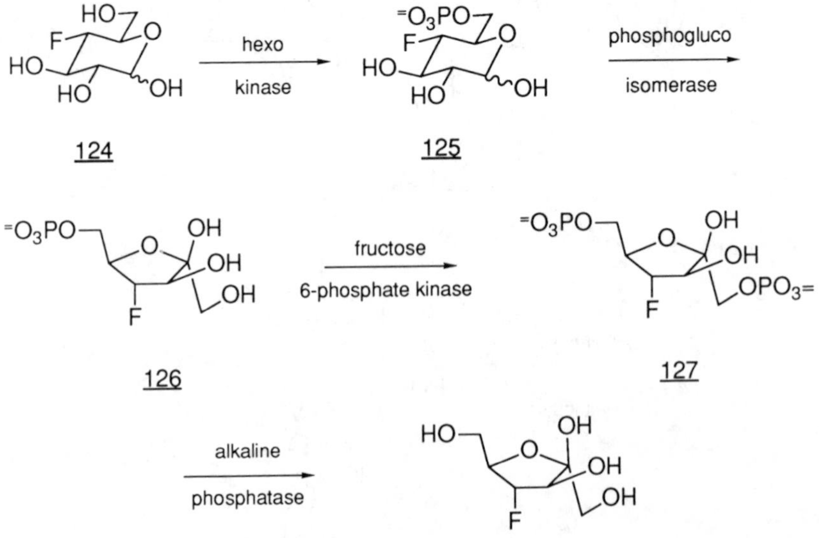

8.6.2 Aldolase- and Isomerase-Catalyzed Reactions

Fructose-1,6-diphosphate aldolase promoted the condensation of 3-fluoro-2-hydroxypropionaldehyde (128) with 1,3-dihydroxyacetonephosphate to form a mixture of 6-deoxy-6-fluoro-D-fructose (129) and 6-deoxy-6-fluoro-D-sorbose (130). 3-Deoxy-3-fluorohydroxyacetone-1-phosphate (131) was not a substrate for this enzyme.[65]

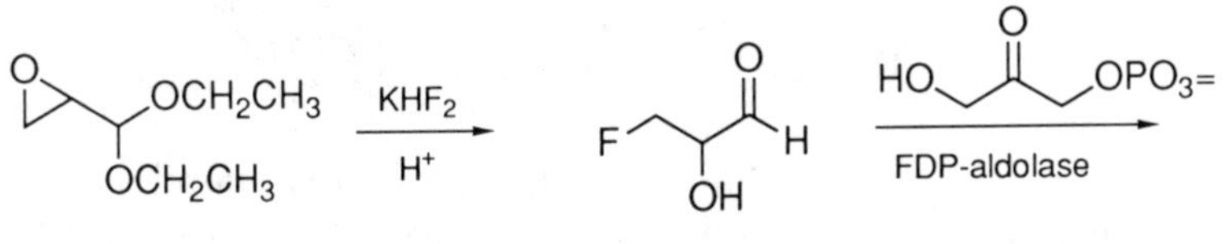

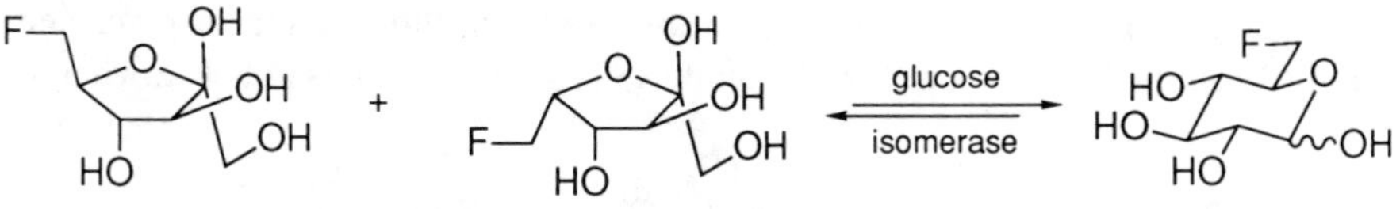

8.7 GLYCOSYL FLUORIDES

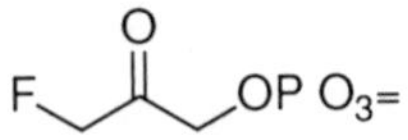

Glycosyl fluorides have found considerable utility as synthetic reagents for glycosylation reactions[66] while retaining their intrinsic utility as substrate analogs for enzymatic transformations that involve the anomeric center.

Recent investigations have shown that glycosyl fluorides are synthetically useful in controlling stereochemistry at the anomeric carbon in glycosidation reactions.[67] The anomeric hydroxyl of protected carbohydrates has been converted to glycosyl fluorides in 60–80% yields by treatment with the hydrogen fluoride–pyridine reagent.[68] 1-Acetyl carbohydrates (132) form glycosyl fluorides (133) in good yields, 70–90%, but also with good anomeric selectivity.[69]

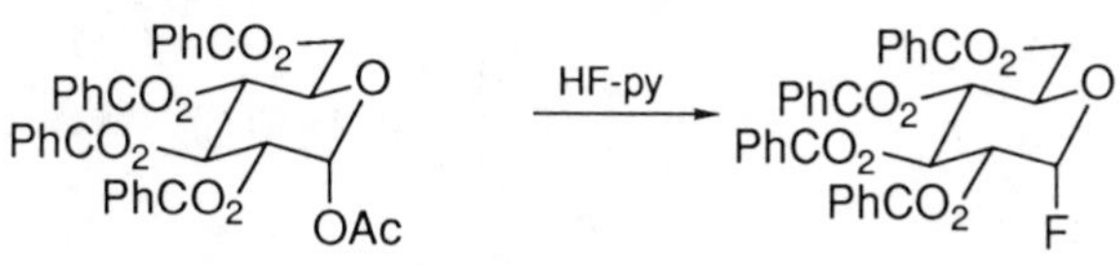

Treatment of protected 1-hydroxy carbohydrates (**134**) with DAST will form the desired fluorides (**135**) in excellent yields, 84–99%.[70] Control of stereochemistry is strongly affected by the choice of the protecting groups.

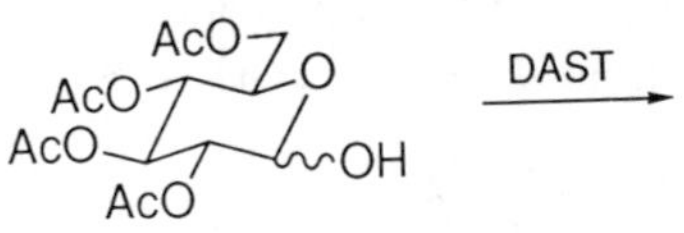

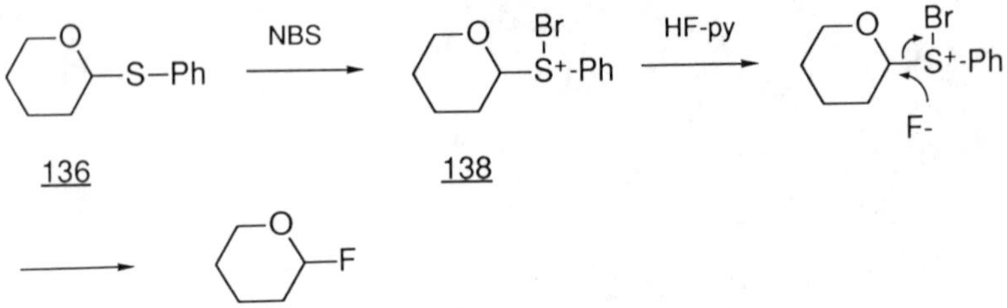

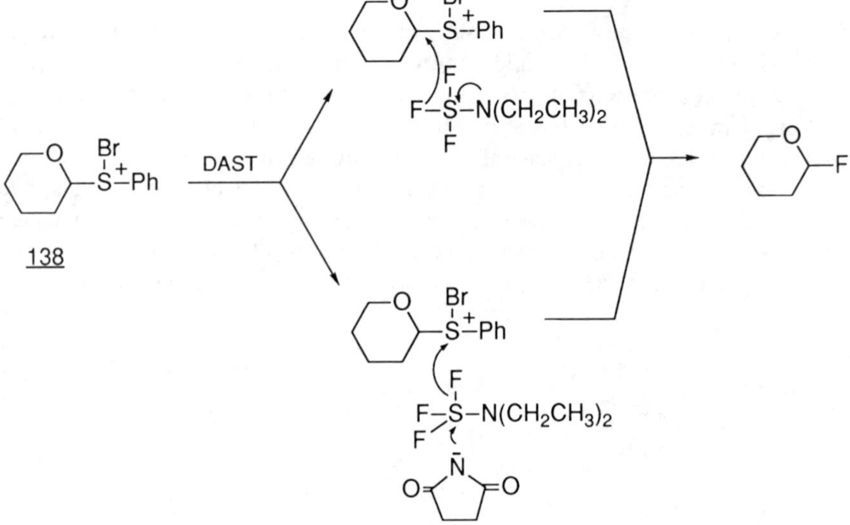

The conversion of phenylthio glycosides to glycosyl fluorides is a gentle, mild procedure tolerant of a variety of protecting groups. Treatment of phenylthio glycoside (**136**) with DAST or hydrogen fluoride–pyridine and N-bromosuccinimide in methylene chloride formed glycosyl fluoride (**137**) in good to excellent yield.[71] Fluorination proceeds via fluoride ion displacement of the bromosulfonium ion (**138**). When DAST is employed it is not clear whether participation by succinimide is necessary. However, when the carbohydrate has an unprotected group trans to the anomeric substituent, a 1,2-migration reaction occurs on treatment with DAST:

RO $\cdots$SPh   RO'''   OH   OR   DAST   RO   F   RO'''   '''SPh   OR

This apparently is a consequence of DAST reacting first with the hydroxyl group to form a good leaving group. Neighboring group participation leads directly to the 1,2-migration:[72]

RO   SPh   RO'''   OSF$_2$N(CH$_2$CH$_3$)$_2$ OR

RO   '''S+Ph   RO'''   OR

RO   O+   RO'''   '''SPh   OR

RO   F   RO'''   '''SPh   OR

REFERENCES

1. (a) N. F. Taylor, ed., *Fluorinated Carbohydrates: Chemical and Biochemical Aspects*, American Chemical Society, Washington, D.C., 1988; (b) *J. Carbohydr. Res.* **4**, 1985 (special fluorocarbohydrate issue); (c) A. A. E. Penglis, *Adv. Carbohydr. Chem. Biochem.* **38**, 195, 1981; (d) A. B. Foster and J. H. Westwood, *Pure Appl. Chem.* **35**, 147, 1973; (e) P. W. Kent, *Chem. Ind.* **1969**, 1128.
2. A. Zamojski, A. Banaszek, and G. Grynkiewcz, *Adv. Carbohydr. Chem. Biochem.* **40**, 1–20, 1982.
3. (a) W. A. Szarek, *Adv. Carbohydr. Chem. Biochem.* **28**, 225, 1973; (b) J. E. G. Barnett, *Adv. Carbohydr. Chem. Biochem.* **22**, 177, 1967; (c) S. Hanessian, *Adv. Chem.* (Ser. 74), **1968**, 159; (d) A. H. Hains, *Adv. Carbohydr. Chem. Biochem.* **33**, 72, 1976.
4. (a) M. E. Phelps, J. C. Mazziota, and S. C. Huang, *Cereb. Blood Flow Metab.* **2**, 113, 1982; (b) M. E. Phelps, E. J. Hoffman, C. Selin, S. C. Huang, G. Robinson, N. MacDonald, H. Schelbert, and D. E. Kuhl, *J. Nucl. Med.* **19**, 1311, 1982.
5. S. Levy, E. Livni, D. Elmaleh, and W. Curatolo, *J. Chem. Soc. Chem. Commun.* **1982**, 972.
6. S. Levy, E. Livni, D. R. Elmaleh, D. A. Varnum, and G. L. Brownell, *Int. J. Appl. Radiat. Isot.* **34**, 1560, 1983.

7. T. Haradahira, M. Maeda, Y. Kai, H. Omae, and M. Kojima, *Chem. Pharm. Bull.* **33**, 165, 1985.

8. T. Haradahira, M. Maeda, Y. Kai, and M. Kojima, *J. Chem. Soc. Chem. Commun.* **1985**, 364.

9. (a) W. J. Middleton, U. S. Pat. 3,940,402 (1976); (b) W. B. Farnham and R. L. Harlow, *J. Am. Chem. Soc.* **103**, 4608, 1981.

10. W. A. Szarek, G. W. Hay, and B. Doboszewski, *J. Chem. Soc. Chem. Commun.* **1985**, 663–664.

11. H. Fukuda, T. Matsuzawa, Y. Abe, S. Endo, K. Yamada, K. Kubota, J. Hatazawa, T. Sato, M. Ito, T. Takahashi, R. Iwata, and T. Ido, *Eur. J. Nucl. Med.* **7**, 294, 1982.

12. T. Ogawa and Y. Takahashi, *J. Carbohydr. Chem.* **2**, 461, 1983.

13. T. Haradahira, M. Maeda, H. Omae, Y. Yano, and M. Kojima, *Chem. Pharm. Bull.* **32**, 4758, 1984.

14. P. J. Card and W. D. Hitz, *J. Am. Chem. Soc.* **106**, 5348, 1984.

15. R. Albert, K. Dax, and A. E. Stuetz, *Tetrahedron Lett.* **24**, 1763, 1983.

16. R. Albert, K. Dax, and A. E. Stuetz, *J. Carbohydr. Chem.* **3**, 267, 1984.

17. F. Latif, A. Malik, and W. Voelter, *Z. Naturforsch.* **40b**, 317, 1985.

18. (a) C. Lopez, K. A. Watanabe, and J. J. Fox, *Antimicrob. Agents Chemother.* **1980**, 803; (b) J. J. Fox, C. Lopez, and K. A. Watanabe, in *Antiviral Chemotherapy* (ed. K. K. Gauri), Academic, New York, 1981, p. 219; (c) C. Marck, B. Lesyng, and W. Saenger, *J. Mol. Struc.* **82**, 77, 1982; (d) J. J. Fox, K. A. Watanabe, C. Lopez, F. S. Philips, and B. Leyland-Jones, in *Herpesvirus, Clinical, Pharmacological and Basic Aspects* (eds. H. Shiota, Y. C. Cheng, and W. H. Prusoff), Excerpta Medica, Amsterdam, 1982, p. 135; (e) J. A. Montgomery, *Med. Res. Rev.* **2**, 271, 1982; (f) K. A. Watanabe, T. -L. Su, R. S. Klein, C. K. Chu, A. Matsuda, M. W. Chun, C. Lopez, and J. J. Fox, *J. Med. Chem.* **26**, 152, 1983; (g) M. P. Fanucchi, B. Leyland-Jones, C. W. Young, J. H. Burchenal, K. A. Watanabe, and J. J. Fox, *Cancer Treat. Rep.* **69**, 55, 1985; (h) J. E. Swigor, K. A. Pittman, and J. Label, *Comp. Radiopharm.* **22**, 931, 1985; (i) K. A. Watanabe, T. -L. Su, U. Reichman, N. Greenberg, C. Lopez, and J. J. Fox, *J. Med. Chem.* **27**, 91, 1984.

19. M. E. Perlman, K. A. Watanabe, R. F. Schinazi, and J. J. Fox, *J. Med. Chem.* **28**, 741, 1985.

20. T. -L. Su, K. A. Watanabe, R. F. Schinazi, and J. J. Fox, *J. Med. Chem.* **29**, 151, 1986.

21. M. G. Ambrose and R. W. Binkley, *J. Org. Chem.* **48**, 674, 1983.

22. S. Hannessian and J.-M. Vatele, *Tetrahedron Lett.* **22**, 3579, 1981.

23. C. H. Tann, P. R. Brodfuehrer, S. P. Brundidge, S. Sapino, Jr., and H. G. Howell, *J. Org. Chem.* **50**, 3644, 1985.

24. T. J. Tewson, *J. Org. Chem.* **48**, 3507, 1983.

25. T. J. Tewson, *J. Nucl. Med.* **24**, 718, 1983.

26. P. Kovac and C. P. J. Glaudemans, *J. Carbohydr. Chem.* **2**, 313, 1983.

27. P. Kovac and C. P. J. Glaudemans, *Carbohydr. Res.* **123**, 326, 1983.

28. L. Hough, A. K. M. S. Kabir, and A. C. Richardson, *Carbohydr. Res.* **125**, 247, 1984.

29. N. F. Taylor, R. F. Childs, and R. V. Brant, *Chem. Ind.* **1964**, 928.

30. S. Cohen, D. Levy, and E. D. Bergmann, *Chem. Ind.* **1964**, 1802.

31. W. A. Szarek, G. W. Hay, and M. M. Perlmutter, *J. Chem. Soc. Chem. Commun.* **1982**, 1253.

32. P. A. Beeley, W. A. Szarek, G. W. Hay, and M. M. Perlmutter, *Can. J. Chem.* **62**, 2709, 1984.

33. D. Picq, D. Anker, C. Rousset, and A. Laurent, *Tetrahedron Lett.* **24**, 5619, 1983.

34. D. Picq and D. Anker, *J. Carbohydr. Chem.* **4**, 113, 1985.

35. L. Hough, A. A. E. Penglis, and A. C. Richardson, *Carbohydr. Res.* **83**, 142, 1980.

36. M. K. Gurjar, V. J. Patil, J. S. Yadav, and A. V. Rama Rao, *Carbohydr. Res.* **135**, 174, 1984.

37. L. Helena, B. Baptistella, A. J. Marsaioli, J. D. de Souza Filho, G. G. de Oliveira, A. B. de Oliveira, A. Dessinges, S. Castillon, A. Olesker, T. T. Thang, and G. Lukacs, *Carbohydr. Res.* **140**, 51, 1985.

38. J. Adamson, A. B. Foster, L. D. Hall, and R. H. Hesse, *J. Chem. Soc. Chem. Commun.*, **1969**, 309.

39. T. Ido, C. -H. Wan, J. S. Fowler, and A. P. Wolf, *J. Org. Chem.* **42**, 2341, 1977.

40. C. -Y. Shiue, C. D. Arnett, and A. P. Wolf, *Eur. J. Nucl. Med.* **9**, 77, 1984.

41. M. Diksic and D. Jolly, *J. Carbohydr. Chem.* **4**, 265, 1985.

42. M. J. Adam, *J. Chem. Soc. Chem. Commun.* **1982**, 730.

43. W. Korytnyk and S. Valentekovic-Horvat, *Tetrahed*ron L ett. **21**, 1493, 1980.

44. K. Bischofberger, A. J. Brink, and A. Jordaan, *J. Chem. Soc. Perkin (I)* **1975**, 2457.

45. W. J. Middleton, *J. Org. Chem.* **40**, 547, 1975.

46. M. Sharma and W. Korytnyk, *Tetrahedron Lett*, **1977**, 573.

47. (a) C. W. Somawardhana and E. G. Brunngraber, *Carbohydr. Res.* **94**, C14, 1981; (b) G. H. Klemm, R. J. Kaufmann, and R. S. Sidhu, *Tetrahedron Lett.* **23**, 2927, 1982.

48. P. J. Card, *J. Org. Chem.* **48**, 393, 1983.

49. (a) C. W. Somawardhana and E. G. Brunngraber, *Carbohydr. Res.* **121**, 51, 1983; (b) P. J. Card and G. S. Reddy, *J. Org. Chem.* **48**, 4734, 1983.

50. P. Herdewlin, J. Balzarini, E. DeClercq, R. Pauwells, M. Baba, S. Bruder, and H. Vanderhaege, *J. Med. Chem.* **30**, 1270, 1987.

51. R. Faigh, F. C. Escribano, S. Castillon, J. Garcia, G. Lukacs, A. Olesker, and T. T. Thang, *J. Org. Chem.* **51**, 4556–4564, 1986.

52. T. Tsuchiya, T. Torii, Y. Suzuki, and S. Umezawa, *Carbohydr. Res.* **116**, 277, 1983.

53. T. Tsuchiya, T. Torii, and S. Umezawa, *J. Antibiot.* **35**, 1245–1247, 1982.

54. S. Castillon, A. Dessinges, R. Faghih, G. Lukacs, A. Olesker, and T. T. Thang, *J. Org. Chem.* **50**, 4913, 1985.

55 A. Dessinges, F. C. Escribano, G. Lukacs, A. Olesker, and T. T. Thang, *J. Org. Chem.* **52**, 1633, 1987.

56. J. E. G. Barnett and P. W. Kent, *J. Chem. Soc.* **1963**, 2743.

57. P. W. Kent and J. E. G. Barnett, *J. Chem. Soc.* **1964**, 2497.

58. J. T. Welch and S. Eswarakrishnan, *J. Chem. Soc. Chem. Commun.* **1985**, 186.

59. L. W. Hertel, J. S. Kroin, J. W. Misner, and J. M. Tustin, *J. Org. Chem.* **53**, 2406, 1988.

60. J. T. Welch and J. S. Samartino, *J. Org. Chem.* **50**, 3663, 1985.

61. J. T. Welch and J. S. Plummer, *J. Carbohydr. Chem.* **1988**, (accepted) unpublished results.

62. J. T. Welch and S. Eswarakrishnan, *J. Am. Chem. Soc.* **109**, 6716, 1987.

63. J. T. Welch, T. Yamazaki, J. S. Plummer, and R. Gimi, *J. Am. Chem. Soc.,* **1990**. (submitted)

64. P. J. Card, W. D. Hitz, and K. G. Ripp, *J. Am. Chem. Soc.* **108**, 155, 1986.

65. J. R. Durrwachter, D. G. Drueckhammer, K. Nozaki, H. M. Sweers, and C. -H. Wong, *J. Am. Chem. Soc.* **108**, 7812, 1986.

66. (a) T. Mukaiyama, Y. Hashimoto, and S. -C. Shoda, *Chem. Lett.* **1983**, 935; (b) S. Hashimoto, M. Hayashi, and R. Noyori, *Tetrahedron Lett.* **25**, 1379, 1984; (c) Y. Araki, K. Watanabe, F. -H. Kuan, K. Itoh, N. Kobayashi, and Y. Ishido, *Carbohydr. Res.,* **127**, C5, 1984.

67. K. C. Nicolau, R. E. Dolle, D. P. Papahatjis, and J. L. Randall, *J. Am. Chem. Soc.* **106**, 4189, 1984.

68. W. A. Szarek, G. Grynkiewicz, B. Doboszewski, and G. W. Hay, *Chem. Lett.* **1984**, 1751.

69. M. Hayashi, S. -I. Hashimoto, and R. Noyori, *Chem. Lett.* **1984**, 1747.

70. (a) G. H. Posner and S. R. Haines, *Tetrahedron Lett.* **26**, 5, 1985; (b) W. Rosenbrook, Jr., D. A. Riley, and P. A. Lartey, *Tetrahedron Lett.* **26**, 3, 1985.

71. K. C. Nicolau, R. E. Dolle, D. P. Papahatjis, and J. L. Randall, *J. Am. Chem. Soc.* **106**, 4189, 1984.

72. (a) K. C. Nicolau, T. Ladduwahetty, J. L. Randall, and A. Chucholowski, *J. Am. Chem. Soc.* **108**, 2466, 1986; (b) A. Hasegawa, M. Goto, and M. Kiso, *J. Carbohydr. Chem.* **4**, 627, 1985.

9

FLUORINATED PROSTAGLANDINS

9.1 INTRODUCTION

Fluorination of prostaglandins, prostacyclins, and thromboxanes had led to exciting and useful modifications of activity, as much as fluorination effectively modified the activity of steroids. The biosynthetic pathway of these compounds begins with arachidonic acid. To better understand the structural relationship between one prostaglandin and another as well as the relation between the three classes of compounds discussed here—prostaglandins, prostacyclins, thromboxanes, which are derived from arachidonic acid via prostaglandin endoperoxide, PGH_2—see Fig. 9.1.

Selective fluorination of prostanoids has been effected on the cyclopentane nucleus and on both side chains. A variety of techniques have been employed that

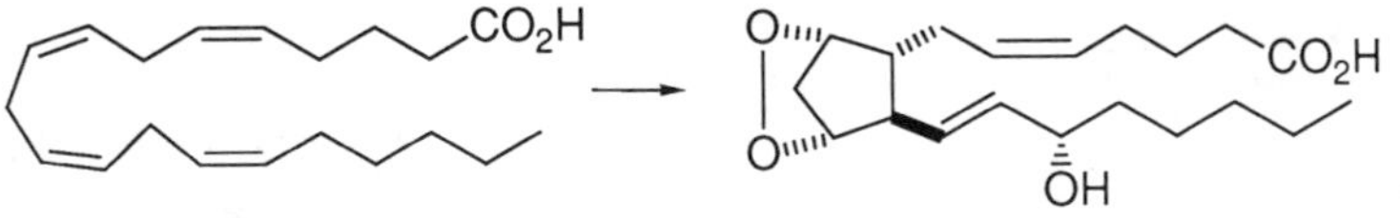

arachidonic acid

PGH_2

PGH_2 ⟶

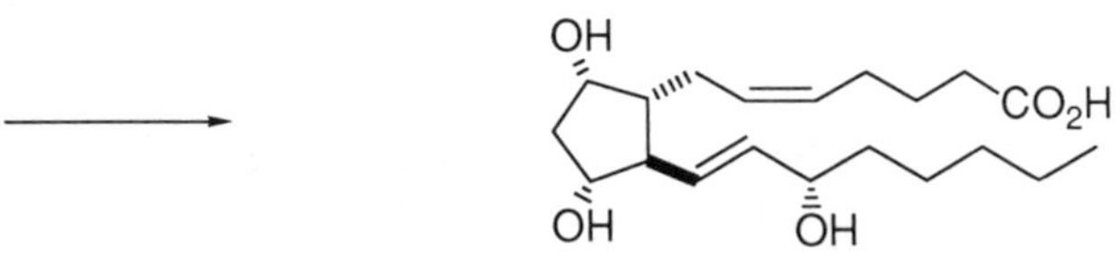

PGF_2

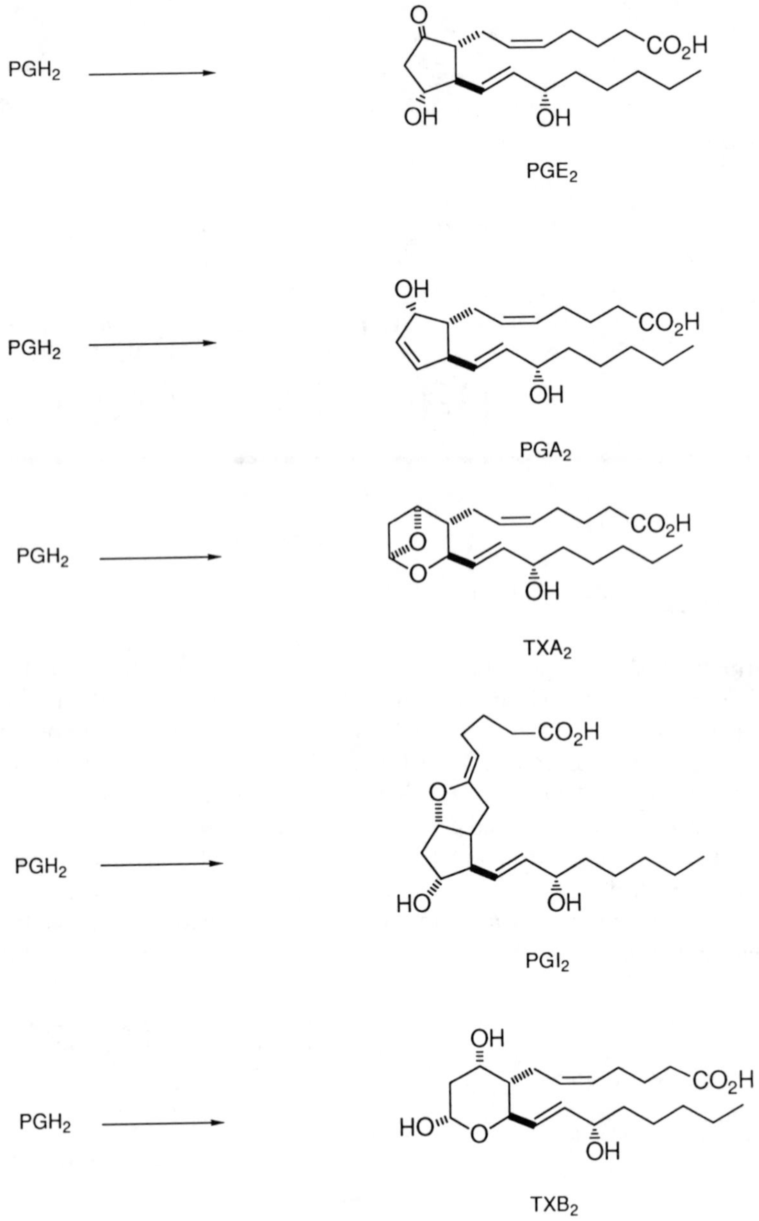

FIGURE 9.1

have been arbitrarily divided into the fluorination reactions of intermediates and the use of fluorinated building blocks to prepare the target compounds. The chemistry and biology of fluorinated prostaglandins, prostacyclins, and thromboxanes through 1981 has been reviewed.[1]

9.2 FLUORINATIVE DEHYDROXYLATION

Fluorinative dehydroxylation is often a particularly attractive approach as it is based on the preparation of an alcohol as starting material, a process for which there are often several methods, including stereoselective transformations. The fluorination step may be effected on the alcohol itself or on conversion of the alcohol to a good leaving group such as a sulfonate ester. Since a displacement is involved, the reaction is potentially stereoselective, proceeding with inversion of stereochemistry. However, the reaction may be accompanied by the usual side reactions such as rearrangements and eliminations found in displacement reactions.

9.2.1 Fluoroalkylamine Reagents

In early work, replacement of hydroxyl by fluorine was effected using diethyl(2-chloro-1,1,2-trifluoroethyl)amine. Methyl 9α-hydroxy-11α, 15α-bis(tetrahydropyranyl)-prosta-5-*cis*-13-*trans*-dienoate (**1**) was treated with diethyl (2-chloro-1,1,2-trifluoroethyl)amine at −10 to −15°C in methylene chloride for 3.5 hours to form the 9β-fluoro material (**2**) in 30–40% yield following chromatography.[2] The procedure was quite general, proceeding in nearly the

same yield independently of the stereochemistry of the 9-hydroxy or the 11α-ether. The preparation of 9α, 15α-diacetoxy-11β-fluoro-prosta-5-*cis*-13-*trans*-dienoate (**4**) was possible utilizing the appropriately protected starting material (**3**).

The 9β-fluoro-11α-hydroxy compound was particularly interesting in that it was three to four times more effective than PGE_2. The other compound 9α-fluoro, 11α-fluoro, and 11β-fluoro PGF_2 analogs were less active.

9.2.2 Diethylaminosulfur Trifluoride

The utility of diethylaminosulfur trifluoride in the preparation of organo-fluorine compounds by replacement of a hydroxyl group by fluorine is well known.[3] However, side reactions can occur with neighboring functionality. For instance, when (7S)-7-hydroxy-PGI_2-11,15-O-bis(t-butyldimethylsilyl) methyl ester (5) was treated with DAST, the product was not the desired fluoride but rather (7R, 11R)-11-deoxy-7, 11-epoxy-PGI_2 ring-15-O-t-butyl dimethylsilyl methyl ester (6)[4]. However, when the diacetyl-protected PGI_2 methyl ester (7) was

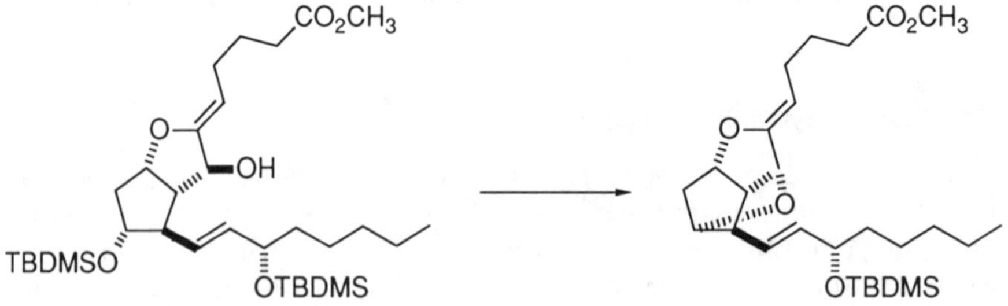

employed, the reaction with DAST formed the (7S)-7-fluoro-PGI_2 diacetylated methyl ester (8) in 22% yield and 5-fluoro-Δ^6-PGI_2 diacetyl methyl ester (9) in

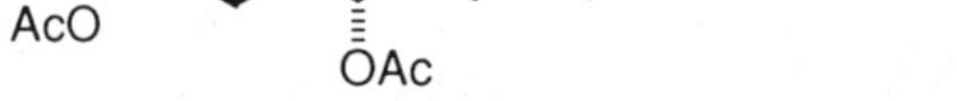

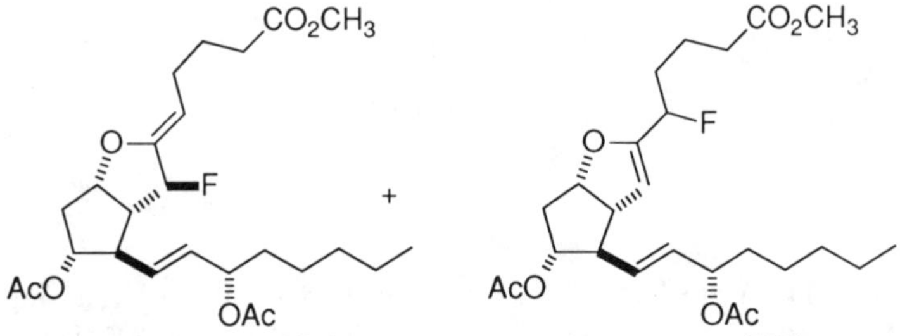

32% yield. It is likely that the 5-fluoro product results from preferential trapping of an intermediate allyl cation at the 5 position, as it is relatively unencumbered.

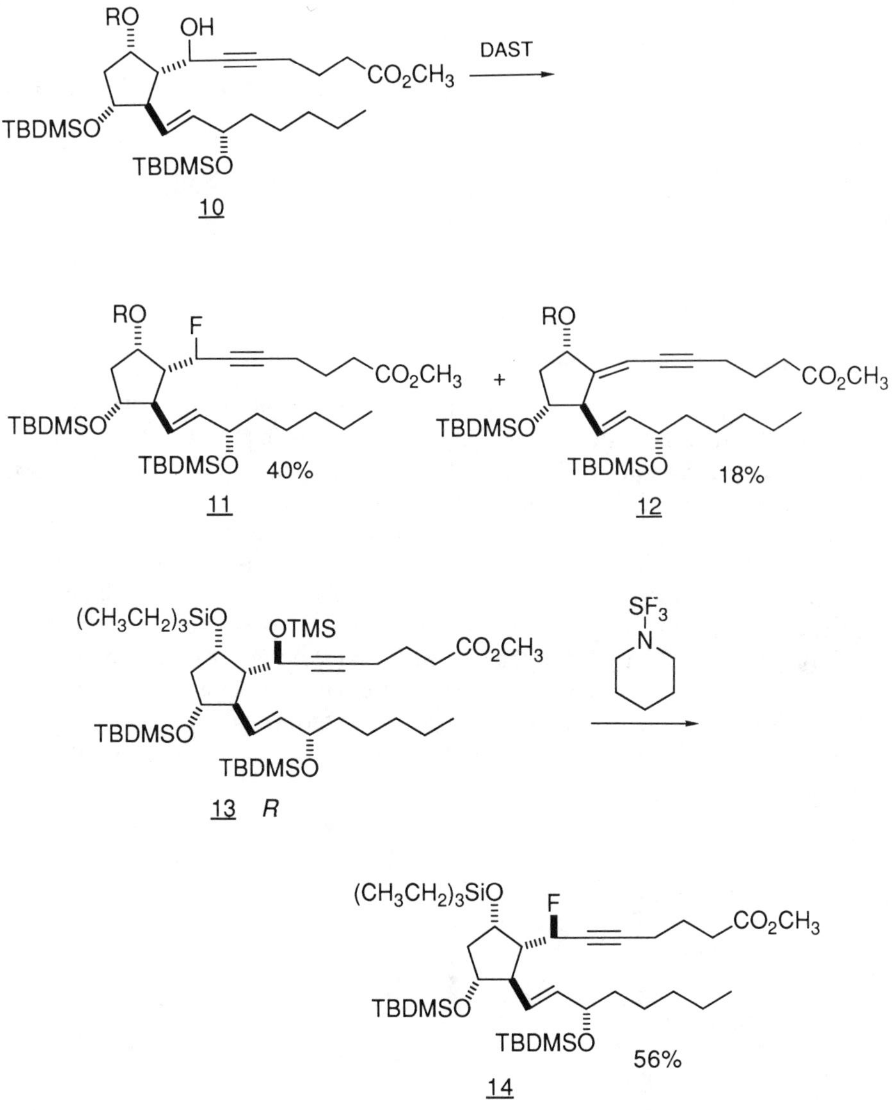

In a different approach to (7R)-7-fluoro-PGI$_2$ (**14**) it was observed that DAST treatment of the readily available acetylenic alcohol (**10**) afforded the desired fluoride (**11**) in only 40% yield and was accompanied by an 18% yield of dehydrated product (**12**)[5]. This side reaction was suppressed by treating the 7-trimethylsilyl ether (**13**) with piperidinosulfur trifluoride in 1,1,2-trichloro-1,2,2-trifluoroethane at 0°C. The desired (7R)-7-fluoro product (**14**) was obtained from either the (7R)- or (7S)-TMS ether in 56% and 72% yield with no detectable (7S)-fluoro material being formed. Deblocking and cyclization formed (7R)-7-fluoro-PGI$_2$ in greater than 30% overall yield.

The use of morpholinosulfur trifluoride was reported to result in an 83% yield of the fluoroprostaglandin (16) when the 11-hydroxy material (15) was allowed to react at $-78°C$ for one hour. The fluorination proceeded stereospecifically with retention of configuration.[6]

15   16

9.2.3 Sulfonate Displacement Reactions

A stepwise process is sometimes preferable for the replacement of hydroxyl by fluorine. Formation of the sulfonate leaving group as the tosylate, mesylate, or trifluoromethanesulfonate is possible. In the preparation of 10α-fluoro-PGF$_{2\alpha}$ a key intermediate was the lactone triflate (17) prepared in situ by treatment of 3,3aα, 4α, 5β, 6α, 6aα-hexahydro-6-hydroxy-5 (phenylmethoxy)-4-[(β-methoxyethoxymethoxy)methyl]-2H-cyclopenta[b]furan-2-one with triflic anhydride. The facile displacement of triflate in five-membered rings is well known; however, potassium fluoride in acetonitrile in the presence of 18-crown-6 led to formation of only 2% of the desired fluoro material (18). Cesium fluoride in hexamethylphosphoramide or dimethylformamide yields 16–18% of the fluorinated material respectively. However, tetra-n-butylammonium fluoride in THF at reflux reproducibly formed the desired product (18) in 55% yields.[7]

17   18

Introduction of fluorine into a 12-methyl substituent as in **19** illustrates the difficulty that frequently accompanies fluorination of multifunctional molecules.

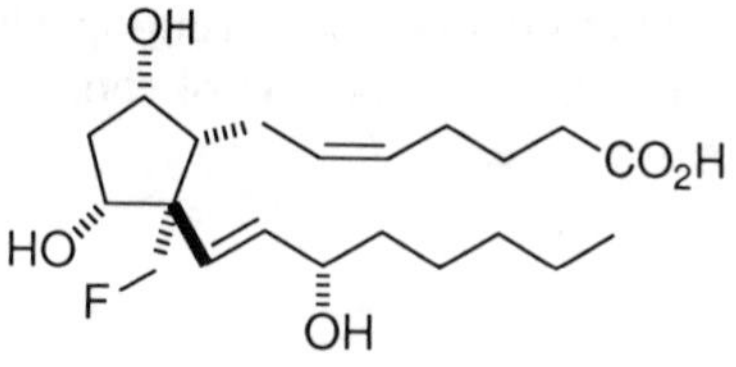

19

Although fluoride ion displacement of the primary methanesulfonate (**20**) provided a 78% yield of the desired fluoride (**21**), all attempts to reduce the carbomethoxy group resulted in loss of fluorine.[8] Reduction of the carbomethoxy

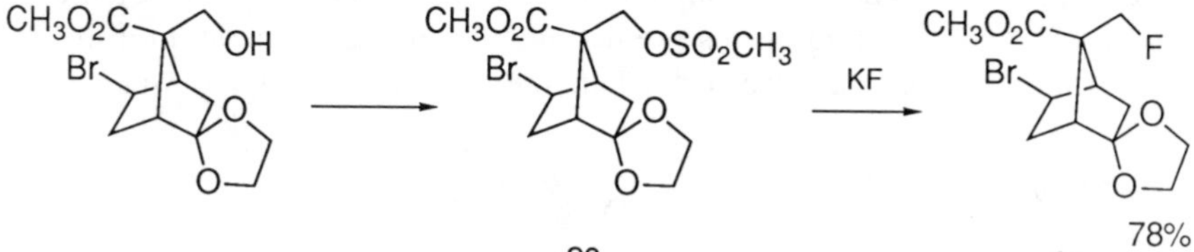

20   **21**

group and protection prior to formation of the methanesulfonate and displacement by potassium fluoride in ethylene glycol avoided this difficulty. However, the protected alcohol (**22**) inhibited the displacement reaction. The recovery of fluoride (**23**) was only 43% as well as 21% of the starting methanesulfonate. Variation in the displacement conditions offered no improvement. Neither the use of tetra-*n*-butylammonium fluoride in THF nor the use of potassium fluoride in dimethylsulfoxide was successful. Elaboration to the prostaglandin proceeded uneventfully. Unfortunately, the compounds did not affect fertility or possess smooth muscle activity.

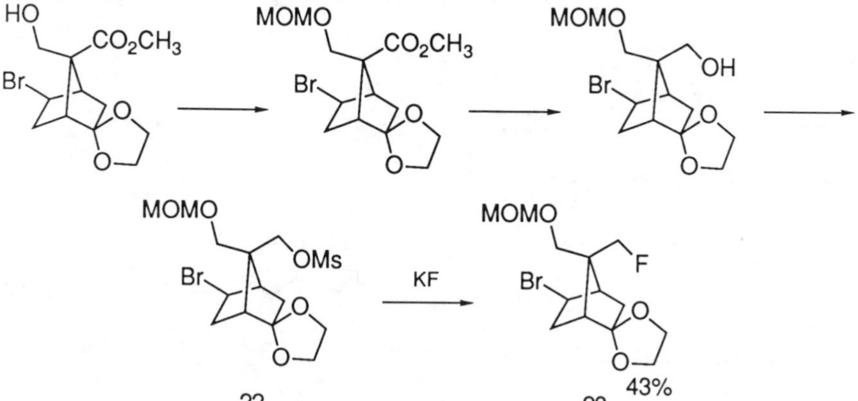

22   **23**

9.3   EPOXIDE RING-OPENING REACTIONS

The development of methods for the stereoselective synthesis of epoxides has dramatically enhanced the utility of epoxide ring-opening reactions by fluoride ion to form fluorohydrins. The ring opening of epoxides is itself often a regioselective and stereoselective process.

9.3.1   Potassium Bifluoride

The ready availability of the epoxide **24** suggested the utility of an epoxide ringopening reaction for the preparation of 10β-fluoroprostaglandin $F_{2\alpha}$ methyl ester, **27**.[9] Unfortunately, hydrogen fluoride, hydrogen fluoride–pyridine, and potassium fluoride–18-crown-6 in acetonitrile all failed to open the epoxide. However, the use of potassium bifluoride in hot ethylene glycol did result in formation of a 40% yield of the 10β-compound (**25**). The 10β-isomer was predicted because the C-12 bearing the benzyloxy methyl group was oriented β-axially. However, in spite of this prediction, 30% of the isomeric fluorohydrin with 11β-fluoro (**26**) was also formed. If the 10β-fluoro material was carried

forward in the synthesis, it was found that it was very sensitive and thus susceptible to side reaction during attempts to elaborate the prostaglandin structure. However, careful reduction and oxidation did in fact permit preparation of 10β-fluoro-PGF$_{2\alpha}$ (**27**). It was also found that if the ring-opening reaction was postponed, the desired 10β-fluoro-PGF$_{2\alpha}$ (**27**) could be prepared in 25% yield.

Although, it was possible to prepare the 10β-fluoro material by ring-opening reaction, the accompanying formation of up to 30% of the undesired isomeric 11β-fluoro compound could not be avoided. Opening of the β-epoxide (**28**) with potassium bifluoride did not form the desired fluoride (**29**). Only the lactone formed by initial saponification of the ester and lactonization was isolated.[10] The 10α-fluoro compound was successfully prepared by displacement of 10β-

28   KHF$_2$ →//   **29**

trifluoromethanesulfonate (**30**). Tetra-*n*-butylammonium fluoride in tetrahydrofuran formed **31** in 55% yield. Upon successful fluorination the remaining elaborations were conventional.

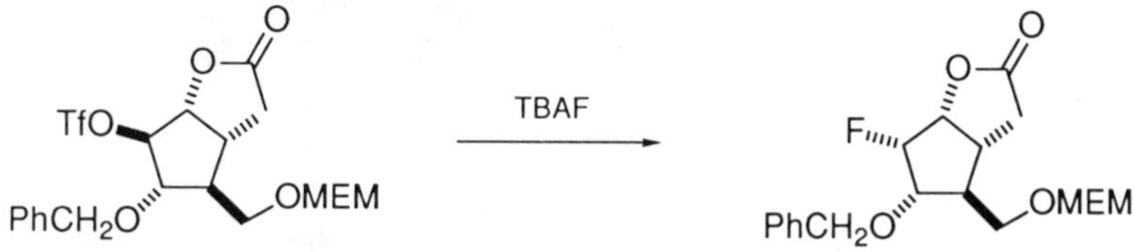

30   TBAF   **31**

9.3.2 Hydrogen Fluoride–Pyridine

The epoxide opening reaction may be used to prepare a fluorinated building block for the preparation of 16-F–9α-carboprostacyclin. A cyano-substituted oxirane (**32**) available by condensation of chloroacetonitrile with a carbonyl compound was opened by treatment with hydrogen fluoride–pyridine at 0°C to give fluorohydrin (**33**) in 86% yield accompanied by a small amount ($< 10\%$) of fluorocyclohexanone carboxaldehyde, **34**.

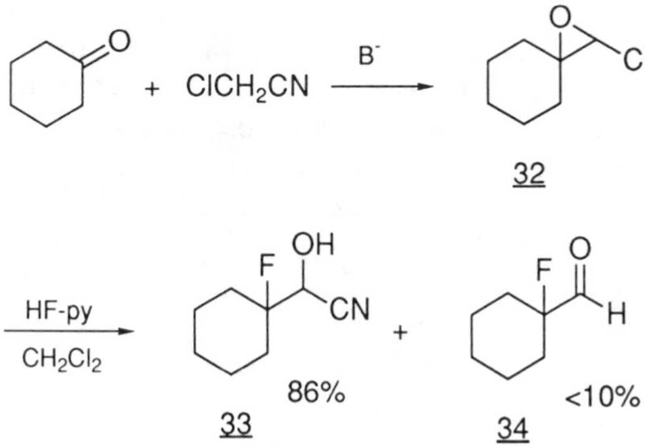

32

33 86%   +   **34** <10%

Further elaboration via methyl 1-fluorocyclohexane carboxylate (**35**) and conversion to the phosphonate (**36**) led to preparation of the reagent required for condensation with the protected aldehyde. The fluorinated olefin was converted in two steps to the target compound, 16-F–9α-carboprostacyclin (**37**).[11] The

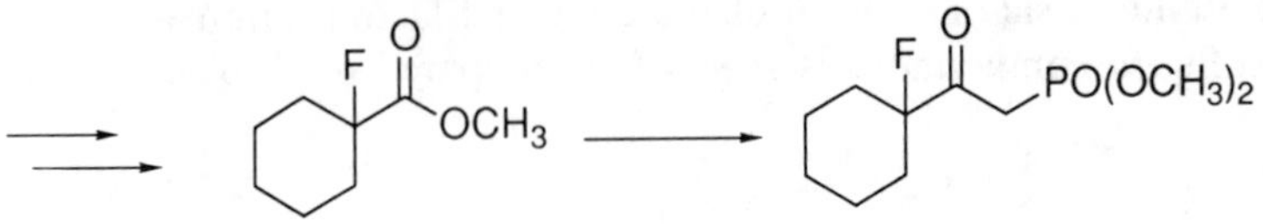

35   **36**

36

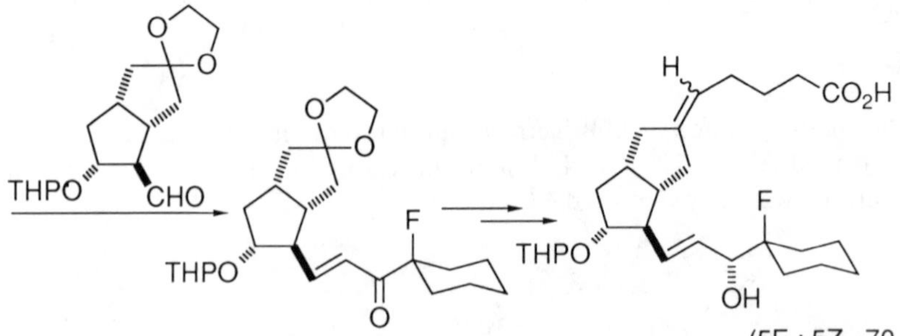

37   (5E : 5Z, 70 : 30)

compound possessed "good" activity, but fluorination did not lead to increased activity or better selectivity.

9.4   DIFLUOROMETHYLENE GROUPS

Direct conversion of a ketone or an aldehyde to a difluoromethylene group is generally a difficult reaction. As relatively few reagents are available to accomplish this transformation, only a limited number of approaches can be employed when materials containing difluoromethylene groups are required.

9.4.1   Molybdenum Hexafluoride

The preparation of ethyl 2,2-difluorohexanoate was required as a key step in the synthesis of the phosphonate ester used for introduction of the prostaglandin side chain. Difluorination of methyl 2-oxohexanoate (**38**) with molybdenum hexafluoride–boron trifluoride formed methyl 2,2-difluorohexanoate (**39**).[12]

$$CH_3CH_2CH_2CH_2COCO_2CH_3 \xrightarrow[BF_3]{MoF_6} CH_3CH_2CH_2CH_2CF_2CO_2CH_3$$

38   **39**

$$\longrightarrow CH_3CH_2CH_2CH_2CF_2COCH_2PO(OCH_3)_2$$

40

Deprotonation and condensation of dimethyl methylphosphonate with the methyl difluorohexanoate formed the reactive Horner-Emmons reagent (**40**), which was subsequently transformed into 16,16-difluoro-PGF$_{2\alpha}$ methyl ester

(**41**).[13]. This compound had very good antifertility activity with diminished smooth muscle activity.

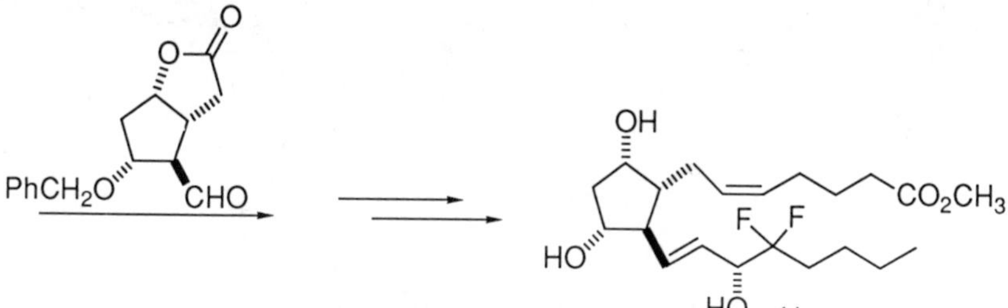

9.4.2 Sulfur Tetrafluoride

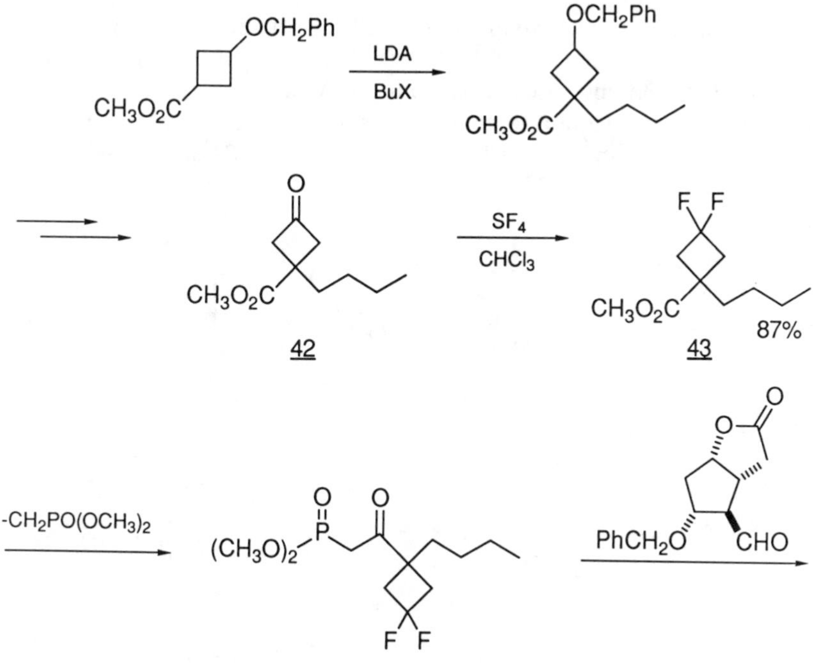

16,16-Trimethylene prostaglandins, PGE$_2$ and PGE$_1$, predicted to be resistant to 15-hydroxy-prostaglandin dehydrogenase, were sought. The methyl ester of the known cyclobutane carboxylate was butylated and oxidized to form the synthon that was to become the analog of the lower side chain of PGE$_2$. Fluorination of cyclobutanone (**42**) with sulfur tetrafluoride in chloroform in the presence of a trace of ethanol formed difluorocyclobutanone (**43**) in 87% yield.[14] Subsequent conventional manipulations were used to prepare the target 16,16-trimethylene-PGE_2 and PGE_1 analogs (**44**, **45**). These materials exhibited potent uterine

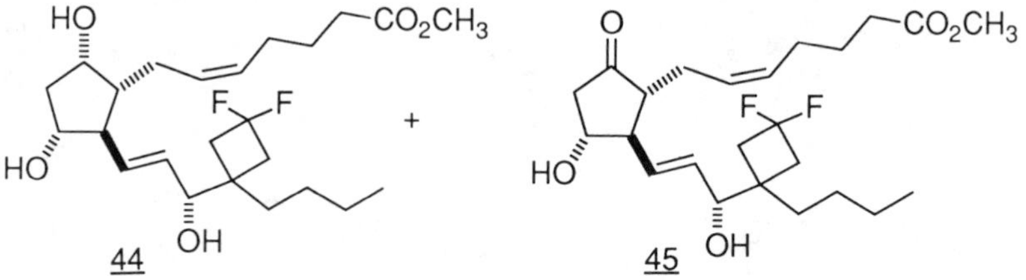

contractile activity in pregnant rats but less severe diarrhea in mice relative to natural PGE_2.

9.5 ELECTROPHILIC FLUORINATION OF ENOLS AND ENOLATES

Reagents for electrophilic fluorination are relatively uncommon, but there has been an increasing research effort aimed at the development of such reagents due to their potential utility. Such reagents are particularly useful in reactions of ubiquitous enols and enolates which have found such broad utility in organic synthesis.

9.5.1 Perchlorylfluoride

The natural prostacyclins contain an unusually acid-labile enol ether grouping, which makes the study of these materials difficult. It was proposed that fluorinated analogs would exhibit increased acid stability. 4-Allyl-1, 3-pentanedione (**46**) was treated with perchlorylfluoride to form the difluoroketone (**47**), which was reduced with potassium sec-butyl borohydride to the cis diol (**48**) in 36% yield

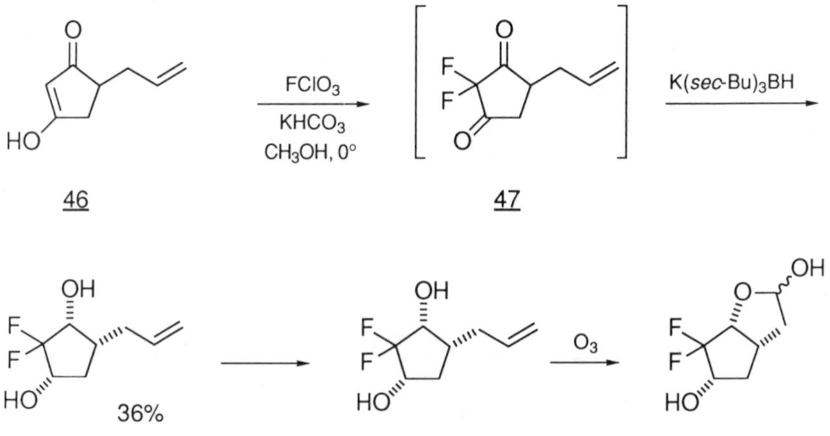

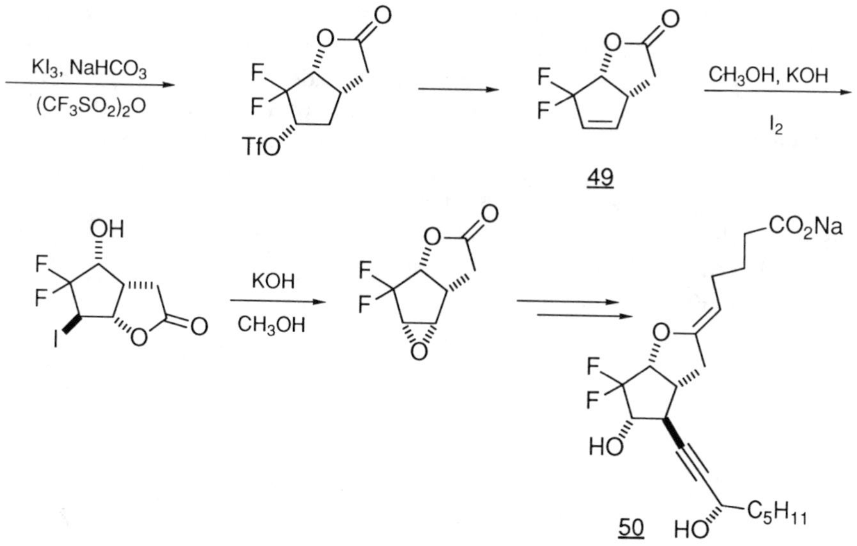

49

50

after purification[15]. Ozonolysis, periodate oxidation, and dehydration via the triflate, formed **49**. Epoxidation was effected via saponification and iodolactonization, as the allylically fluorinated olefin was unreactive toward direct epoxidation. Following conversion to the prostacyclin, biological evaluation showed **50** mimicked PGI$_2$ yet had 150 times greater half-life under the same conditions.

In a related study, the attempted direct fluorination of PGI$_2$ with perchlorylfluoride was generally unsuccessful. However, with the addition of methanol

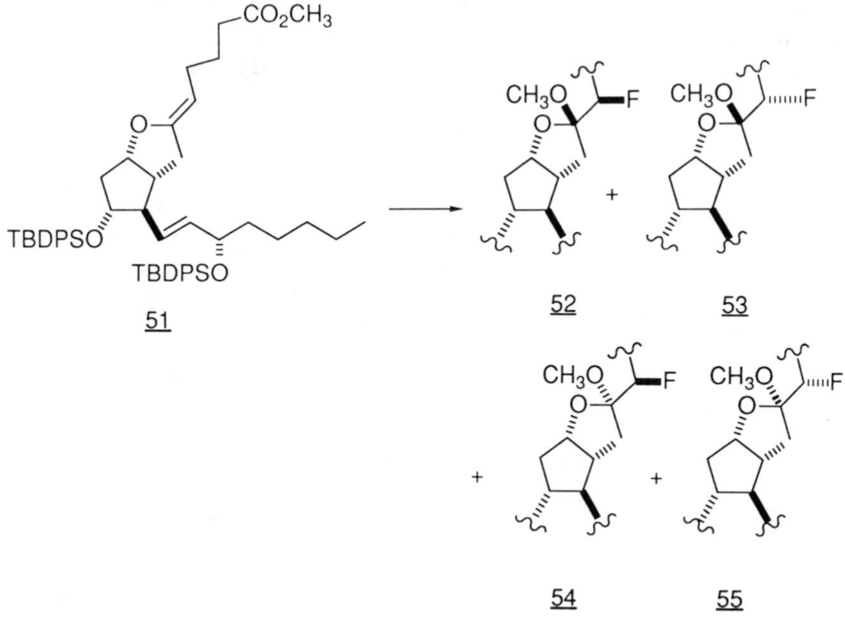

51

52   53

54   55

the reaction did proceed. Treatment of PGI_2 with the 11- and 15-hydroxy groups protected as t-butyldiphenyl silyl ethers as in **51** was preferable for isolation and further transformations.[16] When perchlorylfluoride was added in the presence of cesium carbonate in methanol, **52**, **53**, **54** and **55** were formed in a 6:2:3:1 ratio and the formation of side products from cationic intermediates, such as the 6-fluoro compound, was suppressed. Both **52** and **55** will eliminate methanol to form 5-fluoroprostacyclins, with a 14% yield of methyl 11α, 15(S)-bis[(triethylsilyl)oxy]-6,9α-epoxy-5-fluoro-prosta-5 (E), 13 (E)-dien-1-oate (**56**) and a 70% yield of methyl 11α, 15(S)-bis[(triethylsilyl)oxy]-6,9α-epoxy-5(R)-fluoro-prosta-6,13 (E)-dien-1-oate (**57**).

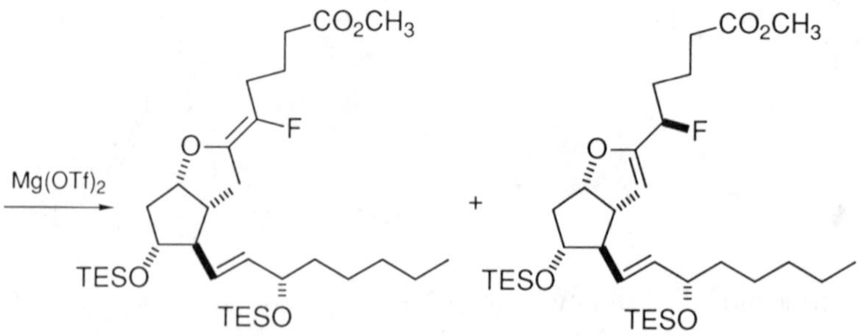

Preparation of the 13-fluoro-$PGF_{2\alpha}$ (**64**) employed fluorination of a phosphorus-stabilized anion.[17] Following addition to **58** of the bicyclic Wittig reagent (**59**) at low temperature, prior to collapse of the betaine (**60**), a second equivalent of butyl lithium was added to the reaction mixture to form the deprotonated betaine (**61**). Perchlorylfluoride was passed into the solution, which on warming formed a 12% yield of Z fluoroolefin (**62**) and a 41% yield of E

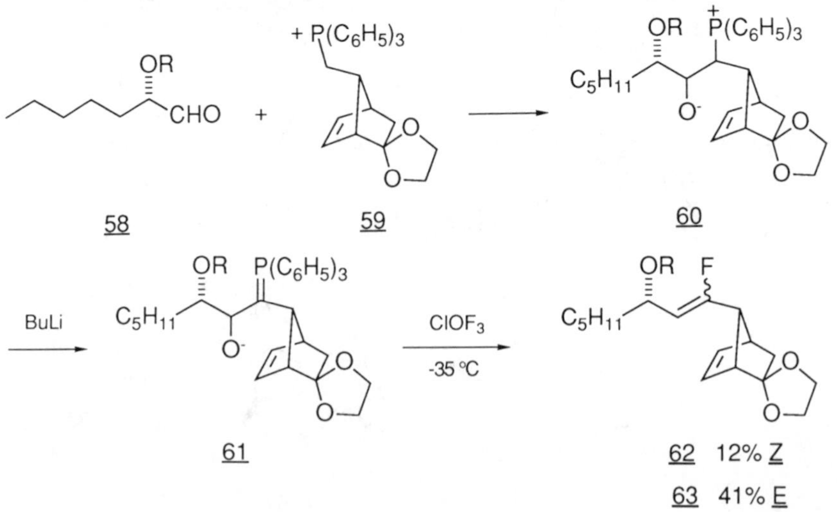

64

fluoroolefin **(63)**. The 13-fluoroprostaglandin **(64)** prepared from this material possessed slightly improved antifertility activity but undiminished smooth muscle response.

9.6 OLEFINATION REACTIONS

Although the syntheses of fluoroolefins were discussed earlier, this section considers those transformations where formation of the fluorinated double bond is the key step of preparation. Fluorinated olefins have been introduced into prostanoid compounds to replace an existing double bond and as additional functional groups or substituents.

9.6.1 Horner-Emmons–Type Reagents

Both the 13 (E)- and 13 (Z)-14-fluoro-PGF$_{2\alpha}$ **(65, 66)** have been prepared via a fluorinated Horner-Emmons reagent. Dimethyl 2-ketoheptyl phosphonate **(67)**

65

66

was deprotonated and then treated with perchlorylfluoride to form a 29% yield of dimethyl 1-fluoroheptyl phosphonate **(68)**[18] in a stepwise process related to that discussed in the previous section. Treatment of the fluorinated phosphonate with sodium hydride generated the reactive ylide which was added to the aldehyde to form a 64% yield of the undesired E stereoisomer **(69)** and a 9% yield of the desired Z isomer **(70)**. Both materials were elaborated by conventional manipulations to the desired 14-fluoro-PGF$_{2\alpha}$ compounds. These compounds retained antifertility activity but exhibited diminished smooth muscle activity.

$$(CH_3O)_2POCH_2CO(CH_2)_4CH_3 \xrightarrow[\text{ClO}_3\text{F}]{B^-} (CH_3O)_2POCHFCO(CH_2)_4CH_3$$

67   68

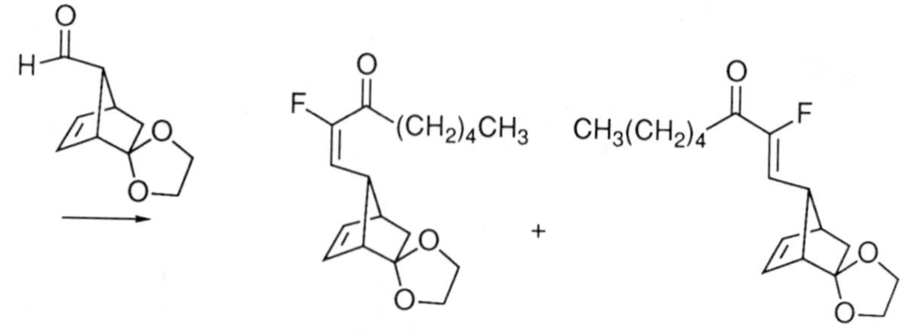

64% **69**   +   9% **70**

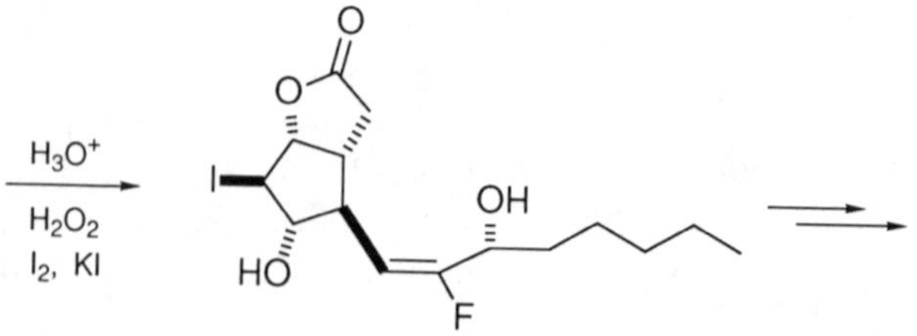

$$\xrightarrow[\text{I}_2,\ \text{KI}]{\substack{\text{H}_3\text{O}^+ \\ \text{H}_2\text{O}_2}}$$

$\longrightarrow$ **66**

An alternative approach to the synthesis of a fluorinated Horner-Emmons reagent has been employed in the synthesis of methyl 4-fluoro-11α,15-dihydroxy-15-methyl-9-oxo-prosta-4 (E), 13 (E)-dien-1-oate, 4-fluoro-enisoprost (**77**).[19] The necessary fluorinated phosphonate (**73**) was prepared from ethyl bromofluoro-acetate (**72**) and triethylphosphonate (**71**). Condensation with the appropriate hydroxyaldehyde (**74**) yielded the E ester (**75**) in 84% after chromatography. Furan rearrangement to the key cyclopentanone intermediate (**76**) occurred in 99% yield. A sequence of standard transformations led directly to the formation of the target compound (**77**). The introduction of vinyllic fluoride was proposed to enhance activity and duration of the analog by interfering with the enzymatic reduction of the C-4, C-5 double bond. The target compound was significantly less active than enisoprost. This diminution in activity was attributed to a hydrogen bond interaction between the 15-hydroxyl and fluorine, which affected the conformation of the substrate.

$$(CH_3CH_2O)_3PO\ +\ CHBrFCO_2CH_2CH_3\ \longrightarrow\ (CH_3CH_2O)_2PO\ CHFCO_2CH_2CH_3$$

71   **72**   **73**

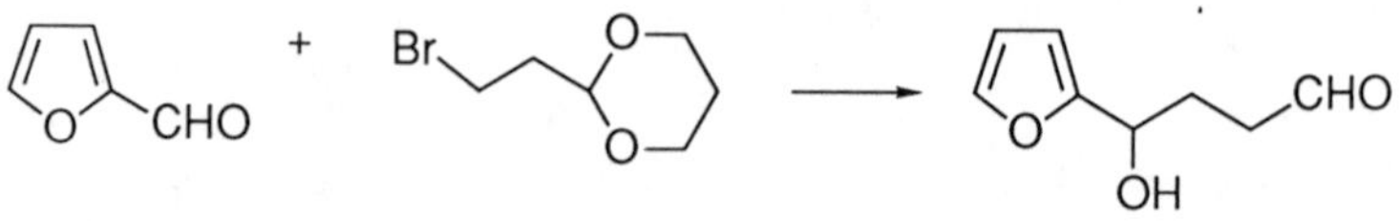

74

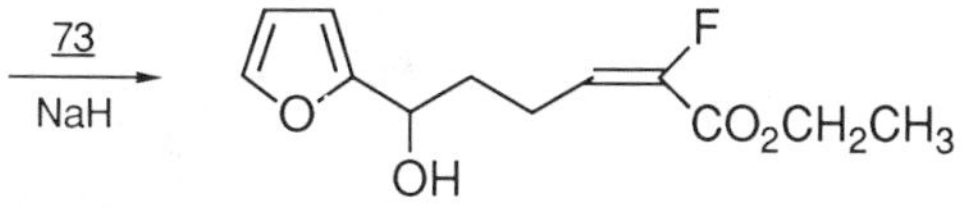

9.6.2 Chlorodifluoromethane

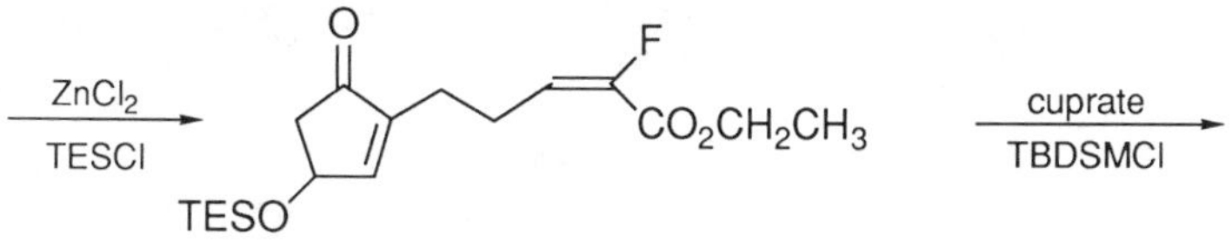

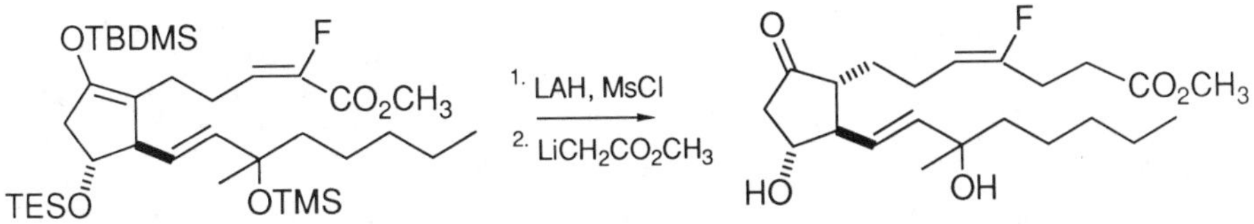

Phosphonate reagents have also been prepared which contain a fluoromethylene unit remote to where the new double bond will be prepared upon olefination. A difluoromethylene group was introduced by treating the sodium salt of diethyl *n*-butylmalonate (**78**) with chlorodifluoromethane in dimethoxyethane at $-40°$C. The difluoromethylenated material was formed in 93% yield.[20] Following partial saponification, decarboxylative elimination in dimethoxyethane with anhydrous potassium fluoride regioselectively formed α-fluoromethylene ester (**79**) in 98% yield. Further elaboration to the phosphonate (**80**) and olefination with an appropriate aldehyde (**81**) yielded the desired product (**82**). While 16-fluoromethylene-PGE$_1$ (**83**) retained biological activity relative to the parent compound, there was no significant reduction of the diarrhea-producing side-effect of the parent material.

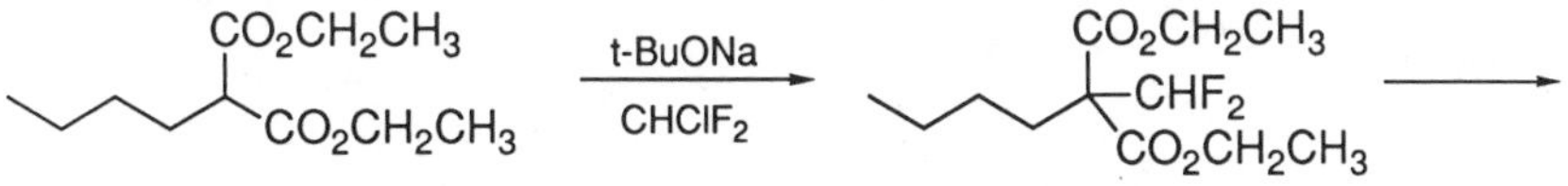

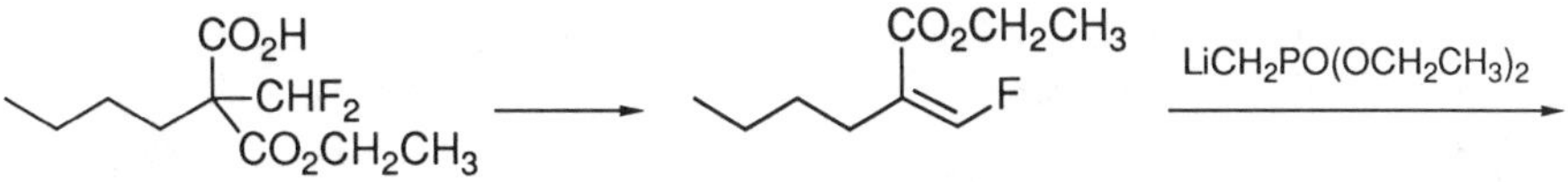

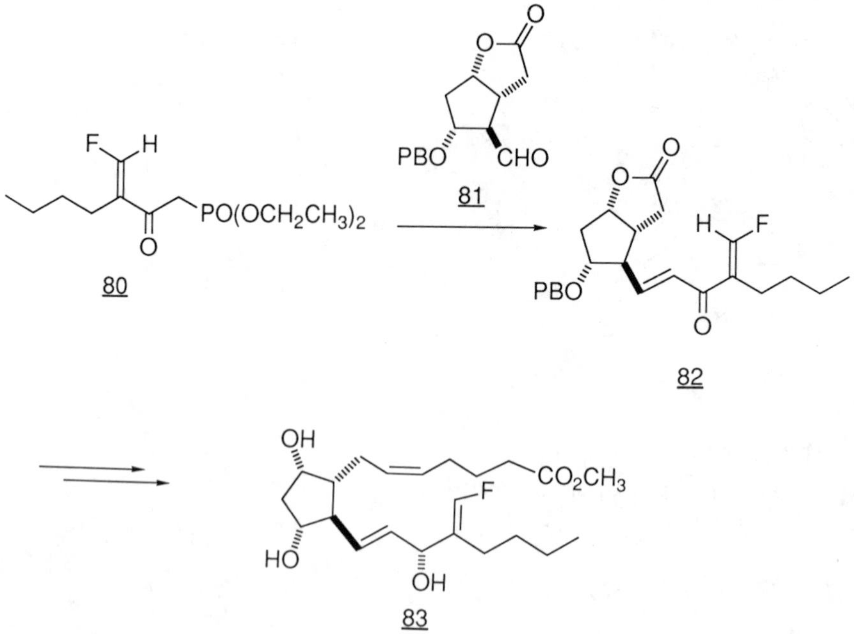

80

81

82

83

9.6.3 Fluorosulfoximines

The development of a new reagent has made possible direct reaction on protected PGE$_2$ itself.[21] A fluoromethylphenyl sulfoximine reagent was prepared by treatment of methylphenyl sulfoxide (84) with DAST. Oxidation of the resultant fluoromethylphenyl sulfide (85) followed by amination yielded the desired fluoromethyl sulfoximine (86). The deprotonated sulfoxime (87) was added to a solution of the protected PGE$_2$ forming the intermediate alcohol 88. Treatment with aluminum amalgam yielded on deprotection the 9 (E) compound (89) in 37% and the 9(Z) material (90) in 40% yield.

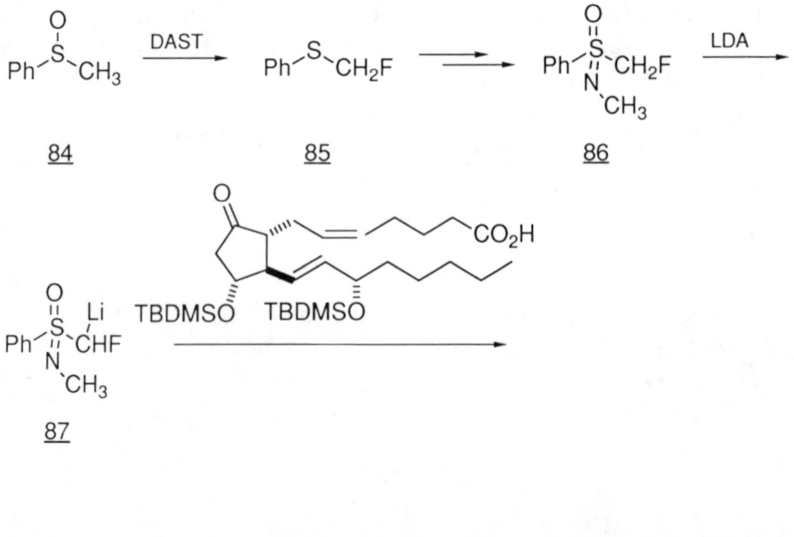

84

85

86

87

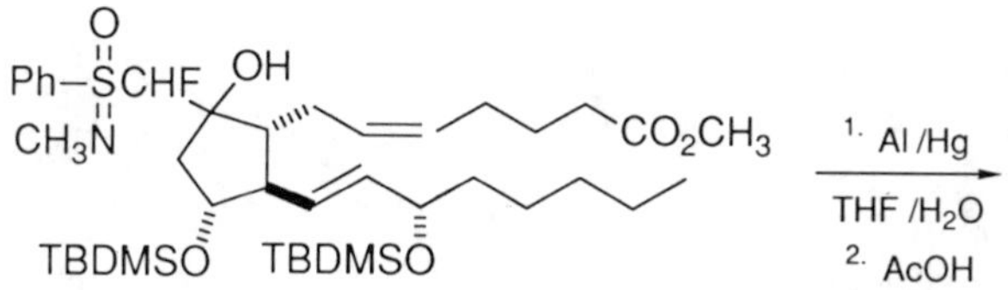

88

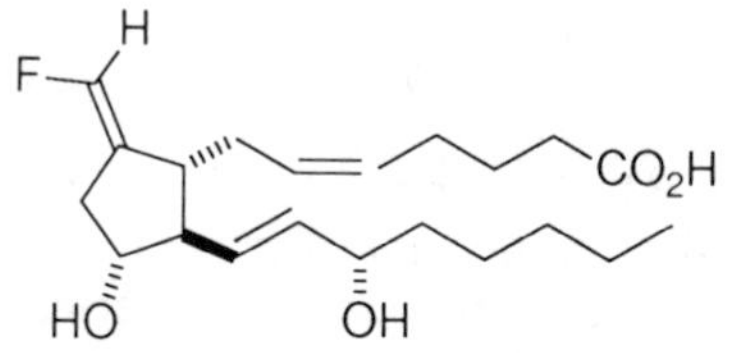

89

9 E 37%

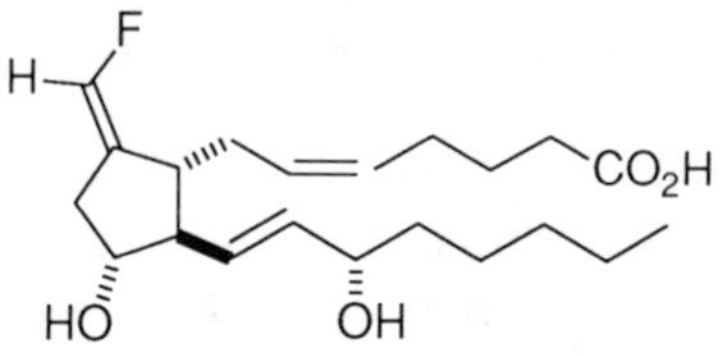

90

9 Z 40%

9.6.4 Chlorodifluoroacetic Acid

The commonly recognized relation between methylene groups and cyclopropanes has been explored by other workers who introduced a fluorinated cyclopropane as a replacement for a fluoroolefin unit in prostaglandins. In an early report, addition of difluorocarbene, generated by the pyrolysis of the sodium salt of chlorodifluoroacetic acid, to the unsaturated lactone (**91**) formed a mixture of the α-(**92**) and β-(**93**) compounds.[22] These materials were further elaborated to the desired difluoromethylene prostaglandin analog (**94**).

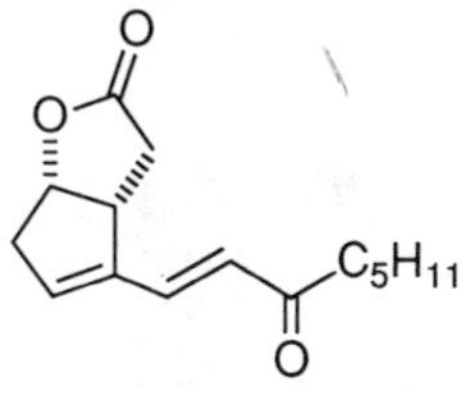

91

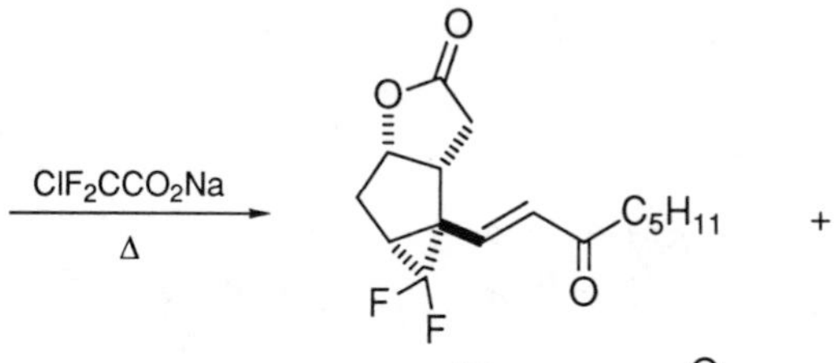

CIF₂CCO₂Na
Δ

92

+

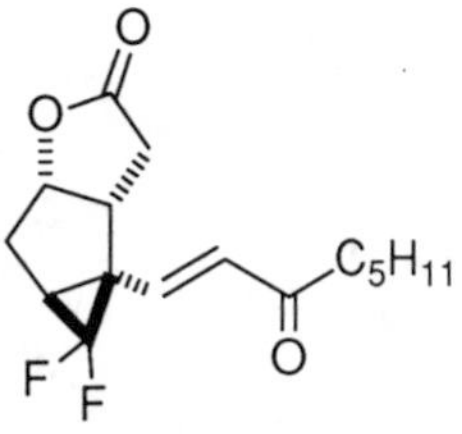

93

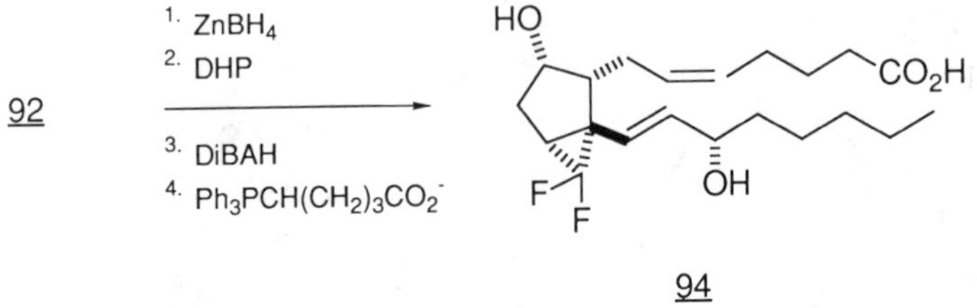

94

9.7 REFORMATSKY-TYPE REACTIONS

The use of readily available fluorinated Reformatsky reagents as well as fluorinated lithium enolates in the preparation of fluorinated prostanoids has also been reported. Generally these reagents react with poor diastereoselectivity in contrast to the excellent selectivity possible with nonfluorinated materials. Additionally, the use of fluorinated acetylenic aluminum reagents is discussed.

9.7.1 Reformatsky Reagents

The preparation of 7,7-difluoro derivative **96** of 2,6-dioxo (3.1.1)bicyclo-heptanes as model compounds for thromboxanes was proposed to improve the acid stability of these materials. Thromboxane (**95**), with a half-life of 32 seconds under physiological conditions, pose an extraordinary challenge to synthetic chemists. In the proposed model compounds, the β-cation (**97**) formed on ring cleavage would be strongly destabilized by the two adjacent fluorines.[23] Reformatsky reaction of ethyl bromodifluoroacetate and **98** led to a 2:1 mixture of diastereomers **99** and **100**. The 3β-trifluoromethanesulfonate (**102**) prepared after reduction to the lactol (**101**) was deprotonated with lithium hexa-methyldisilazide to form the target oxetane (**103**). The half-life of this material under physiological conditions was improved to 86 minutes.

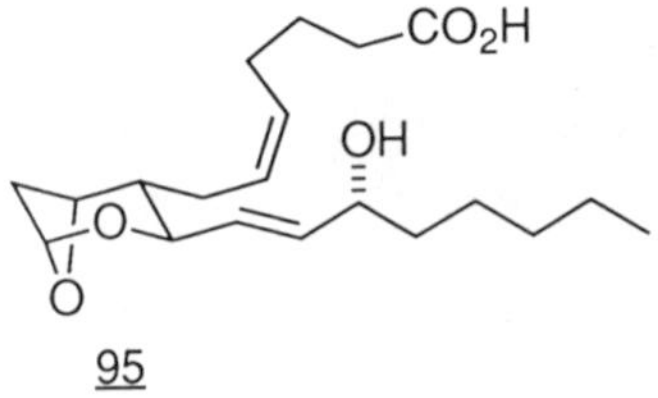

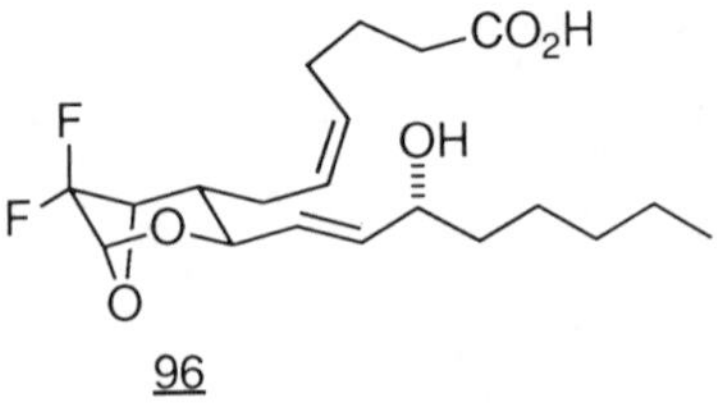

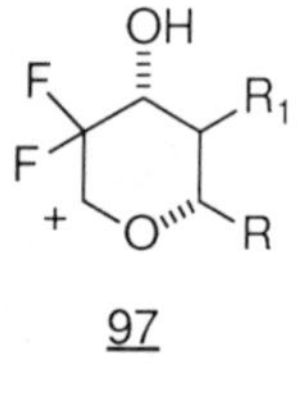

95

96

97

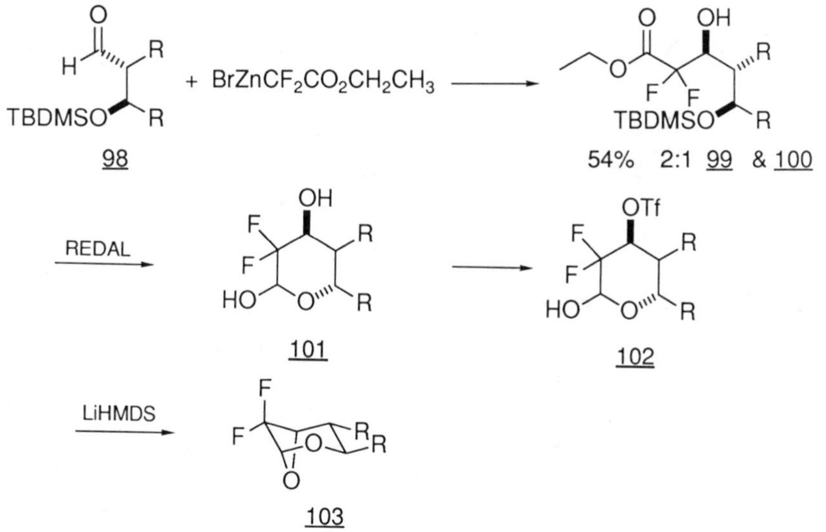

98

+ BrZnCF$_2$CO$_2$CH$_2$CH$_3$ ⟶

54%   2:1 99 & 100

REDAL ⟶

101   102

LiHMDS ⟶

103

9.7.2 Fluorinated Lithium Enolates

Both the E and Z isomers of 5-fluoro-PGF$_{2\alpha}$ were accessible from the aldol condensation of a functionalized methyl ester of 2-fluorohexanoate (**104**).[24] The lithium enolate was prepared by deprotonation and was condensed with the required aldehyde to afford a 70% yield of a mixture of diastereomers **105**. Decarboxylative elimination formed a mixture of the E and Z isomers, **106** which separated on esterification. Oxidation of (**105**) followed by decarboxylation with aqueous dimethylsulfoxide formed the oxidized and protected aldol adduct (**107**) which lead to preparation of 5-fluoro-6-keto-PGE$_1$ methyl ester (**108**). 5-Fluoro-6-keto-PGE$_1$ methyl ester was 10 times more potent than PGE$_1$ in inhibition of stress ulcers in rats and 10 times more potent in uterine contractile activity.

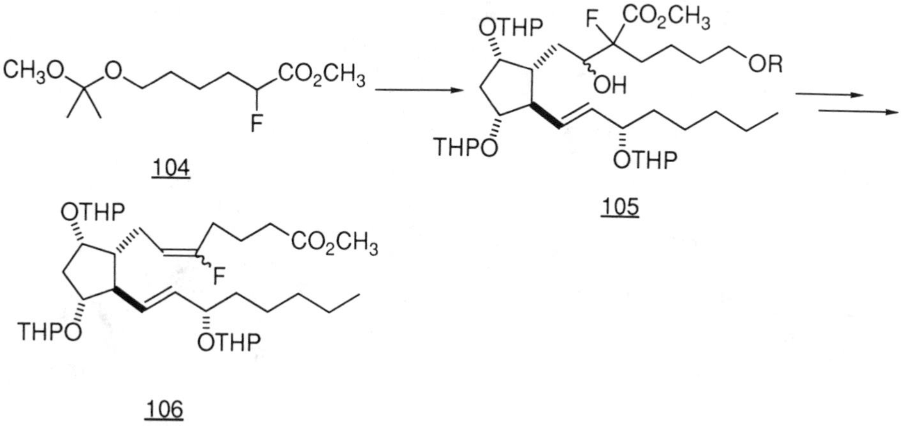

104   105

106

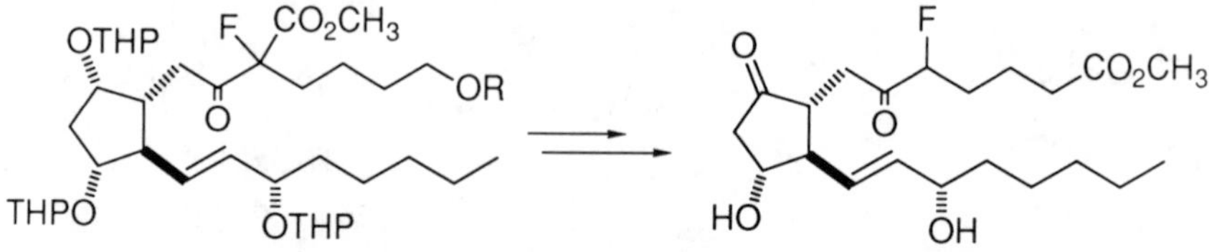

107 → 108

9.7.3 Fluorinated Alane Reagents

Reduction of ethyl 2-fluorohexanoate (109) with diisobutylaluminum hydride, gave 2-fluorohexanal (110) which was condensed with acetylene magnesium bromide. The erythro (111) and threo (112) diastereomers, formed in a 7:3 ratio, were separated by chromatography. Following protection of the alcohol as a *t*-butyl ether (113), the acetylene was converted into the alane reagent (114). Addition of 114 to cyclopentene epoxide (115) formed the triol *t*-butyl ether (116) with the correct stereochemical relationships. Conversion to the lactol, Wittig olefination and deprotection formed the target compound (117).[25] These compounds possessed potent antifertility activity but one-fifth lower smooth muscle activity than the nonfluorinated parent dehydro compound.

$C_4H_9CHFCO_2CH_2CH_3$
109
$\xrightarrow{\text{DIBAL}}$
$C_4H_9CHFCHO$
110
$\xrightarrow{\equiv-MgBr}$

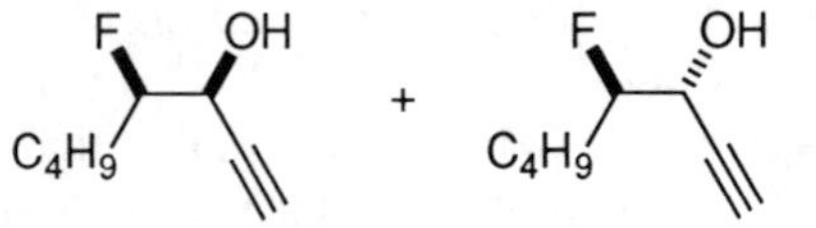

111 + 112 → $C_4H_9CHFCHC\equiv CH$ with $OC(CH_3)_3$
113

$C_4H_9CHFCHC\equiv CAlR_2$ with $OC(CH_3)_3$
114

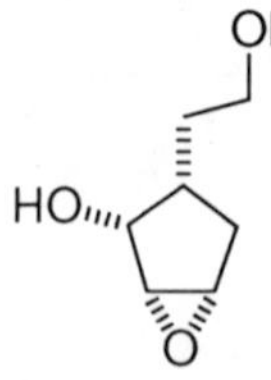

115 →

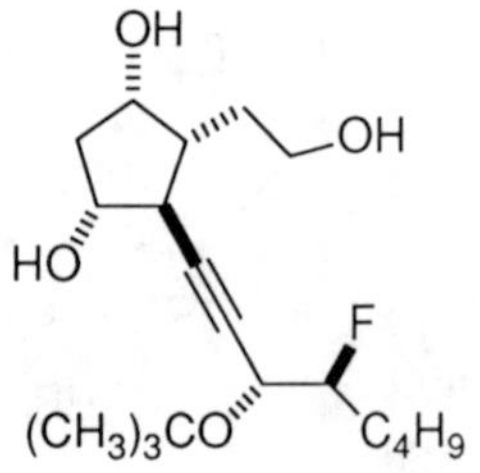

116

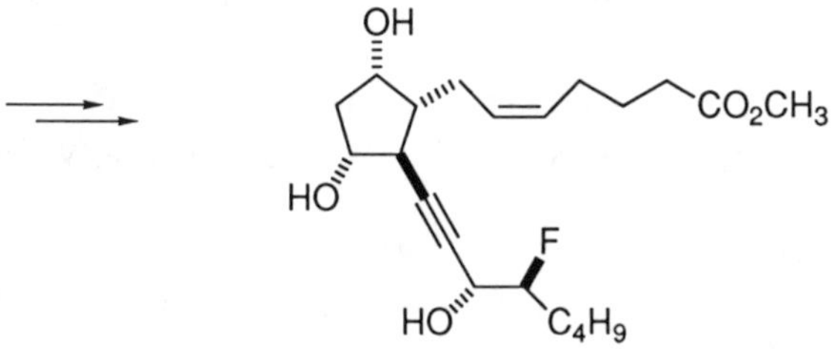

117

9.7.4 α-Fluorinated Carbonyls as Electrophilic Building Blocks

Previously α-fluorinated carbonyls were utilized as nucleophilic reagents. However, electrophilic addition to the carbonyl carbon has also been effectively employed in synthesis. 1-Fluoro, 1,1-difluoro, and 1,1,1-trifluoro-2-hexanone (**118**) have been condensed with propargyl magnesium bromide to form a hydroxy alkyne (**119**) in 86, 51, and 65% yield respectively.[26] The alkyne products were reacted with tributyl stannane to form the E vinyl tin compounds, which were in turn converted to the E vinyl lithium reagents and were used to prepare the required copper lithium reagents (**120**). Conjugate addition of these reagents to an appropriately functionalized cyclopentenone (**121**) formed 16-fluoromethyl-15-deoxy-16-hydroxy-PGE$_2$ (**122**).

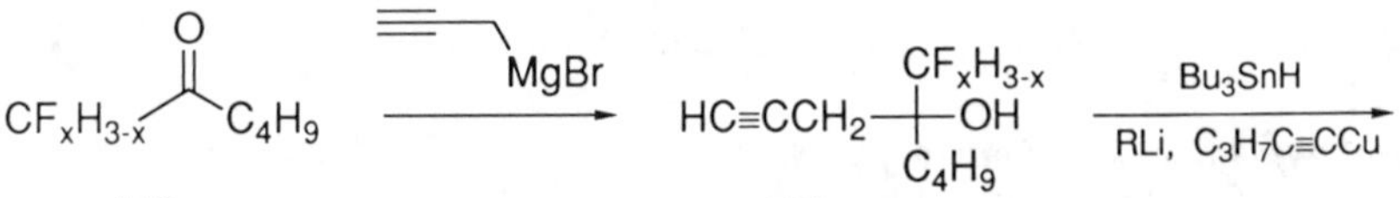

118

a	x = 1
b	x = 2
c	x = 3

119

a	x = 1	86%
b	x = 2	51%
c	x = 3	65%

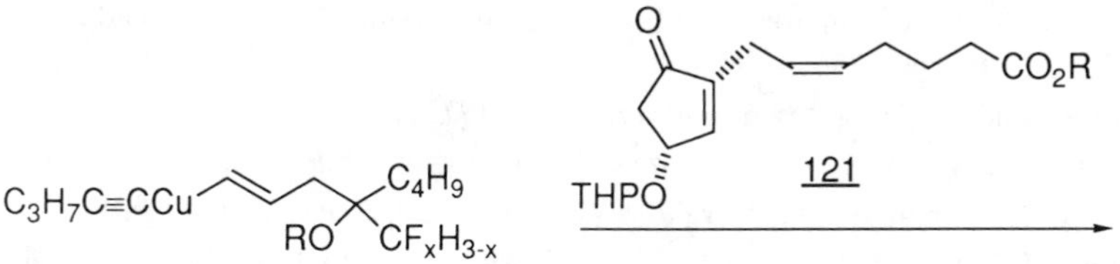

120

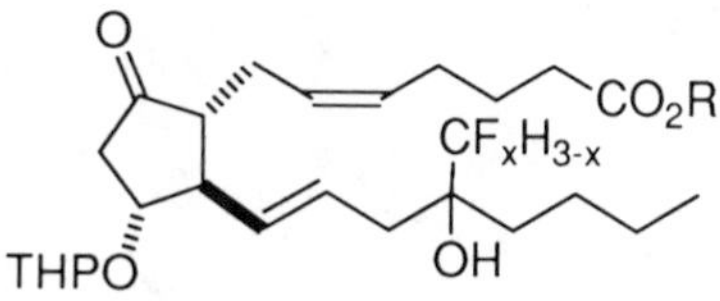

121

122

REFERENCES

1. W. E. Barnette, *Crit. Rev. Biochem.* **15**, 201, 1984.

2. C. E. Arroniz., J. Gallina, E. Martinez, J. M. Muchowski, E. Velarde, and W. H. Rooks, *Prostaglandins* **16**, 47, 1978.

3. M. Hudlicky, *Org. React.* **35**, 513–637, 1987.

4. K. Bannai, T. Toru, T. Oba, T. Tanaka, N. Okamura, K. Watanabe, A. Hazato, and S. Kurozumi, *Tetrahedron* **39**, 3807, 1983.

5. A. Yasuda, T. Asai, M. Kato, K. Uchida, and M. Yamabe, personal communication, 1986.

6. V. V. Bezuglov, I. V. Serkov, R. G. Gafurov, and L. D. Bergel'son, *Dokl. Akad. Nauk,* **277**, 1400–1402, 1984.

7. P. A. Grieco, E. Williams, and T. Sugahara, *J. Org. Chem.* **44**, 2194, 1979.

8. P. A. Grieco and T. R. Vedananda, *J. Org. Chem.* **48**, 3497, 1983.

9. P. A. Grieco, E. Williams, and T. Sugahara, *J. Org. Chem.* **44**, 2194, 1979.

10. P. A. Grieco, T. Sugahara, Y. Yokoyama, and E. Williams, *J. Org. Chem.* **44**, 2189, 1979.

11. N. Mongelli, F. Animati, R. D'Alessio, L. Zuliani, and C. Gandolfi, *Synthesis* **1988**, 310–313.

12. F. Mathey and J. Bensoam, *Tetrahedron* **27**, 3965, 1971.

13. B. J. Magerlein and W. L. Miller, *Prostaglandins* **9**, 527, 1975.

14. H. Nakai, N. Hamanaka, and M. Kurno, *Chem. Lett.* **1979**, 63–66.

15. J. Fried, D. K. Mitra, M. Nagarajan, and M. M. Mehrotra, *J. Med. Chem.* **23**, 234, 1980.

16. S. W. Djuric, R. B. Garland, L. N. Nysted, R. Pappo, G. Plume, and L. Swenton, *J. Org. Chem.* **52**, 978–990, 1987.

17. P A. Grieco, T. Takigawa, and T. R. Vedananda, *J. Org. Chem.* **50**, 3111, 1985.

18. P. A. Grieco, W. J. Schillinger, and Y. Yokoyama, *J. Med. Chem.* **23**, 1077, 1980.

19. P. W. Collins, S. W. Kramer, and G. W. Gullikson, *J. Med. Chem.* **30**, 1952, 1987.

20. S. Kosuge, H. Nakai, and M. Kurono, *Prostaglandins* **18**, 737–743, 1979.

21. M. L. Boys, E. W. Collington, H. Finch, S. Swanson, and J. F. Whitehead, *Tetrahedron Lett.* **29**, 3365–3368, 1988.

22. P. Crabbe and A. Cervantes, *Tetrahedron Lett.* 1973, 1319.

23. J. Fried, E. A. Hallinan, and M. J. Szewedo, Jr., *J. Am. Chem. Soc.* **106**, 3871, 1984.

24. H. Hakai, N. Hamanaka, H. Miyake, and M. Hayashi, *Chem. Lett.* **1979**, 1499.

25. J. Fried, M.-S. Lee, B. Gaede, J. C. Sih, Y. Yoshikawa, and J. A. McCracken, *Adv. Prostaglandin Thromboxane Res.* **1**, 183, 1976.

26. S.-M. L. Chen and C. V. Grudzinskas, *J. Org. Chem.* **45**, 2278, 1980.

10

FLUORINATED STEROIDS

10.1 INTRODUCTION

The fluorination of steroids has been known to have profound effects on biological activity since the early work of Fried.[1] Fluorinated steroids have been described in other reviews.[2] The utility of fluorination as a tool to enhance selectivity and to improve the utility of biologically active materials has been clearly demonstrated in studies of vitamin D_3.

The preparation of fluorinated analogs of vitamin D_3 has been reviewed.[3] Fluorination has been employed to block metabolic hydroxylation and to modify the reactivity of hydroxyl groups. Vitamin D_3 (1) is activated by hydroxylation at C-1 and C-25 to yield an active steroid hormone (2). Fluorination at C-25 (3), C-1

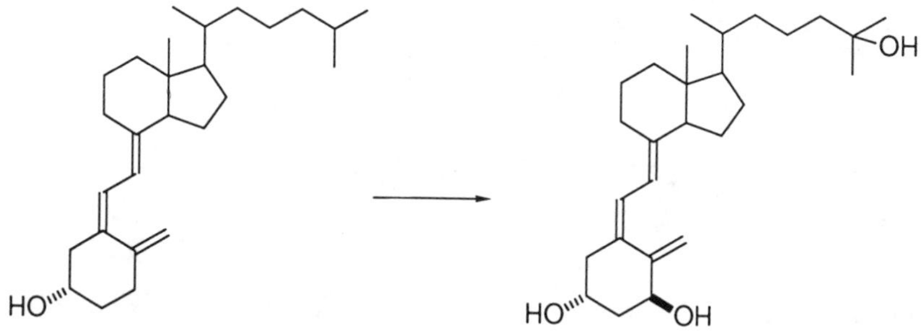

1   2

(4), or C-3 (5) prohibited the hydroxylation reaction that was essential for activation while facilitating study of the role each of these hydroxyl groups played in the various aspects of the activity of vitamin D_3. Compounds which were fluorinated at C-24 (6), C-26 (7), or C-23 (8) blocked the hydroxylation reactions that lead to metabolic deactivation of vitamin D_3. Further, fluorination at C-24

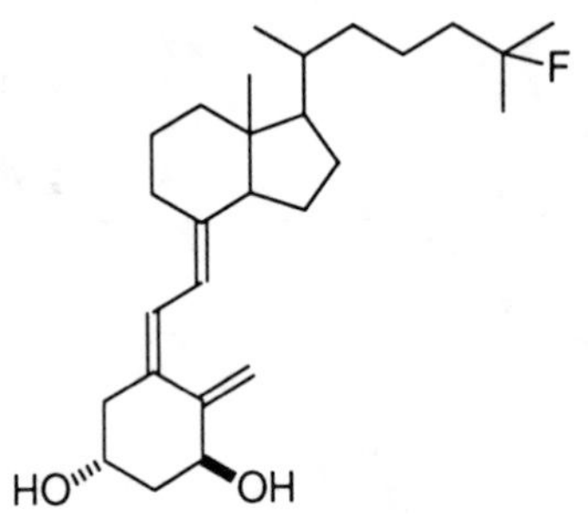

3

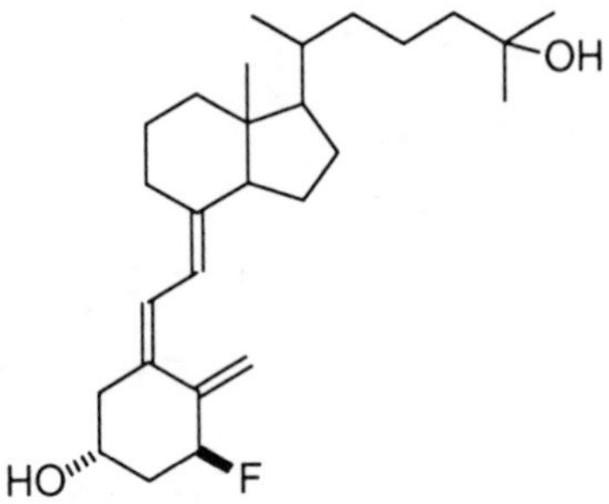

4

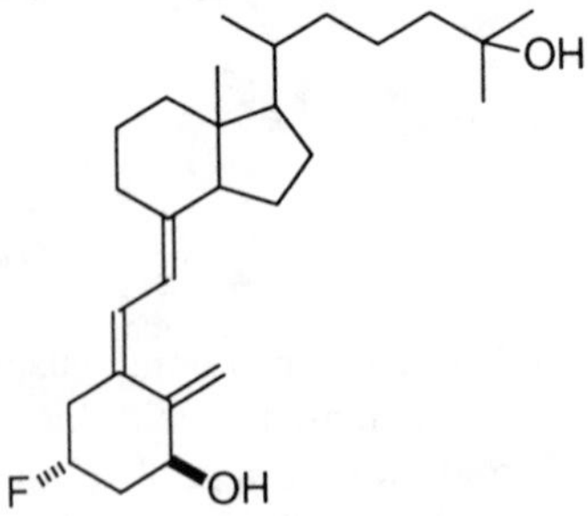

5

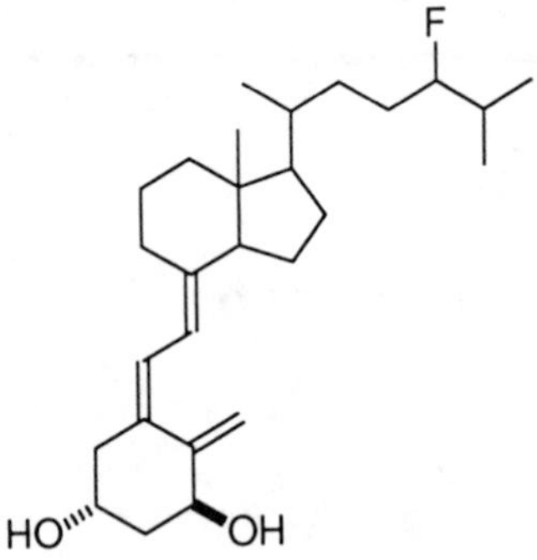

6

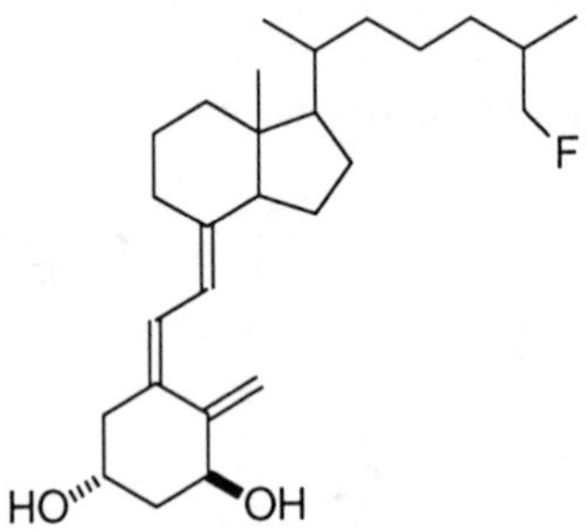

7

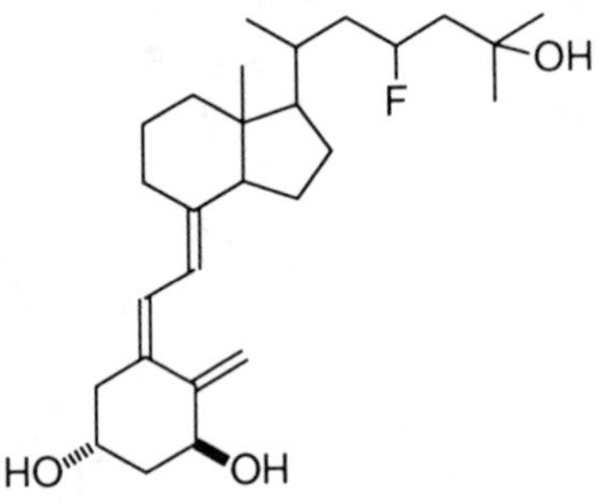

8

(9), C-26 (10), or C-2 (11), where the fluorine substitution would be adjacent to the 25- or 1-hydroxyls essential for activity, was useful in determining the role of those hydroxyls in hydrogen bonding, either as hydrogen bonding donors or acceptors. Additionally, fluorination was used to study the photochemical reaction which forms vitamin D_3. Where substitution of the reactive olefin system was made, both the 6-fluoro and 19, 19-difluoro analogs (12, 13) were prepared.

Although the scope of the utility of fluorination in modifying activity is clearly illustrated in the preceding examples, this utility is not limited to vitamin D

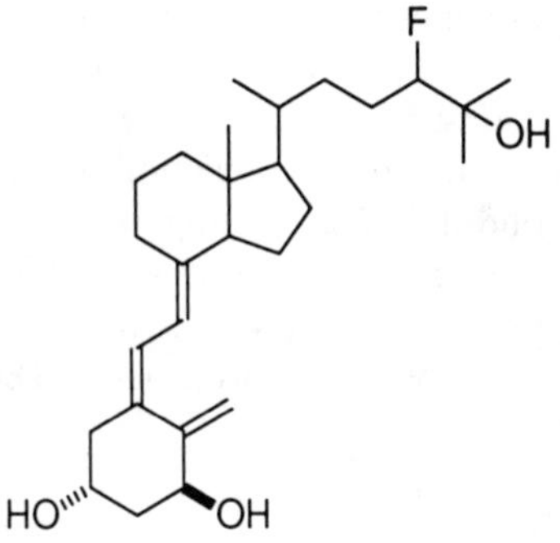

9

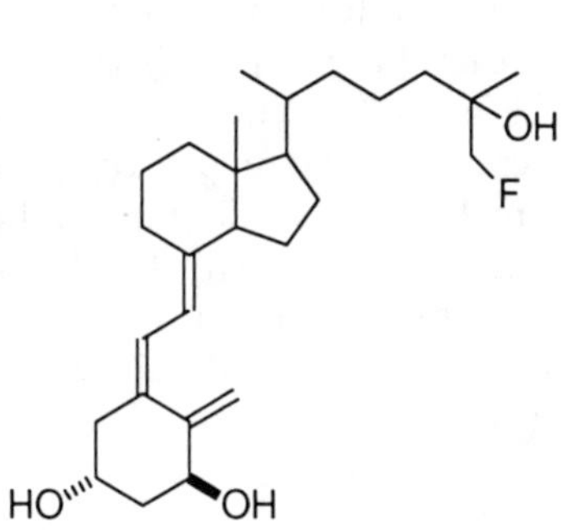

10

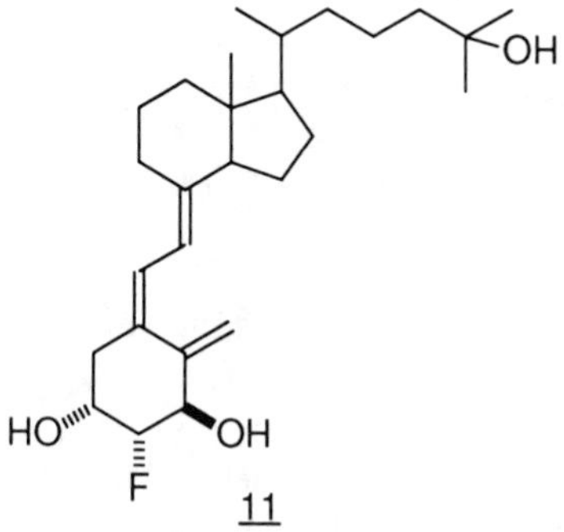

11

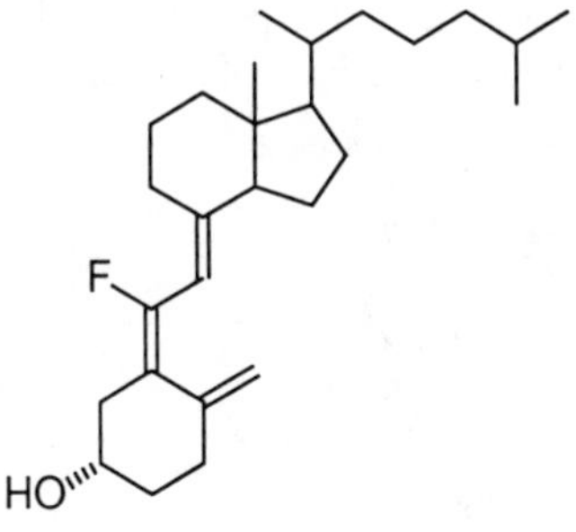

12

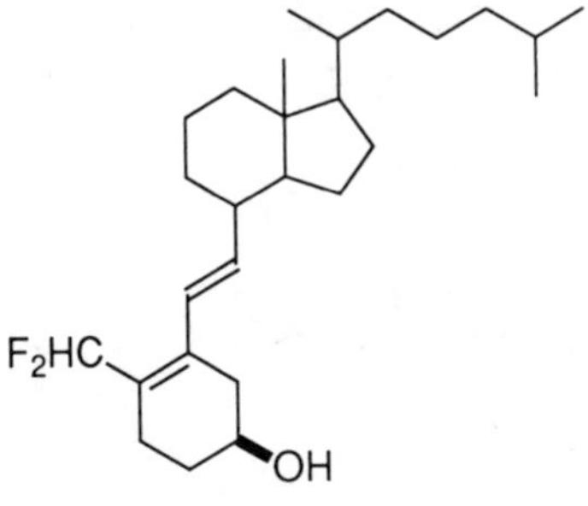

13

derivatives. Therefore, the use of the various reagents and strategies employed for preparation of a variety of steroids of biological interest is described.

10.2 DIRECT FLUORINATION

Molecular fluorine, diluted with nitrogen or another inert gas, has been used as a reagent for the electrophilic substitution of an sp^3 hybridized carbon. Reactions at such saturated, unactivated centers are rare; however, molecular fluorine not only reacts but reacts with selectivity. The crucial feature of the reaction is the choice of the solvent. In a nonpolar solvent such as $CFCl_3$ the reaction is indiscriminate, although when $CHCl_3$ is added to the reaction mixture it apparently polarizes the fluorine–fluorine bond, as in **14**, allowing the reaction to exhibit more of the character of an ionic process. Successful reaction occurs with those hydrogens that have relatively high ionic character yet at the same time have a substantial *p*-orbital contribution in their bonding with carbon. Tertiary C–H bonds are therefore the most reactive.

14

Selectivity can be illustrated by the reaction at $-78°C$, of 17-oxo-5β-androsten-3α-ol acetate (**15**) with fluorine diluted with nitrogen. 17-Oxo-5β-fluoroandrosten-3α-ol acetate (**16**) was formed in 35% yield and 17-oxo-14α-5β-androsten-3α-ol acetate (**17**), in 25% yield.[4] No 9-fluorination was observed, presumably as a result of steric obstruction to reaction by the A–B cis ring junction, even though the 9-C–H has the highest *p*-orbital contribution relative to other C–H bonds. The stereospecificity and regiospecificity of product formation are remarkable considering the alleged reactivity of elemental fluorine.

10.3   FLUORINATIVE DEHYDROXYLATION

10.3.1   Diethylaminosulfur Trifluoride

Diethylaminosulfur trifluoride (DAST) is one of the best known and most convenient reagents for fluorinative dehydroxylation.[5] Although DAST readily reacts with sterols, the product stereochemistry depends on the structure of the steroid.[6] Cholesterol (**18a**), pregnenolone (**18b**), and androstenolone (**18c**) produce 3β-fluoro-5-cholestene (**19a**) in 95% yield, 3β-fluoro-5-pregnen-20-one (**19b**) in 82% yield, and 3β-fluoro-5-androsten-17-one (**19c**) in 81% yield. The

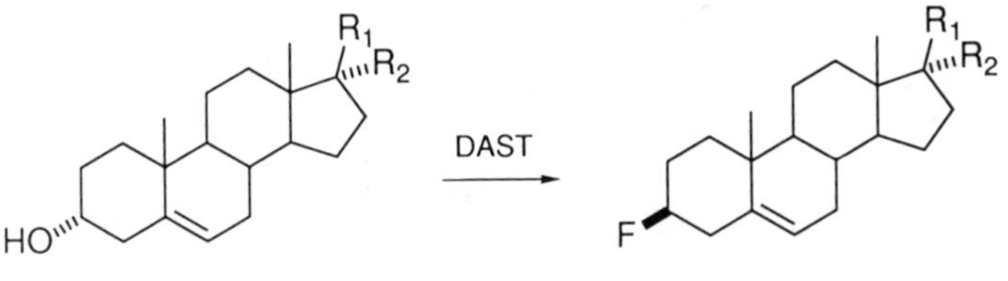

$$18 \quad a \quad R_1 = C_8H_{17} \quad R_2 = H \qquad\qquad 19$$
$$b \quad R_1 = AcO \quad R_2 = H$$
$$c \quad R_1 = R_2 = O$$

yields and stereoselectivity are excellent as a result of the interaction of the δ^5 double bond as in **20**. In the absence of the double bond, elimination was an

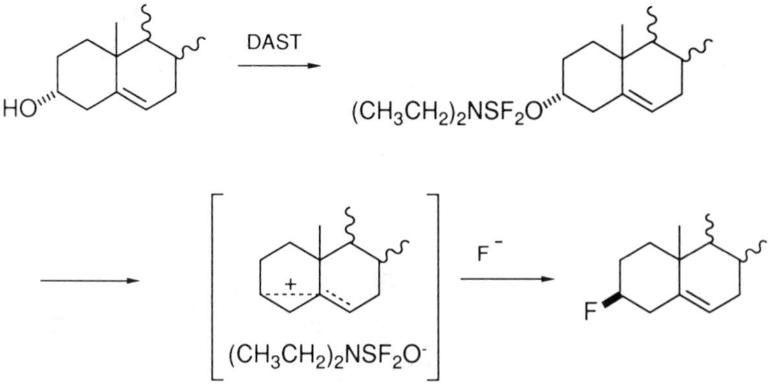

20

important side reaction. When cholestanol (**21**) was treated with DAST, 3α-fluorocholestan (**22**) was formed in only 43% yield but 2-cholestene (**23**) was isolated in 32% yield.

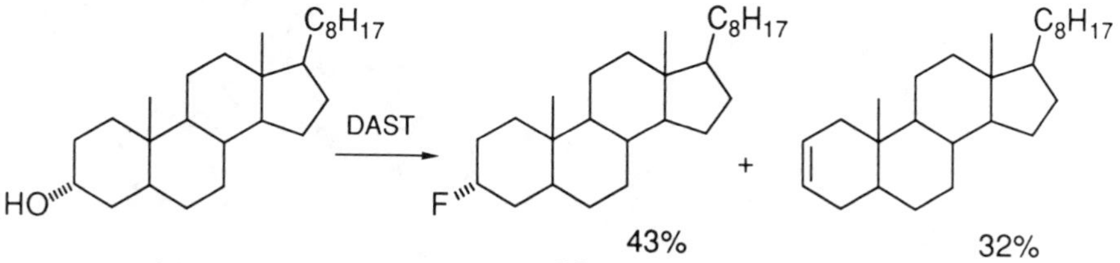

When cholesta-5,7-dien-3α-ol, protected as 5α,8α-(4-phenyl-1,2-urazolo)-cholesta-6-ene (**24**) (the Diels-Alder adduct of 4-phenyl-1, 2, 4-triazoline-3,5-

dione), was treated with DAST, reaction of the free hydroxyl resulted in a 35% yield of the protected 3β-fluoride (**25**). On deprotection and photolysis, 3β-fluoro-vitamin D$_3$ (**26**) was formed.[7] The 3β-fluoro-vitamin D$_3$ was less active than vitamin D$_3$ but more active than 3-dehydroxy-vitamin D$_3$.

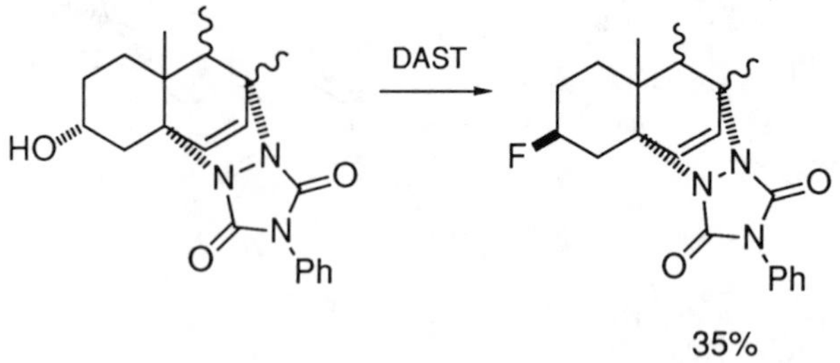

24   **25**

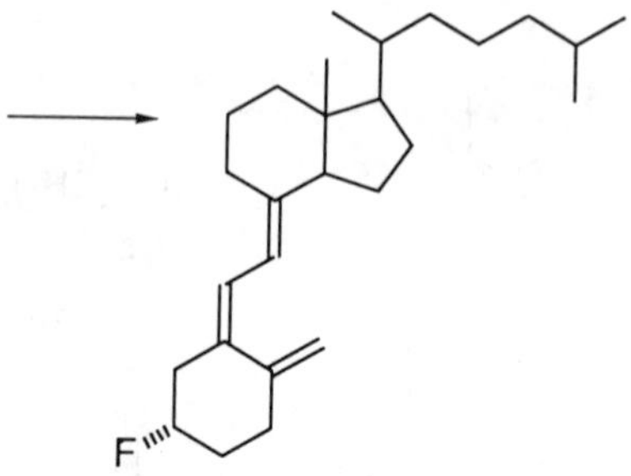

26

The utility of DAST is not limited to fluorination of the steroid phenanthrene nucleus. The transformation of clerosterol to 25-fluoroclionosterol (**27**) required the fluorination of the tertiary alcohol with DAST.[8] Even though this reaction

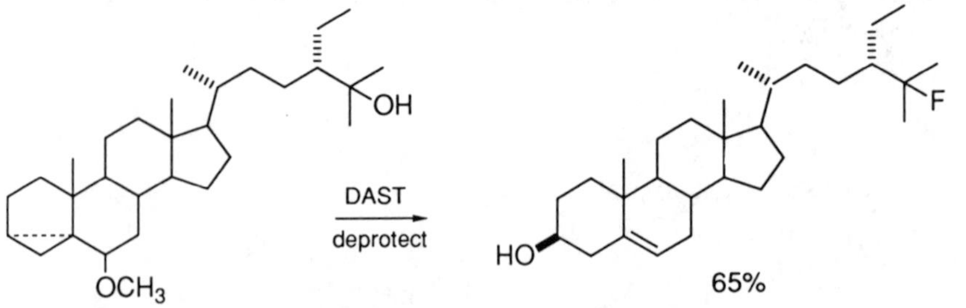

27

successfully proceeded in 65% yield, the preparation of 26-fluorositosterol (**28**) required the fluorination of a secondary alcohol by DAST, which failed. The

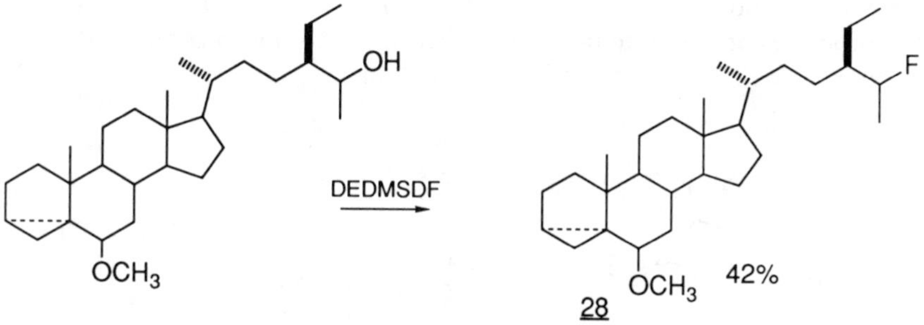

42%

<u>28</u>

fluorination was successfully effected by modifying the reactivity of DAST by the addition of another equivalent of dimethylaminotrimethylsilane to the DAST solution prior to adding the sterol. Application of the new reagent, diethylaminodimethylamino–sulfur difluoride, efficiently formed the desired fluorinated sterol in 42% yield.

10.3.2 Chlorotrifluoroethyl(diethyl)amine

Chlorotrifluoroethyl(diethyl)amine (fluoroalkyl amine reagent, FAR) was introduced in 1962 for the fluorination of steroids.[9] The fluorination of 3β-hydroxypregn-5-en-20-one (**29**) with FAR was dependent on the solvent. Both THF and acetonitrile gave yields of less than 30% of the desired 3β-fluorosteroid (**30**), whereas with methylene chloride or chloroform as solvent yields of the 3β-fluorosteroid could be as high as 90%.[10]

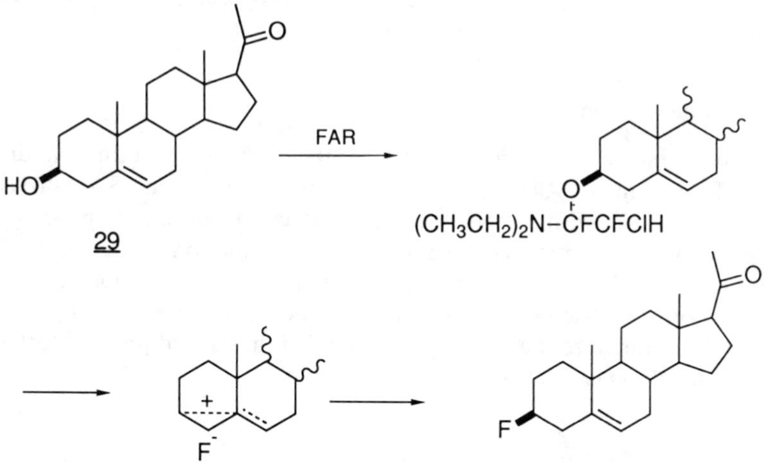

10.3.3 Phenyltetrafluorophosphine

Phenyltetrafluorophosphine was introduced as a fluorinating reagent by Kobayashi,[11] but its use has largely been supplanted by DAST and more modern reagents. Reaction with 24-hydroxycholesteryl acetate (31) failed to form 24-fluorocholesteryl acetate but rather formed the 25-fluoro compound (32) in 30% yield as well as demonsteryl acetate (33) in 33% yield.[12]

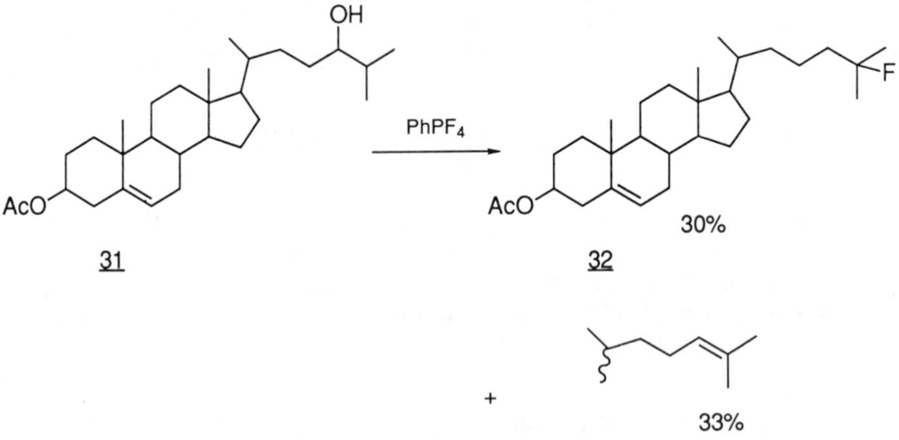

10.4 DISPLACEMENT REACTIONS

Under some conditions the use of one of the previously described fluorinating reagents may be accompanied by side reactions such as elimination. Introduction of a leaving group, such as a sulfonate ester, prior to attempted displacement by fluoride ion very effectively avoids these problems.

10.4.1 Toluenesulfonates

This result is nicely illustrated in a study of the preparation of fluoroandrosta-nones. Treatment of the hydroxyl compounds, such as 3α-hydroxy-5α-androstanone (34), with DAST yielded olefinic side products in 83%. This reaction could be substantially suppressed by formation of 3α-tosylate (35) followed by treatment with tetra-n-butylammonium fluoride in N-methylpyrrolidine. The desired 3β-fluoro-5α-androstanone (36) was formed in 33% yield in contrast to the less than 10% yield of the desired product formed in one-step transformations.[13]

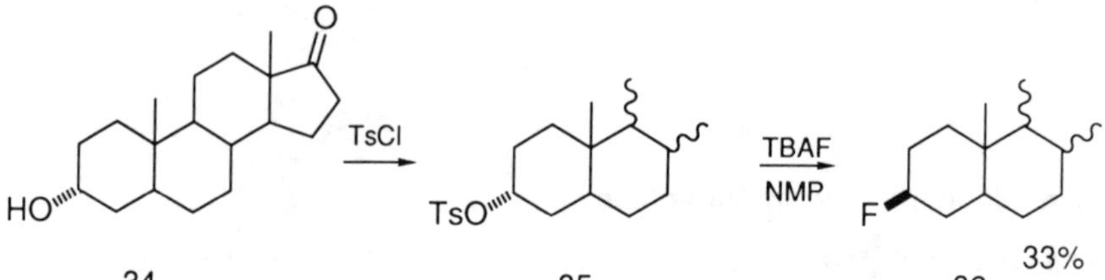

Methyl 3β-tetrahydropyranyloxy-24-tosyloxychol-5-ene-24-carboxylate (**37**) was treated with potassium fluoride in dimethylformamide in the presence of 18-C-6 to form methyl 24-fluoro-3β-tetrahydropyranylchol-5-ene-24-carboxylate (**38**) in 73% yield.[12] The same 24-hydroxy starting material had failed to yield any of the desired 24-fluoro product on treatment with phenyltetrafluorophosphine.

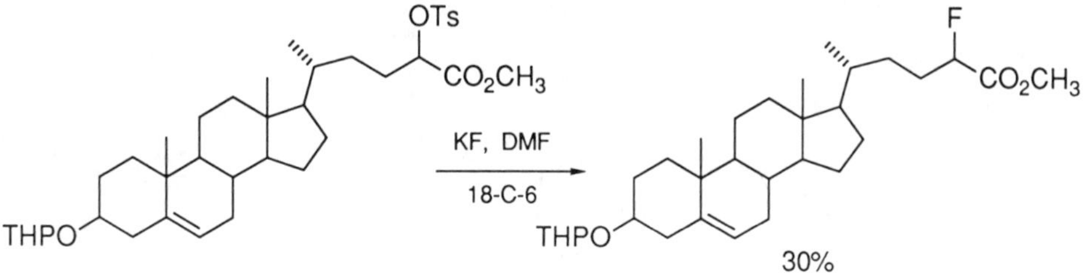

10.4.2   Trifluoromethanesulfonates

The 16α-trifluoromethanesulfonate ester (**39**) prepared in four steps from estrone was treated with tetra-n-butylammonium fluoride in tetrahydrofuran to afford the 16β-fluoroestrone 3-trifluoromethanesulfonate (**40**) in 82% yield.[14] The reaction does not proceed with clean inversion of configuration if excess tetra-n-butylammonium fluoride is used. The excess tetra-n-butylammonium fluoride epimerizes the newly created 16-fluoride. Reduction with lithium aluminum hydride yielded 16β-fluoro-17β-estradiol (**41**) in 78% yield, presumably as a result of the accessibility of the β face of the carbonyl. The 16α-fluoro compound (**43**) was prepared in 62% yield by tetra-n-butylammonium fluoride displacement of

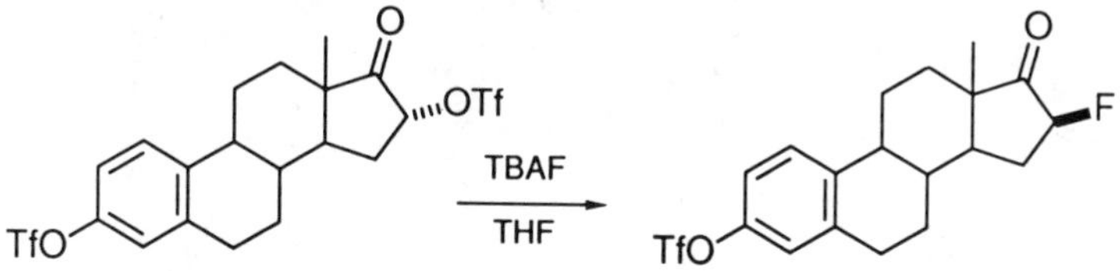

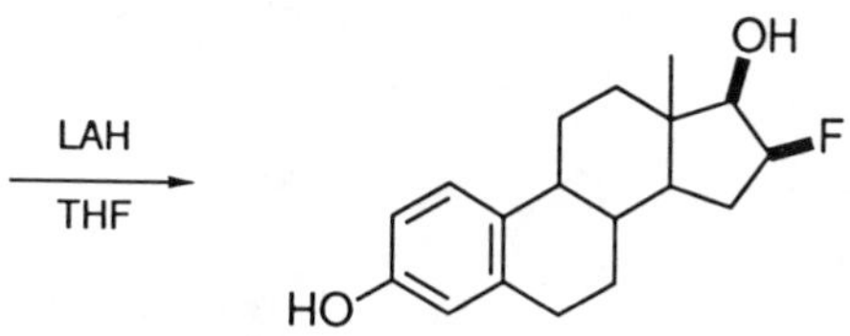

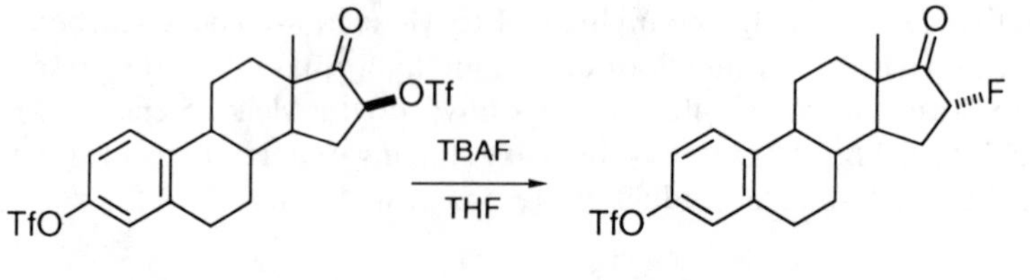

16β-trifluoromethanesulfonate (**42**). The reduction with lithium aluminum hydride was somewhat less selective, affording the 16α-fluoro-17β-estradiol (**44**) in 65% yield, reflecting a 4:1 selectivity of reduction from the β face. Yields on displacement of the 21-triflate (**45**) with excess tetra-*n*-butylammonium fluoride are between 50 and 60%. When fluoride is taken as the limiting reagent, yields fall to 30%. Isolation is simplified by hydrolysis of the unreacted triflate to the alcohol.[15]

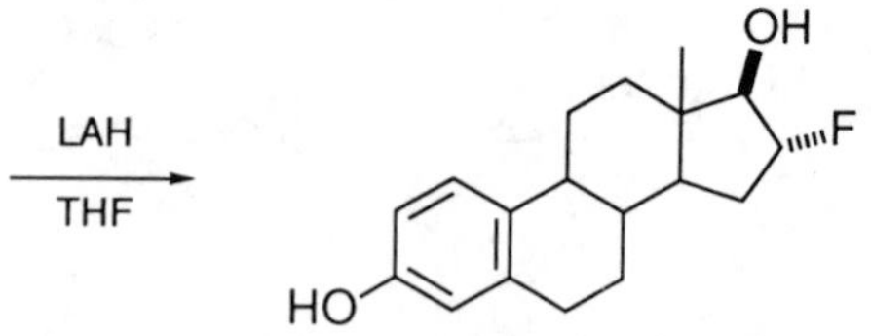

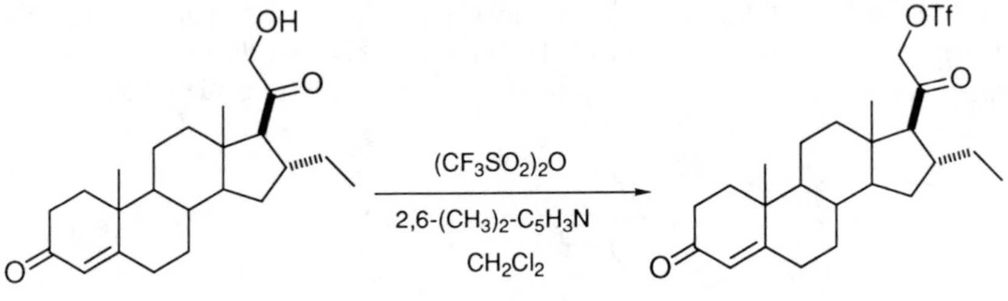

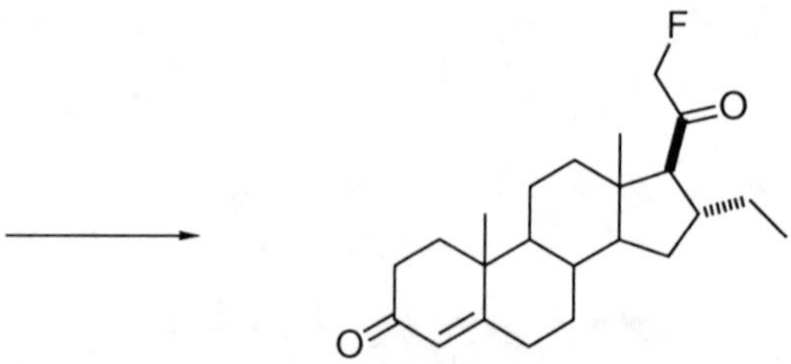

60% (xs TBAF)

30% (1 x TBAF)

Nonsteroidal estrogen receptor binding agents have also been prepared.[16] (2R, 3S)-1-Fluoro-2,3-bis(4-hydroxyphenyl)pentane (**48**) was prepared by tetra-n-butylammonium fluoride displacement of the (2R, 3S)-1-trifluoro-methanesulfonate-2,3-bis (4-phenyl-trifluoromethanesulfonate)pentane (**46**). The (2R, 3S)-1-fluoro-2, 3-bis(4-phenyl-trifluoromethanesulfonate)pentane (**47**) was not isolated but the phenolic trifluoromethanesulfonates were reduced directly with lithium aluminum hydride to form the desired compound (**48**) in 57% yield. The nonsteroidal material had a very high affinity for estrogen receptors. As a consequence, the radiolabeled material was also prepared.

10.5 BALZ-SCHIEMANN REACTIONS

The Balz-Schiemann reaction has generally been employed for the nuclear fluorination of aromatic compounds. When ring A fluorinated estrogens were required for a study of estrogen sulfotransferase they were prepared by this process. 4-Fluoroestrone and 2-fluoroestrone were proposed to be accessible via the Balz-Schiemann reaction or by the electrophilic fluorination of a 19-nortestosterone intermediate or estrone 3-O-methyl ether by perchlorylfluoride or xenon difluoride.[17]

The desired 4-fluoroestrones were finally prepared by diazotization of 4-amino-estra-1, 3, 5 (10)-trien-17β-ol (**49**) in absolute ethanol and 48% fluoboric acid. After stirring at 0°C for 1.5 hours, a 17% yield of 4-fluoroestrone (**51**) was isolated on chromatography. 2-Fluoroestrone was prepared by diazotization and dediazoniation of 2-amino-estra-1,3,5(10)-trien-17β-ol (**52**) in tetrahydro-furan, dioxane, and 48% fluoboric acid at 0°C for 1 hour. The isolated diazonium salt (**53**), suspended in xylene, was heated under reflux for 18 hours yielding on chromatography 16% of the predicted mass of 2-fluoroestrone (**54**).

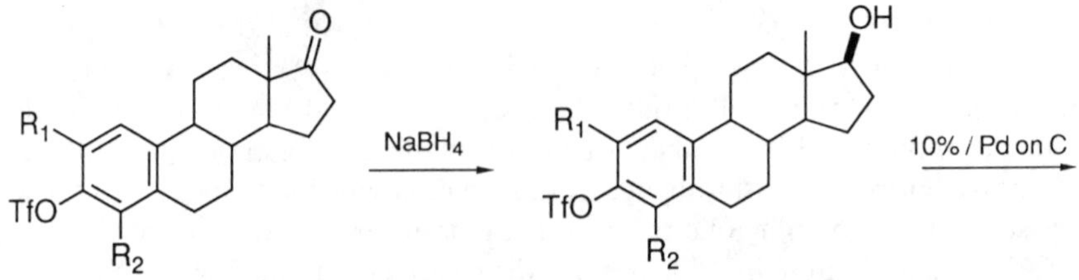

$R_1 = NO_2$, $R_2 = H$

$R_1 = H$, $R_2 = NO_2$

<u>49</u> $R_1 = H$, $R_2 = NH_2$

<u>52</u> $R_1 = NH_2$, $R_2 = H$

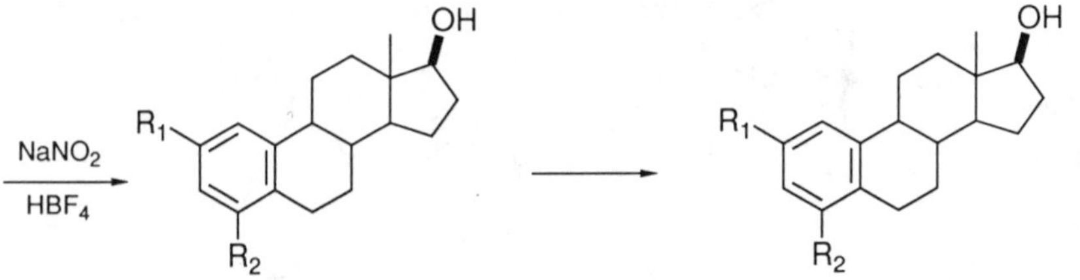

<u>50</u> $R_1 = H$, $R_2 = N_2^+$

<u>53</u> $R_1 = N_2^+$, $R_2 = H$

<u>51</u> $R_1 = H$, $R_2 = F$

<u>54</u> $R_1 = F$, $R_2 = H$

10.6 EPOXIDE OPENING REACTIONS

10.6.1 Hydrogen Fluoride

Solutions of hydrogen fluoride, stabilized by a Lewis base, have been very effectively employed to regioselectively open epoxides. 21-Chloro-9β,11β-epoxy-16α,17α-isopropylidendioxy-4-pregnen-3,20-dion (**55**) was treated with a solution of hydrogen fluoride in pyridine at $-40°C$ and was then allowed to stir at 15°C for one hour. Following chromatographic workup, 21-chloro-9α-fluoro-11β-hydroxy-16α,17α-isopropylidendioxy-4-pregnen-3,20-dion (**56**), the topical anti-inflammatory corticosteroid Halcinonid, was isolated in 70% yield.[18]

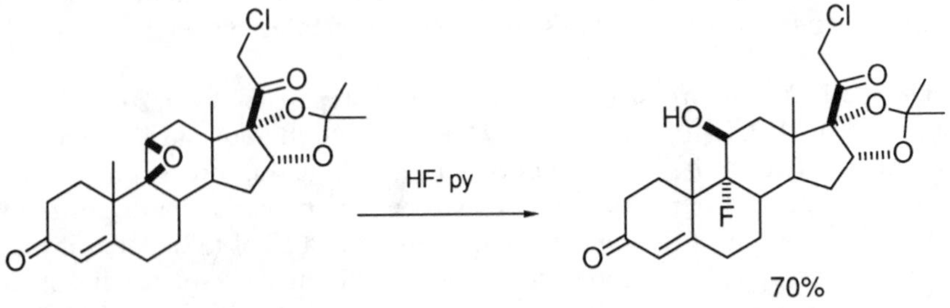

55

56

70%

In a related epoxide-opening reaction, 5α,6α-epoxy-16α,17α-dimethylpregnen-20-yn-3β-yl acetate (**57**) was treated with boron trifluoride etherate and a 9.82 *M* solution of hydrogen fluoride in diglyme for only 15 minutes. A 99% yield of the pharmacologically interesting 3β-acetoxy-6β-fluoro-16α, 17α-dimethyl-pregnen-20-yn-5α-ol **58** was recovered.[19]

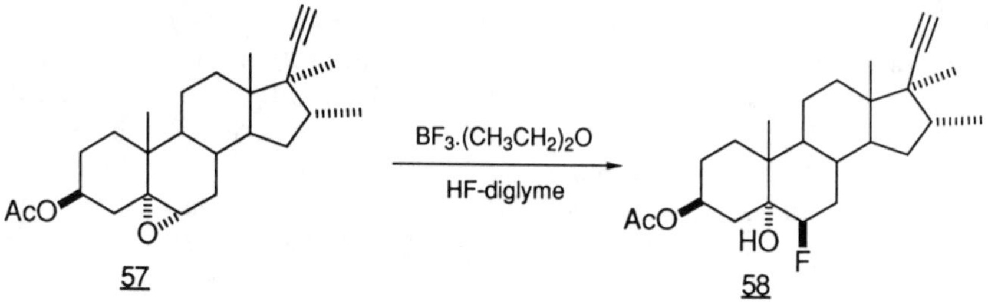

10.6.2 **Potassium Fluoride–Potassium Hydrogen Difluoride**

2β-Fluoro-1α-hydroxy-vitamin D$_3$ was prepared as part of a research program to determine the effect of the introduction of fluorine at carbon-2 of vitamin D$_3$ on the hydrogen bonding interactions of the 1α-hydroxyl with the appropriate receptor.[20] Treatment of 1α,2α-epoxy-5α-cholesten-3β-ol (**59**) with tetra-*n*-butylammonium fluoride or potassium fluoride in the presence of dicyclohexyl 18-C-6 was fruitless. However, treatment of 1α,2α-epoxy-5α-cholesten-3β-ol (**59**) with potassium bifluoride in ethylene glycol at 170°C for 1.5 hours formed 2β-fluoro-cholesten-1α,3β-diol (**60**) in 48% yield.

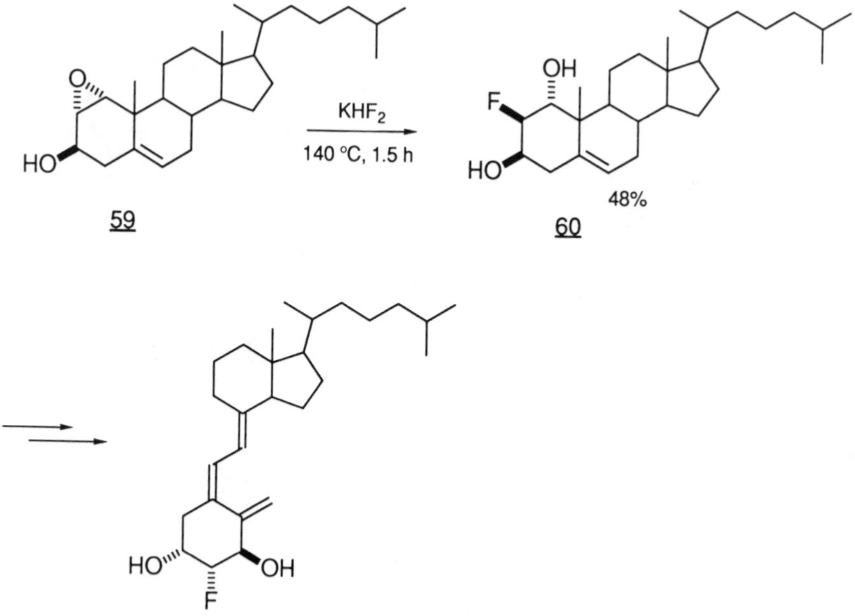

Previously 1α-Fluoro-vitamin D$_3$ was prepared by direct fluorinative dehydroxylation of 1α-hydroxy-vitamin D$_3$ with DAST,[21] but the stereochemistry of fluorination has not been determined definitively. More recently a regioselective and stereoselective synthesis based on the potassium hydrogen fluoride opening of 6β-acetoxy-1β,2β-epoxy-5α-cholesten-3β-ol (61) in ethylene glycol at 160°C over 7 hours has been reported.[22] Workup with acetone and acid formed acetonide (62) in 48% yield. The acetate was saponified and the 6-hydroxy group dehydrated with phosphorus oxychloride. Deprotection of the acetonide, followed by selective protection of the 3β-hydroxyl with triethylsilylchloride, facilitated conversion of the 2β-hydroxyl to a xanthate ester. Reduction with tributyltin hydride, allylic bromination, dehydrobromination formed the diene, which was photolyzed to yield 1α-fluoro-vitamin D$_3$ (63). From these studies it was determined that the earlier work with DAST had formed the 1β-fluoro material.

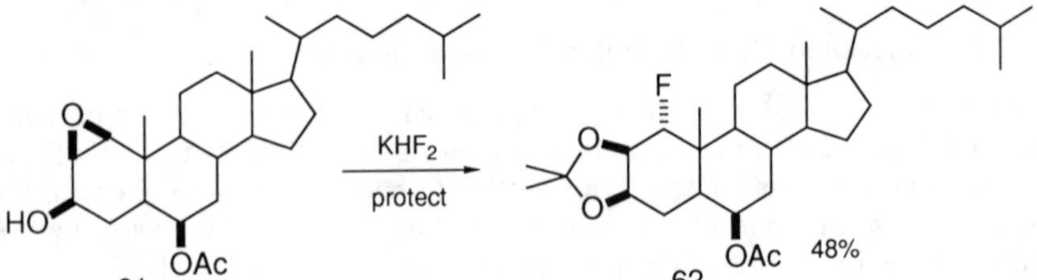

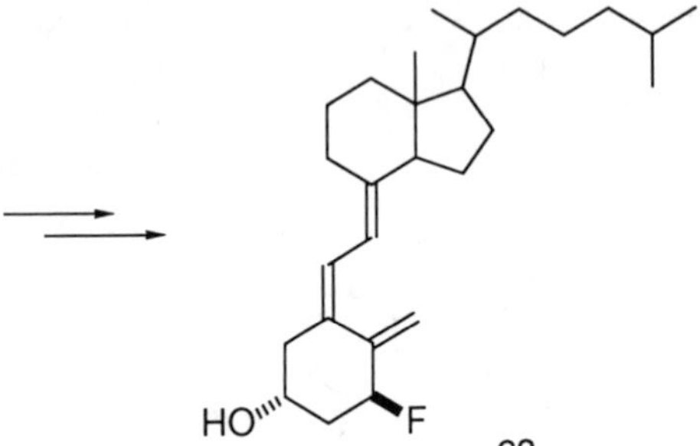

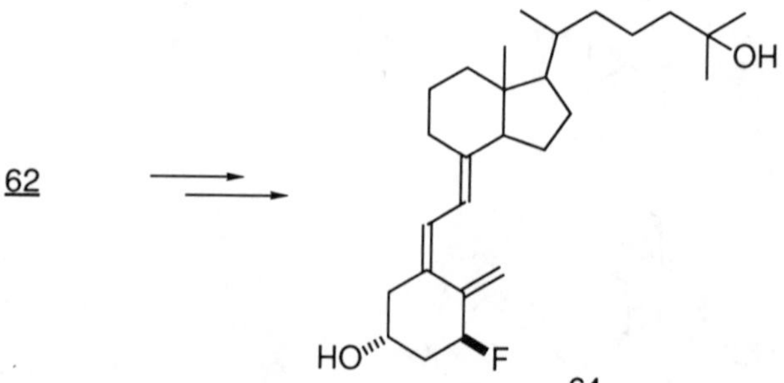

1α-Fluoro-25-hydroxy-vitamin D$_3$ (**64**) can also be prepared by adaptation of this synthetic approach to transformation of protected cholenic acid.[23] Hydroxylation at carbon-1 and carbon-25 produces the hormonally active form of vitamin D$_3$ which mediates calcium and phosphorus metabolism. The 1α-fluoro-25-hydroxy-vitamin D$_3$ analog did not stimulate intestinal calcium transport or bone mobilization of calcium. However, the synthetic material was nearly 30 times more effective at binding than the 25-hydroxy-vitamin D$_3$.

10.7 GEMINAL DIFLUORINATION REACTIONS

The choice of reagents for the generation of geminal difluorides from carbonyl compounds is limited. Difluorination reactions generally require more forcing conditions and result in low yields of the desired product. The popular DAST reagent is most widely employed and its use is illustrated in two examples. Another approach to the synthesis of geminal difluorides which involves the intermediacy of hydrazones is also discussed.

10.7.1 Diethylaminosulfur Trifluoride

Hydroxylation of the side chain of vitamin D$_3$ also has profound effects on the activity of the hormone and on the site specificity of the hormone. Fluorination at carbon-23 was proposed to prohibit the hydroxylation at that position known to occur during the course of vitamin D$_3$ metabolism. Fluorination of the ketoester,

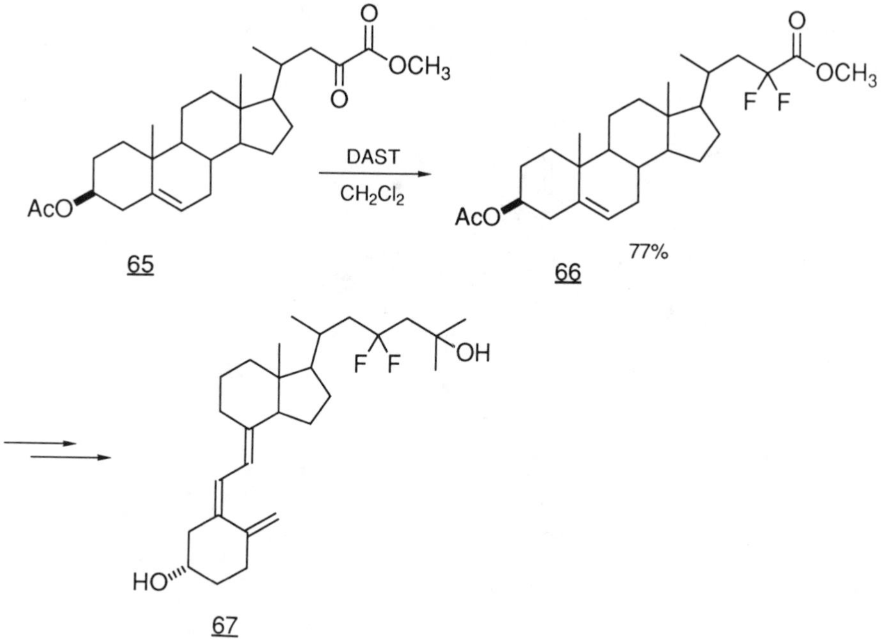

(65) prepared in three steps from 6β-methoxy-3α-5-cyclo-23,24-dinor-5α-cholen-22-ol with DAST in methylene chloride at room temperature for 16 hours yielded difluoro ester (66) in 74% yield.[24] Additional functional group manipulations led to formation of the desired previtamin. Conventional vitamin D_3 transformation followed by photolysis form 23,23-difluoro-25-hydroxy-vitamin D_3 (67). Although the biological activity of the analog was somewhat reduced, apparently the binding was diminished[25] and 1-hydroxylation did occur.

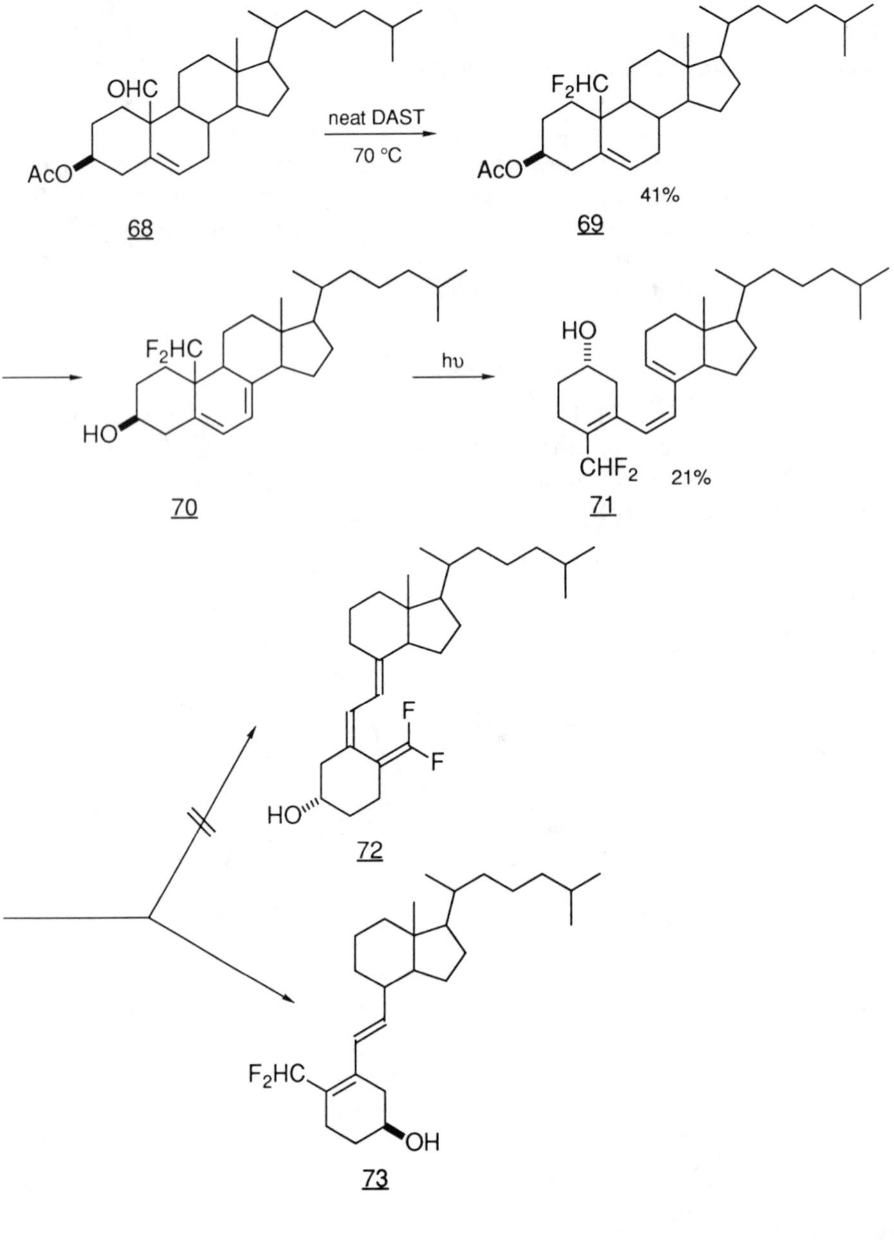

To study the thermal isomerization of previtamin D_3 to vitamin D_3 which has been shown to be sensitive to substituents on the triene portion of the molecule, 19,19-difluoroprevitamin D_3 was prepared.[26] Treatment of 19-oxocholesterol acetate (**68**) with neat DAST at 70°C for 15 hours yielded after chromatography 41% of 19,19-difluorocholesterol acetate (**69**). Conventional manipulations resulted in synthesis of 19,19-difluorocholesta-5,7-dien-3β-ol (**70**), which was irradiated to form 19,19-difluoroprevitamin D_3 (**71**) in 20% yield. Neither thermal nor photochemical transformations led to the desired vitamin D_3 analog (**72**). Only 19,19-difluorotachysterol (**73**) was isolated. The thermally induced (1,7) hydrogen shift is no longer thermodynamically favored when carbon-19 is difluorinated.

10.7.2   Iodine Monofluoride

Another approach to the synthesis of difluorides from carbonyl compounds is based on the in situ reaction of a 3 to 4 molar excess of iodine monofluoride,[27] prepared by the reaction of fluorine and iodine in fluorotrichloromethane at -78°C, with the hydrazone of cholestenone (**74**) to form the difluoride (**75**) in 70% yield.[28] The reaction is not postulated to require the elimination of hydrogen

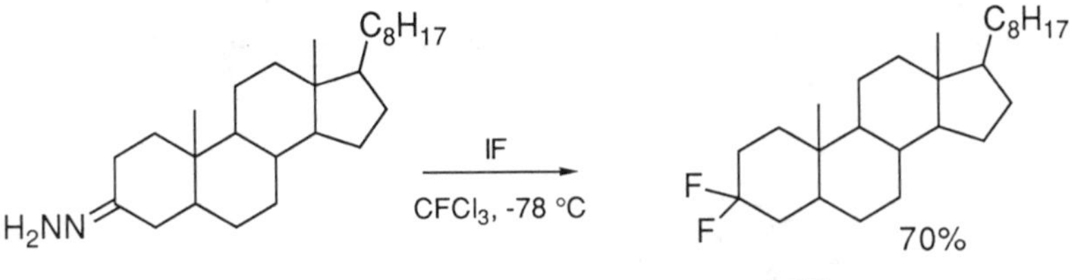

iodide or hydrogen fluoride. The more basic nitrogen of the unsaturated hydrazone is attacked by the electrophilic iodine, which polarizes the hydrazone toward attack by the nucleophilic fluoride ion (Fig. 10.1). After extrusion of iodide, which requires a net two electron transfer to iodine, a second fluoride ion then can react to form the geminal difluoride.

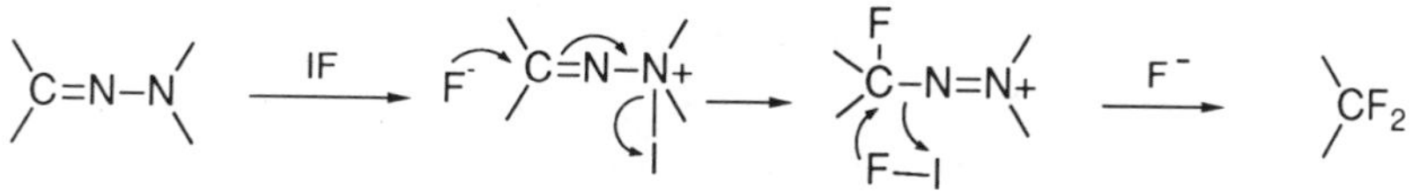

FIGURE 10.1

10.8   ADDITIONS TO OLEFINS

When vicinally difluorinated products are sought, direct addition of molecular fluorine to a multiple bond has not generally been successful. However, vicinal halofluorination is much more common.

10.8.1 Molecular Fluorine

An effective approach to the synthesis of vicinal difluorides is based on the opening of epoxides with hydrogen fluoride under conditions where the initially formed fluorohydrin was converted without isolation to the vicinal difluoride. Introduction of fluorine at C-16 and C-17 of a steroid has been accomplished by opening of the required C-16, C-17 epoxide by hydrogen fluoride.[29]

The addition of bromine fluoride to a C-16, C-17 olefin has been easily effected using hydrogen fluoride containing N-bromosuccinimide.[30] These methods will be discussed in the next section.

Perchlorylfluoride has also been reported to add across an activated olefin, such as steroidal C-17, C-20 enol ether double bond.[31]

As mentioned earlier, the addition of molecular fluorine across double bonds is not a commonly employed transformation; however, such an addition was reported when 3β-acetoxy-5α,6β-dichloropregn-16-en-20-one (**76**) in fluorotrichloromethane was treated with fluorine diluted to 5% by nitrogen at −78°C. Two major products were isolated, 3β-acetoxy-5α,6β-dichloropregn-16α,17α-difluoro-20-one (**77**) in 40% yield and 3β-acetoxy-5α,6β-dichloropregn-13α,16α-difluoro-17β-methyl-18-nor-17α-pregnan-20-one (**78**) in 12% yield.[32] Electrophilic attack from the α face was predicted on steric grounds. The 3β-acetoxy-5α,6β-dichloropregn-16α,17α-difluoro-20-one (**77**) was subsequently transformed to 16α,17α-difluoroprogesterone (**79**).

10.8.2 Bromofluorination

A more modern approach to the addition of bromine fluoride to olefins than that discussed earlier employs the use of N-bromosuccinimide, or bromine and silver nitrate, with pyridinium polyhydrogen fluoride. However convenient the use of these reagents, the lack of regiochemical and stereochemical control associated with their employment severely limits their utility.[33] When 3β-acetoxy-5-androsten-17-one (**80**) was treated with BrF generated by the reaction of aqueous hydrogen fluoride with N-bromosuccinimide, four compounds were isolated, 5α-bromo-6β-fluoro (**81**) in 58% yield, 5β-fluoro-6α-bromo (**82**) in 17% yield, and 5α-fluoro-6β-bromo (**83**) in 9% yield.[34] Theoretically two bromonium cations could

be formed; however, for steric reasons it was suggested that the 5α,6α isomer (**84**) was more probable than the 5β,6β isomer (**85**). Attack at C-5 or C-6 by fluoride ion would lead in the first case to formation of the 5β-fluoro-6α-bromo isomer (**86**) or the 5α-bromo-6β-fluoro isomer (**87**) with the second isomer being preferred as trans diaxial opening is preferred.

10.9 ADDITION TO ENOL ETHERS

The use of activated electron-rich olefins as a substrate for functionalization by fluorine or a fluorinating agent is a logical extension of the reactions of unactivated olefins. It is clear that both the nature of the fluorinating reagent and the enol ether can have a pronounced effect on the yield and the stereochemistry of the reaction. A particularly illustrative example compares the fluorination of estrone enol ethers with cesium fluoroxysulfate, xenon difluoride, fluorine, or trifluoromethyl hypofluorite.[35] When estrone silyl enol ether (**88a**) was treated with cesium fluoroxysulfate, protected 16-fluoroestrone was formed in 15% yield with a 9:1 ratio of α to β fluoride (**89** to **90**). Xenon difluoride gave a 44% yield of exclusively the α-fluorinated material (**89**). Fluorine diluted with nitrogen and trifluoromethyl hypofluorite yielded fluoroestrone in 10–15% yield and in a 7:1 α to β ratio (**89a** to **90a**) and exclusively α fluoride respectively.

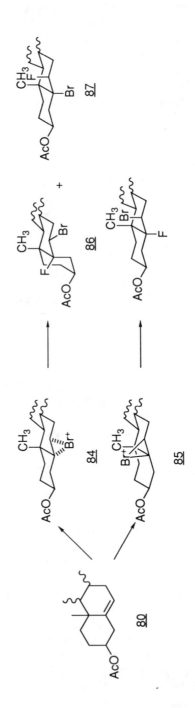

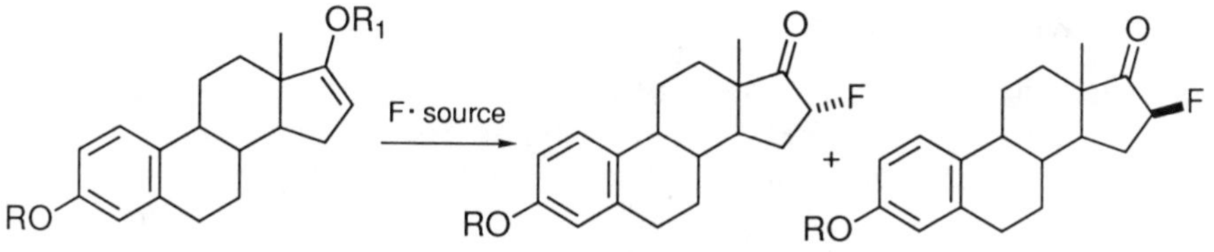

88 a R = CH$_3$ R$_1$ = Si(CH$_3$)$_3$ 89 a   90 a
    b R = R$_1$ = CH$_3$CO

F· source = F$_2$, CF$_3$OF, FSO$_4$Cs, XeF$_2$

In contrast, reaction of the enol acetate of estrone (**88b**) with xenon difluoride showed a degraded selectivity of 9:1 for the α to β (**89b** to **90b**) but the yield was significantly improved to 99%. The stereoselectivity of the reaction with cesium fluoroxysulfate improved to 20:1 with only a modest improvement in the yield to 22%. Both fluorine and trifluoromethyl hypofluorite yielded between 56 and 65% of fluoroestrone with a selectivity favoring the α-fluoroestrone of 12:1.

10.9.1 Perchlorylfluoride

Enol ethers are not the only activated olefins that can be fluorinated; enamines also can be employed in this process. The pyrrolidine enamine of 19-nortestosterone (**91**) was treated with perchlorylfluoride to form 4-fluoronortestosterone (**92**) in 80% yield. Oxidation of the A ring with selenium dioxide yielded the desired 4-fluoroestradiol (**93**).[36] The 4-fluoro-17β-estradiol was an active antitumor agent against rat mammary adrenocarcinoma.[37]

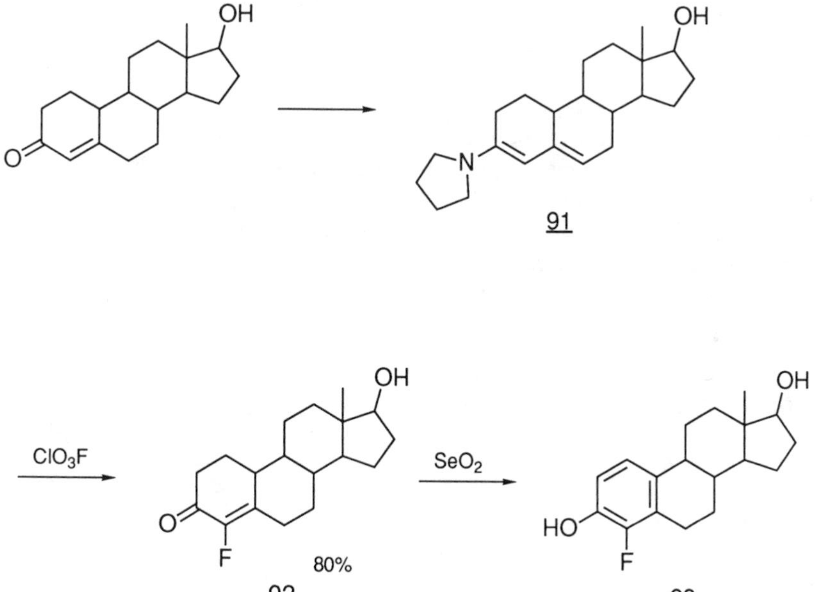

91

92   80%   93

10.9.2 Xenon Difluoride and *p*-Iodotoluene Difluoride

Steroidal silyl enol ethers **94–97** derived from 17β-hydroxy-5α-androstan-3-one acetate, 5-pregnen-3,11,20-trione-3,20-bisethylene ketal, 3β-hydroxy-5α-spirostan-12-one acetate, and 3β-hydroxy-5-androsten-17-one acetate were treated with xenon difluoride or *p*-iodotoluene difluoride. Yields of fluorinated materials with xenon difluoride were between 60 and 70% and strongly favored formation of the α-product (Fig. 10.2).

 p-Iodotoluene difluoride is a particularly attractive fluorinating reagent as it is synthetically accessible without the use of fluorine. Unfortunately, the yields of fluorinated materials when *p*-iodotoluene difluoride was used were substantially lower, between 23 and 30%, and were accompanied by significant amounts of elimination products.[38]

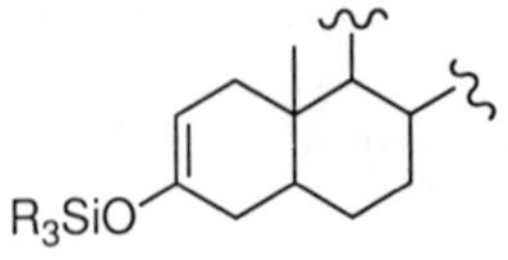

94

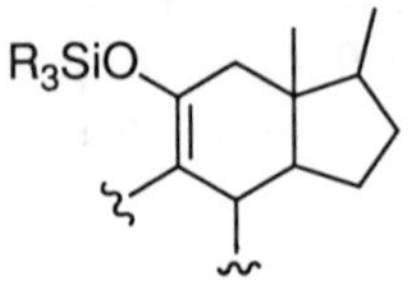

95

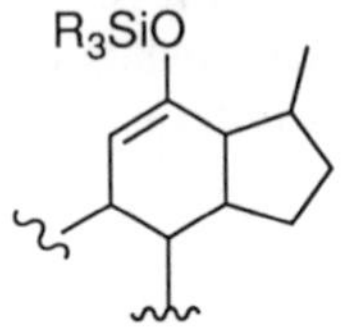

96

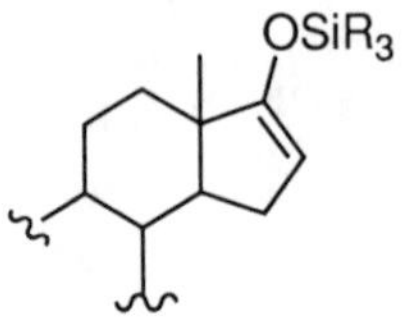

97

Comparison of the mechanism of both reactions is informative. Xenon difluoride is a source of electrophilic fluorine, whereas *p*-iodotoluene difluoride acts as a fluoride ion donor (Fig. 10.3). The reaction with *p*-iodotoluene difluoride does proceed with inversion of configuration. The reagents generally approach the double bond from the less hindered face.

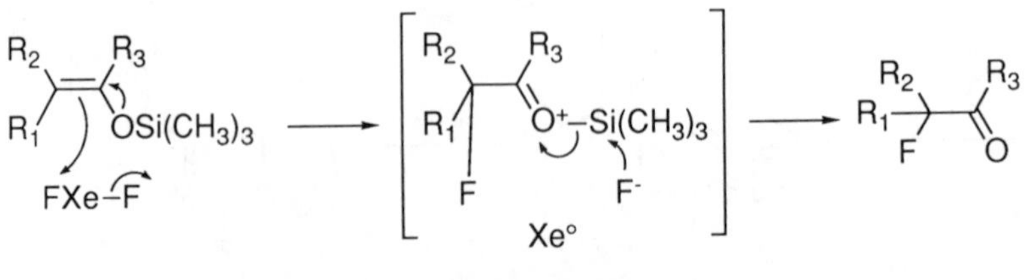

FIGURE 10.2

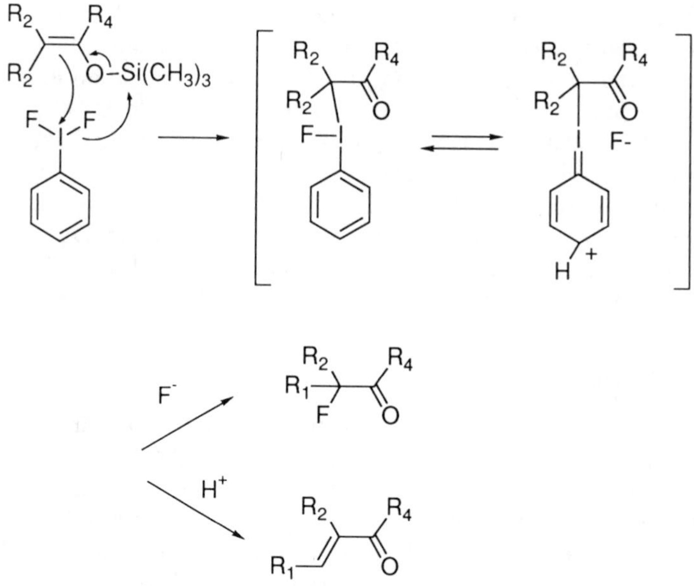

FIGURE 10.3

10.9.3 Diethylaminosulfur Trifluoride

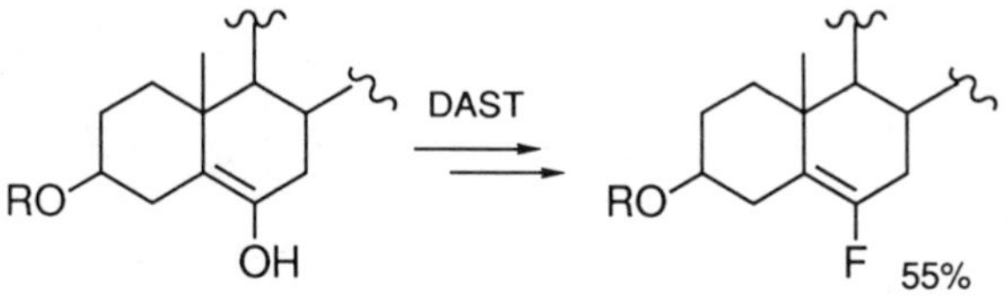

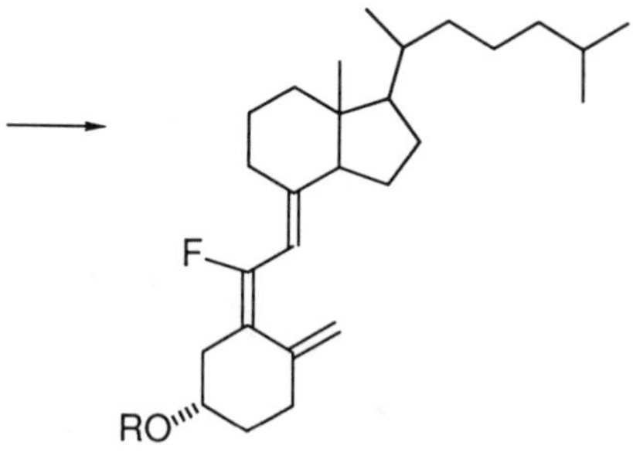

A closely related reaction of enols, the replacement of an enolic hydroxyl by fluorine, is discussed in this section as it is also based on the reactivity of enols. Fluorination of the trienic portion of vitamin D_3 was successfully effected at the 6 position. Treatment of 6-ketocholestanyl acetate (**98**) with DAST led to formation of 6-fluorocholesteryl acetate (**99**) in 55% yield.[39] Although DAST was successfully used to effect this transformation, piperidinosulfur trifluoride simplified the synthesis and enhanced the yield.[40] Allylic bromination and dehydrohalogenation led to 6-fluoro-7-dehydro-cholesterol acetate in 48% yield. On irradiation, the previtamin D_3 analog was formed in 43% yield. Rearrangement to acetyl 6-fluoro-vitamin D_3 (**100**) in 30% yield occurred on heating to 120°C. Fluorination of carbon-6 had relatively little effect on the (1,7)-hydrogen shift; therefore, thermal conversion was successful. 6-Fluoro-vitamin D_3 bound to the receptor but did not stimulate calcium absorption or mobilization.[41] It was the first vitamin D_3 analog reported to antagonize $1\alpha,25$-dihydroxy-vitamin D_3.

10.9.4 Difluorocarbene

It is also possible to add a highly electrophilic fluorinated fragment to an electron-rich olefinic system such as an enol ether. 24,24-Difluoro-25-hydroxy-vitamin D_3 was prepared by conversion of cholenic acid enol acetate **101**, which was then treated with difluorocarbene, generated by thermolysis of sodium chlorodifluoroacetate.[12,42] The difluorocyclopropane (**102**) was formed in 34% yield. Ring opening followed treatment with lithium hydroxide forming both the difluoroketone (**103**) in 9% yield and the fluoroenone (**104**) in 61% yield. Following the addition of methyl magnesium iodide to the difluoroketone, the usual operation formed 24,24-difluoro-25-hydroxy-vitamin D_3. These compounds were equal in activity or slightly more active than the unfluorinated materials[43].

With 24,24-difluoro-25-hydroxy-vitamin D$_3$ (**105**) in hand, it was possible to introduce the 1α-hydroxyl enzymatically.[44] 1α,25-Dihydroxy-24,24-difluoro-vitamin D$_3$ (**106**) was found to be 5–10 times more active than 1α,25-dihydroxy-vitamin D$_3$ in vivo. Although this increased activity was notable, the very high activity associated with 1α,25-dihydroxy-vitamin D$_3$ makes it less significant than the prolonged lifetime of the analog in vivo.[45] The compound also demonstrated potential anticancer activity.

10.10 FLUORINATED BUILDING BLOCKS

Although it is an arbitrary distinction, it is possible to consider the preparation of fluorinated steroids by using fluorinated building blocks. Clearly the previously described application of difluorocarbene could just as well be depicted utilizing this approach. As another example of this type of synthetic strategy, the use of trifluoromethylphenol in steroid synthesis, will be outlined.

10.10.1 3-Trifluoromethylphenol

3-(Trifluoromethyl)phenol (**107**) has been used for the preparation of 2-trifluoromethyl-1,3-cyclopentandione (**110**),[46] the key fluorinated building block required for the Torgov reaction in the preparation of the 18,18,18-trifluoro analog of estrone, estradiol, ethynyl estradiol, and 19-nortestosterone. Oxidation of 3-(trifluoromethyl)phenyl with chlorous acid led to formation of 2-trifluoromethyl-1,4-benzoquinone (**108**) in 32% yield accompanied by formation of the required 2-chloro-2-trifluoromethyl-1,3-cyclopentandione (**109**) in 10% yield. Dechlorination was effected with hydroiodic acid in 78% yield.

2-Trifluoromethyl-1,3-cyclopentanedione was condensed with the required allylic alcohol in presence of triethylamine to form **111** in 44% yield.[47] The diketone was cyclized to **112** in 84% yield. Further transformation yielded the trifluoroestradiol (**113**) in 40% yield over four steps. The deprotected compound manifested antiestrogenic activity and only 1% of the estrogenic activity of estradiol in mice.

107 108 109

109 110

110 111

112 113

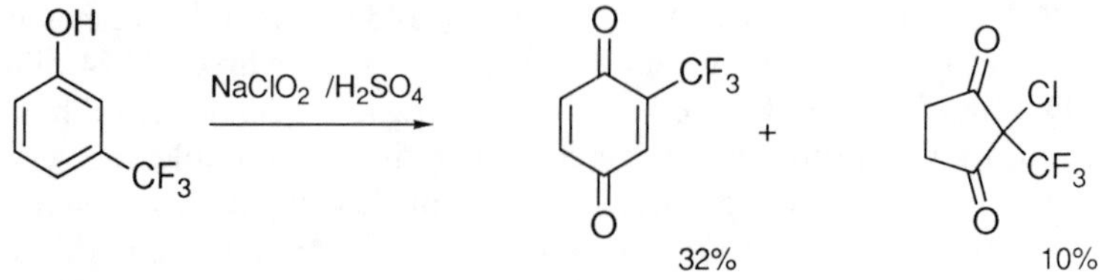

10.10.2 4-Methoxy-α,α,α-trifluoroacetophenone

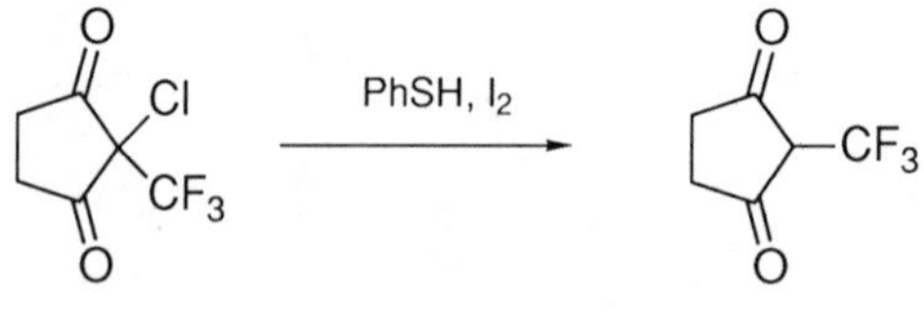

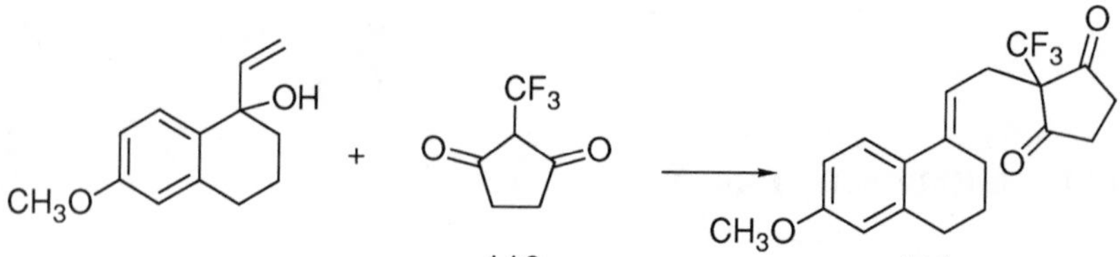

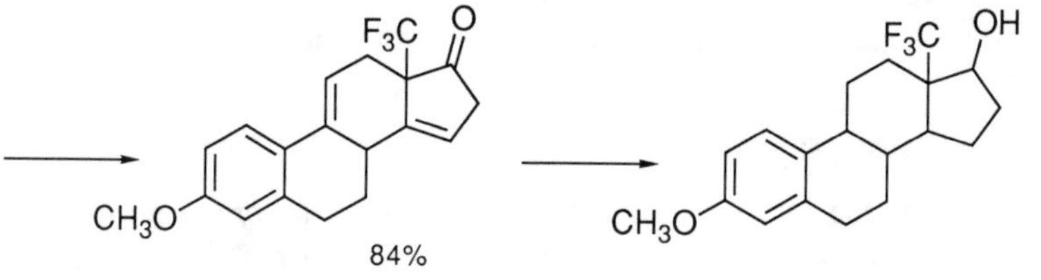

The syntheses of *trans*-1,2-bis(trifluoromethyl)-1,2-bis (4-methoxyphenyl) ethylene (**115**) were effected by reductive coupling with titanium tetrachloride and zinc in 11.4% yield. The corresponding cis compounds were also isolated, but in only 9% yield.[48] The necessary trifluoroacetophenone **114** was prepared by the addition of 4-methoxyphenyl magnesium bromide to trifluoroacetic acid in 59% yield. The synthetic fluorinated materials demonstrated a twofold to tenfold increase in binding affinity for the estradiol receptor and enhanced estrogen reactivity. Additionally, the compounds showed tumor growth inhibitory activity against hormone-dependent mammary tumors.

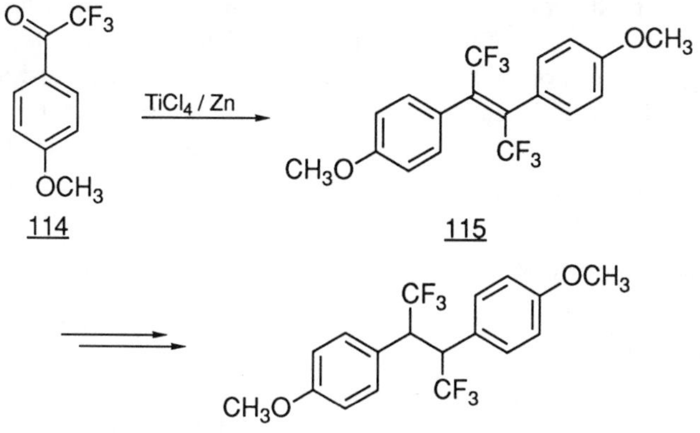

10.10.3 Ethyl Trifluoroacetate

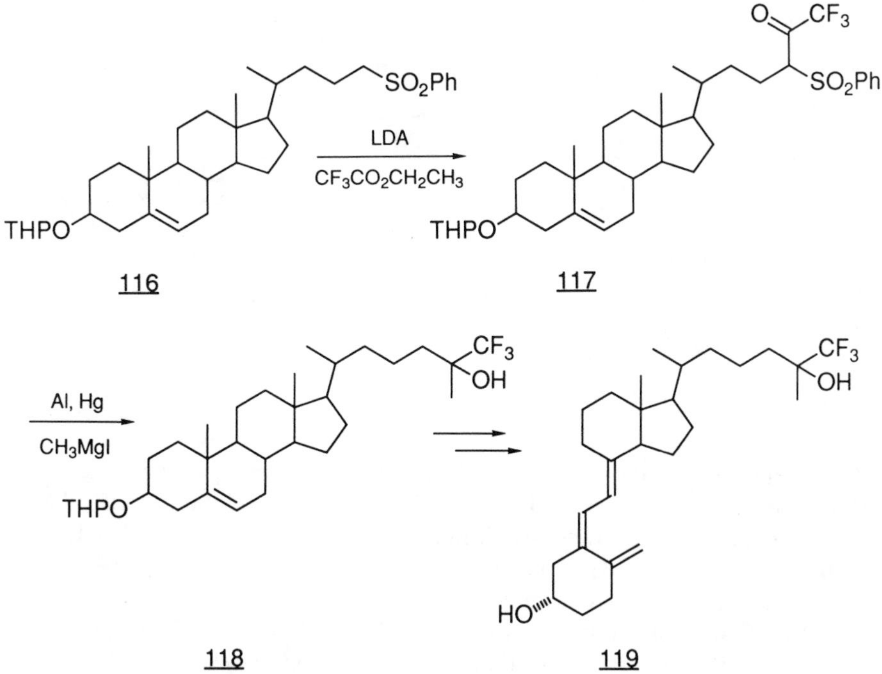

The ethyl ester of trifluoroacetic acid has also been used to introduce a trifluoromethyl group. 24-Phenylsulfonyl-25,26,27- trinorcholest-5-en-3β-yl tetrahydropyranyl ether (**116**) was deprotonated with LDA and condensed with ethyl trifluoroacetate to give **117**. The sulfone was removed with aluminum amalgam and treatment of the ketone with methyl magnesium iodide formed the 25-hydroxy-26,26,26-trifluoro compound (**118**) in 66% yield. Desulfonylation followed by sodium borohydride reduction rather than Grignard addition led to the 25-hydroxy-26,26,26,-trifluoro-27-nor compound in 80% yield. Both materials were elaborated to the vitamin D_3 analogs, e.g., (**119**), following previously described procedures.[49]

10.10.4   Hexafluoroacetone

Deprotonated   phenylsulfonyl-25,26,27-trinorcholest-5-en-3β-yl   tetrahydro-pyranyl ether (**116**) was condensed with hexafluoroacetone in 84% yield. The resultant sulfone (**120**) could be desulfonylated in 64% yield with sodium amalgam.[50]. This material (**121**) was uneventfully converted to 25-hydroxy-26,26,26,27,27,27-hexafluoro-vitamin D_3 (**122**). The intermediate (**121**) was

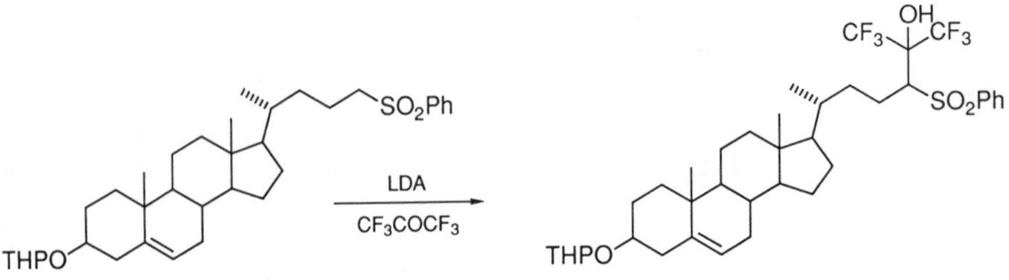

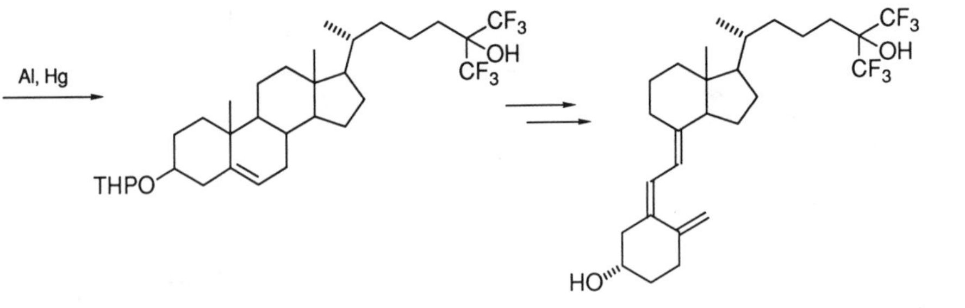

deprotected and oxidized with 2,3-dichloro-5,6-dicyano-1,4-benzoquinone to form the 1,4,6-triene-3-one (**123**) in 55% yield. Epoxidation with alkaline hydrogen peroxide followed by lithium ammonia reduction yielded the 1α-hydroxy compound in 63% yield.[51] The product was converted to the vitamin D_3 analog (**124**) in the usual manner.

It was found that although 1α,25-Dihydroxy-26,26,26,27,27,27-hexafluoro-vitamin D_3 (**124**) is more active than 1α-25-dihydroxy- vitamin D_3, this analog

121 ⟶

123   55%

63%

124

exhibited diminished binding to the receptor.[52] The increase in activity has been attributed to the resistance to metabolism exhibited by the analog. 25-Hydroxy-26,26,26,27,27,27-hexafluoro-vitamin D_3 (**124**) was approximately 20 times more active than the 25-hydroxy-vitamin D_3 itself. The trifluoro analogs 25-hydroxy-26,26,26-trifluoro-27-nor-vitamin D_3 and 25-hydroxy-26,26,26-trifluoro-vitamin D_3 were slightly less active and 8 times more active than 25-hydroxy-vitamin D_3, respectively.

From these studies of variously fluorinated vitamin D_3 compounds it becomes apparent that with the 1α, 25-dihydroxy vitamin D_3 affinity for the receptor is apparently optimized. In the case of 24,24-difluoro-1α,25-dihydroxy-vitamin D_3 as well as 1α,25-dihydroxy-26,26,26,27,27,27-hexafluoro-vitamin D_3, the activity observed has been suggested to be a result of the slower metabolism of the fluorinated analogs.

10.10.5 (*R*)-3-Fluoro-5-iodo-2-methyl-2-pentanol

The evidence that 1α,25-dihydroxy-vitamin D_3 was metabolized by a stereo-specific hydroxylation at carbon-24 to form 1α,24(*R*),25-trihydroxy-vitamin D_3 suggested that a single enantiomer of 24-monofluorinated 1α,25-dihydroxy-vitamin D_3 might be extremely potent. To test this hypothesis 24(*R*)-fluoro-1α,25-dihydroxy-vitamin D_3 was prepared by condensation of (*R*) 3-fluoro-5-iodo-2-methyl-2-pentanol (**126**) with the lithium enolate of functionalized, commercially available dehydroepiandrosterone (**127**). The alkylating agent was prepared stereospecifically by DAST fluorination of 2-hydroxy-butyrolactone (**125**) prepared from 1-malic acid.[53] Reduction, tosylation, hydrogenolysis,

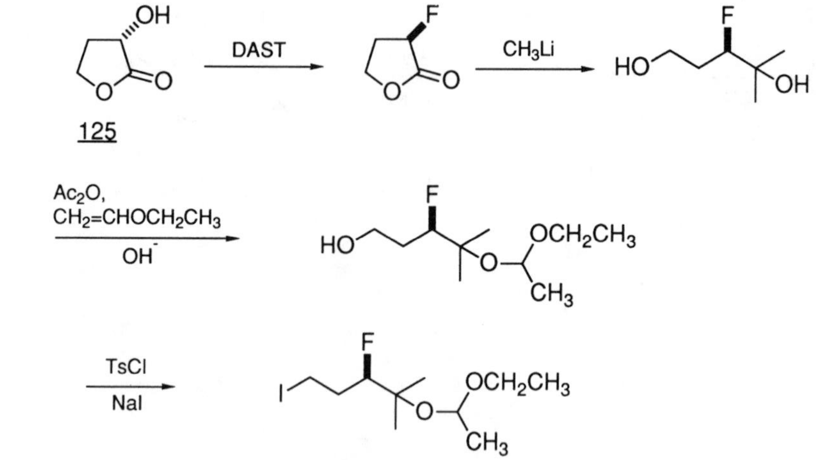

125

126

deprotection, and acetylation formed the acylated 24(R)-fluoro-1α,25-dihydro-xychloesterol (**128**) (Fig. 10.4). Conversion into the 5,7-diene and photolysis were effected under standard conditions to form 24(R)-fluoro-1α,25-dihydroxy-vitamin D_3 (**129**). This synthetic scheme was easily adapted to the preparation of 24,24-difluoro analogs. 2,2-Difluorosuccinic acid was converted into the iodide required for enolate alkylation by standard techniques. Alkylation and elabo-

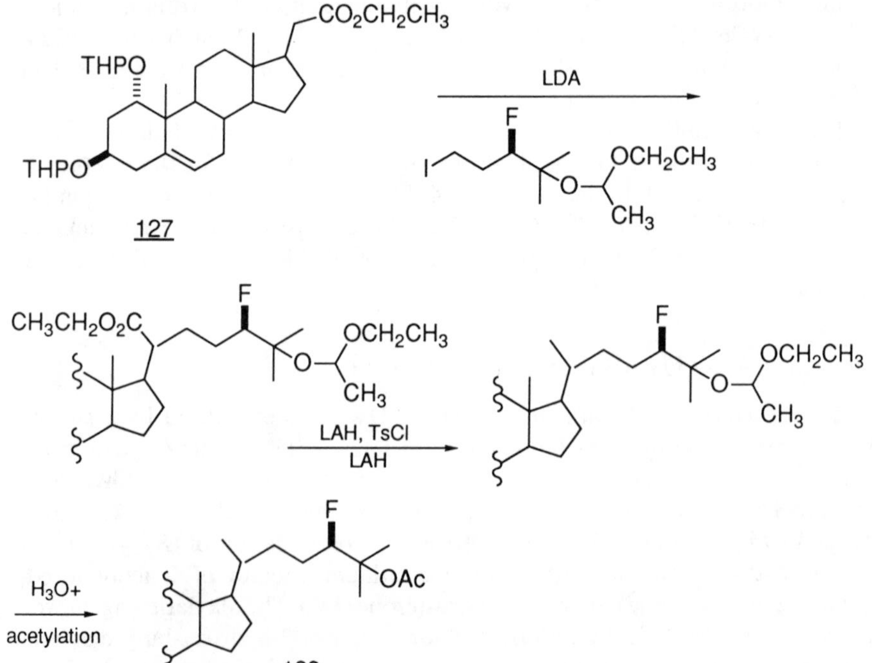

127

128

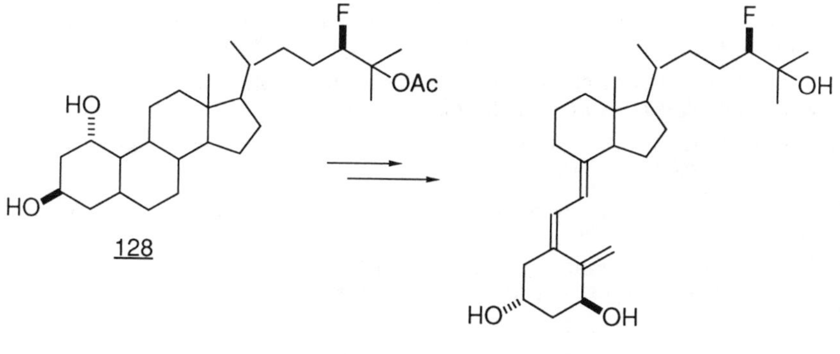

FIGURE 10.4

ration were as described for the monofluorinated material. These compounds inhibited tumor cell proliferation and differentiation. They also had good antirachitogenic activity in concert with prolonged half-lives relative to non-fluorinated materials.

REFERENCES

1. J. Fried and E. F. Sabo, *J. Am. Chem. Soc.* **76**, 1455, 1954.

2. A. Wettstein, in *Ciba Foundation Symposium, Carbon–Fluorine Compounds, Chemistry, Biochemistry and Biological Activities*, Elsevier, New York, 1972.

3. (a) Y. Kobayashi and T. Taguchi, in *Biomedical Aspects of Fluorine Chemistry* (ed. R. Filler and Y. Kobayashi), Kodansha, Tokyo, 1982; (b) Y. Kubayashi and T. Taguchi, *J. Syn. Org. Chem. Jpn.* **43**, 1073, 1985; (c) N. Ikekawa, *Med. Chem. Rev.* **7**, 333–366, 1987.

4. (a) S. Rozen and G. Ben-Shushan, *Tetrahedron Lett.* **25**, 1947, 1984; (b) S. Rozen and G. Ben-Shushan, *J. Org. Chem.* **51**, 3522–3527, 1986.

5. M. Hudlicky, *Org. React.* **35**, 513–637, 1987.

6. S. Rozen, Y. Faust, and H. Ben-Yakov, *Tetrahedron Lett.* **20**, 1823–1826, 1979.

7. (a) R. I. Yakhimovich, N. F. Fursaeua, and V. E. Pashinnik, *Khim. Prir. Soedin.* **4**, 580, 1980; (b) L. K. Revelle, J. M. Londowski, S. B. Kost, R. A. Corradino, and R. Kumar, *J. Steriod Biochem.* **22**, 469, 1985; (c) R. I. Yakhimovich, V. M. Klimashevskii, and G. M. Segal, *Khim.-Farm. Zh.* **10**, 58, 1976.

8. G. D. Prestwich, *Pestic. Sci.* **37**, 430, 1986.

9. (a) D. E. Ayer, *Tetrahedron Lett.* **1962**, 1065; (b) L. H. Know, E. Velarde, S. Berger, D. Cuadriello, and A. D. Cross, *Tetrahedron Lett.* **1962**, 1249; (c) L. H. Knox, E. Velarde, S. Berger, D. Cuadriello, and A. D. Cross, *J. Org. Chem.* **29**, 2187, 1964.

10. A. Néder, A. Uskert, É. Nagy, Zs. Méhesfalvi, and J. Kuszmann, *Acta Chim. Acad. Sci. Hung.* **103**, 231–240, 1980.

11. Y. Kobayashi, I. Kumadaki, A. Ohsawa, M. Honda, and Y. Hanzawa, *Chem. Pharm. Bull.* **23**, 196, 1975.

12. Y. Kobayashi, T. Taguchi, T. Terada, J. -I. Oshida, M. Morisaki, and N. Ikekawa, *J. Chem. Soc. Perkin Trans.* (*I*) **1982**, 85.

13. T. G. C. Bird, G. Felsky, P. M. Fredericks, E. R. H. Jones, and G. D. Meakins, *J. Chem. Res.* **1979**, 4728–4755.

14. (a) D. O. Kiesewetter, J. A. Katzenellenbogen, M. R. Kilbourn, and M. J. Welch, *J. Org. Chem.* **49**, 4900, 1984; (b) D. O. Kiesewetter, M. R. Kilbourn, S. W. Landvatter, D. F. Heimann, J. A. Katzenellenbogen, and M. J. Welch, *J. Nucl. Chem.*, **25**, 1212–1221, 1984; (c) M. A. Mintun, M. J. Welch, C. J. Mathias, J. W. Brodack, B. A. Siegel, and J. A. Katzenellenbogen, *J. Nucl. Med.* **28**, 561, 1987.

15. M. G. Pomper, J. A. Katzenellenbogen, M. J. Welch, J. W. Brodack, and C. J. Mathias, *J. Med. Chem.* **31**, 1360–1363, 1988.

16. S. W. Landvatter, D. O. Kiesewetter, M. R. Kilbourn, J. A. Katzenellenbogen, and M. J. Welch, *Life Sci.* **33**, 1933, 1983.

17. J. P. Horwitz, V. K. Iyer, H. B. Vardhan, J. Corombos, and S. C. Brooks, *J. Med. Chem.* **29**, 692–698, 1986.

18. K. Annen, H. Hofmeister, H. Laurent, and R. Wiechen, *Justus Liebigs Ann. Chem.* **1982**, 966–972.

19. R. T. Logan, R. G. Roy, and G. F. Woods, *J. Chem. Soc. Perkin Trans.* (*I*) **1982**, 1079–1084.

20. J. Oshida, M. Morisaki, and N. Ikekawa, *Tetrahedron Lett.* **21**, 1755, 1980.

21. J. F. Napoli, M. A. Fivizzani, H. K. Schnoes, and H. F. DeLuca, *Biochemistry* **18**, 1641, 1979.

22. E. Ohshima, S. Takatsuto, N. Ikekawa, and H. F. DeLuca, *Chem. Pharm. Bull.* **32**, 3518, 1984.

23. E. Ohshima, H. Sai, S. Takatsuto, N. Ikekawa, Y. Kobayashi, Y. Tanaka, and H. F. DeLuca, *Chem. Pharm. Bull.* **32**, 3525, 1984.

24. T. Taguchi, S. Mitsuhashi, A. Yamanouchi, Y. Kobayashi, H. Sai, and N. Ikekawa, *Tetrahedron Lett.* **25**, 4933, 1984.

25. M. Nakada, Y. Tanaka, H. F. DeLuca, and Y. Kobayashi, *Arch. Chem. Biochem.* **241**, 173, 1985.

26. B. Sialom and Y. Mazur, *J. Org. Chem.* **45**, 2201, 1980.

27. (a) S. Rozen and M. Brand, *J. Org. Chem.*, **51**, 222, 1986; (b) S. Rozen and M. Brand, *J. Org. Chem.* **50**, 3342, 1985.

28. S. Rozen, M. Brand, D. Zamir, and D. Hebel, *J. Am. Chem. Soc.* **109**, 896–897, 1987.

29. H. L. Herzog, M. J. Gentles, H. M. Marshall, and E. B. Hershberg, *J. Am. Chem. Soc.* **82**, 3691, 1960.

30. (a) A. Bowers, *J. Am. Chem. Soc.* **81**, 4107, 1959; (b) A. Bowers, L. C. Ibanez, E. Denot, and R. Becerra, *J. Am. Chem. Soc.* **82**, 4007, 1960.

31. J. Pataki and E. V. Jensen, *Steroids* **28**, 437, 1976.

32. D. H. R. Barton, J. Lister-James, R. H. Hesse, M. M. Pechet, and S. Rozen, *J. Chem. Soc. Perkin Trannans.* (*I*) **1982**, 1105–1110.

33. G. A. Olah, M. Nojima, and I. Kerekes, *Synthesis* **1973**, 780.

34. A. Néder, A. Uskert, Zs. Mehesfalvi, and J. Kuszmann, *Acta Chim. Acad. Sci. Hung.* **104**, 123–140, 1980.

35. T. B. Patrick and R. Mortezania, *J. Org. Chem.* **53**, 5153–5155, 1988.

36. V. C. O. Njar, T. Arunachalam, and E. Caspi, *J. Org. Chem.* **48**, 1007, 1983.

37. M. Neeman, G. Kartha, K. Go, J. P. Santodonato, and O. Dodson-Simmons, *J. Med. Chem.* **26**, 465, 1983.

38. T. Tsushima, K. Kawada, and T. Tsuji, *Tetrahedron Lett.* **23**, 1165–1168, 1982.

39. W. G. Dauben, B. Kohler, and A. Roesle, *J. Org. Chem.* **50**, 2007, 1985.

40. L. N. Markovskij, V. E. Pashinnik, and A. V. Kirsanov, *Synthesis* **1973**, 787.

41. F. Wilhelm, W. G. Dauben, B. Kohler, A. Roesle, and A. W. Norman, *Arch. Biochem. Biophys.* **233**, 127, 1984.

42. Y. Kobayashi, T. Taguchi, T. Terada, J. -I. Oshida, M. Morisaki, and N. Ikekawa, *Tetrahedron Lett.* **20**, 2023, 1979.

43. Y. Tanaka, H. F. DeLuca, Y. Kobayashi, T. Taguchi, N. Ikekawa, and M. Morisaki, *J. Biol. Chem.* **254**, 7163, 1979.

44. Y. Tanaka, H. F. DeLuca, H. K. Schnoes, N. Ikekawa, and Y. Kobayashi, *Arch. Biochem. Biophys.* **199**, 473, 1980.

45. (a) S. Okamoto, Y. Tanaka, H. F. DeLuca, Y. Kobayashi, and N. Ikekawa, *Am. Physiol. Soc.* **1983**, E159; (b) H. P. Koeffler, T. Amatruda, N. Ikekawa, Y. Kobayashi, and H. F. DeLuca, *Cancer Res.* **44**, 5624, 1984.

46. J. -C. Blazejewski, R. Dorme, and C. Wakselman, *Synthesis* **1985**, 1120–1121.

47. (a) J. -C. Blazejewski, R. Dorme, and C. Wakselman, *J. Chem. Soc. Chem. Commun.* **1983**, 1050; (b) J.-C. Blazejewski, R. Dorme, and C. Wakselman, *J. Chem. Soc. Perkin Trans. (I)* **1986**, 337.

48. R. W. Hartmann, A. Heindl, M. R. Schneider, and H. Schoenenberger, *J. Med. Chem.* **29**, 322–328, 1986.

49. Y. Kobayashi, T. Taguchi, N. Kanuma, N. Ikekawa, and J. -I. Oshida, *Tetrahedron Lett.* **22**, 4309, 1981.

50. Y. Kobayashi, T. Taguchi, N. Kanuma, N. Ikekawa, and J. -I. Oshida, *J. Chem. Soc. Chem. Commun.* **1980**, 459.

51. Y. Kobayashi, T. Taguchi, S. Mitsuhashi, T. Eguchi, E. Ohshima, and N. Ikekawa, *Chem. Pharm. Bull.* **30**, 4297, 1982.

52. (a) Y. Tanaka, H. F. DeLuca, Y. Kobayashi, and N. Ikekawa, *Arch. Chem. Biochem.* **229**, 348, 1984; (b) P. H. Stern, T. Mavreas, Y. Tanaka, H. F. DeLuca, N. Ikekawa, and Y. Kobayashi, *J. Pharm. Exp. Ther.* **229**, 9, 1984.

53. J. J. Partridge, Abstracts of Papers, 188th National Meeting of the American Chemical Society, Philadelphia, Pennsylvania, August 26–31, 1984, American Chemical Society, Washington, D.C., 1984, MEDI 1.

11

FLUORINATED PURINES
AND PYRIMIDINES

11.1 INTRODUCTION

Fluorinated analogs of the naturally occurring nucleic acids have become established as antiviral, antitumor, and antifungal agents. A high percentage of fluorinated nucleosides exhibit biological activity. Their structures differ only slightly sterically from the naturally occurring molecules. The electronic effects of the fluorine substituent play a major role in the biological activity of the analogs. Fluorine has been employed as a replacement for both hydroxyl and hydrogen, and difluoromethylene units have been employed as isosteric replacements for oxygen. The difluoromethylene group is only slightly larger than an oxygen atom; in fact, no other functional group that can replace oxygen matches so well the steric and electronic demand of oxygen.[1]

11.2 5-FLUORO PYRIMIDINES

The fluorinated nucleosides 5-fluoro-2'-deoxyuridine and 5-trifluoromethyl-2'deoxyuridine are well established as clinically useful therapeutic agents. The preparation of these materials as well as their biochemistry and pharmacology are currently under investigation. However, other 5-substituted materials under development also show good biological activity.[2]

11.2.1 5-Fluorouracil

One of the best known methods for the preparation of 5-fluorouracil (**2**) is the direct treatment of uracil (**1**) with fluorine in acetic acid or water.[3] There are, however, conflicting reports on the stability of the intermediate formed during the reaction. It has been proposed that the first step in the reaction of fluorine with uracil is the addition of F_2 across the double bond followed by displacement of the 6-fluorine by an acetate to form 5-fluoro-6-acetoxy-5,6-dihydrouracil (**3**).[4]

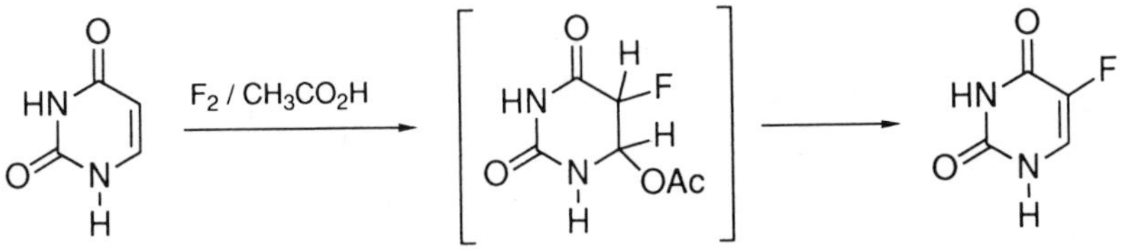

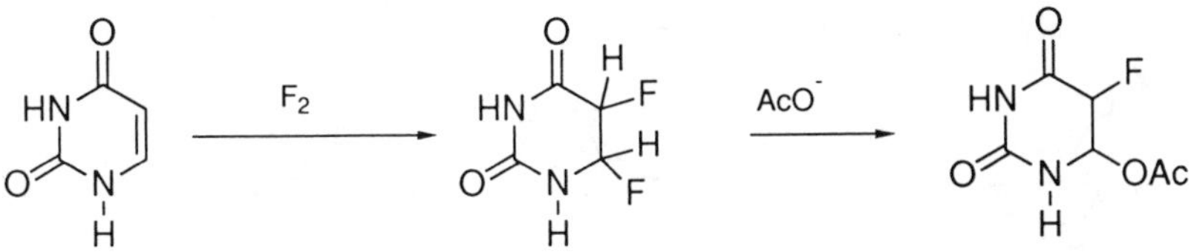

Other workers have proposed the first step in the process to be the addition of CH_3CO_2F (**4**), formed by the reaction of fluorine with acetic acid, across the double bond.[5]

$$F_2 \quad + \quad AcOH \quad \longrightarrow \quad AcOF$$

4

When the fluorination in acetic acid was immediately followed by the addition of an alcohol,[6] the corresponding 5-fluoro-6-alkoxy-5,6-dihydrouracil (**5**) was formed. This material was thermally very sensitive and lost the elements of alcohol to form 5-fluorouracil on warming. This finding clearly illustrated the reactivity of the intermediate 6-acetoxy compound.

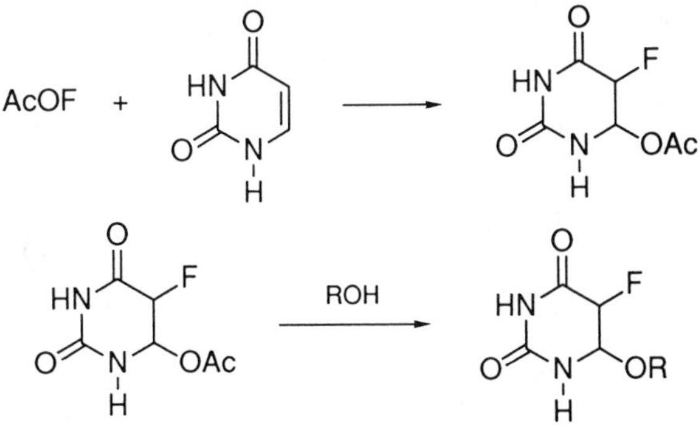

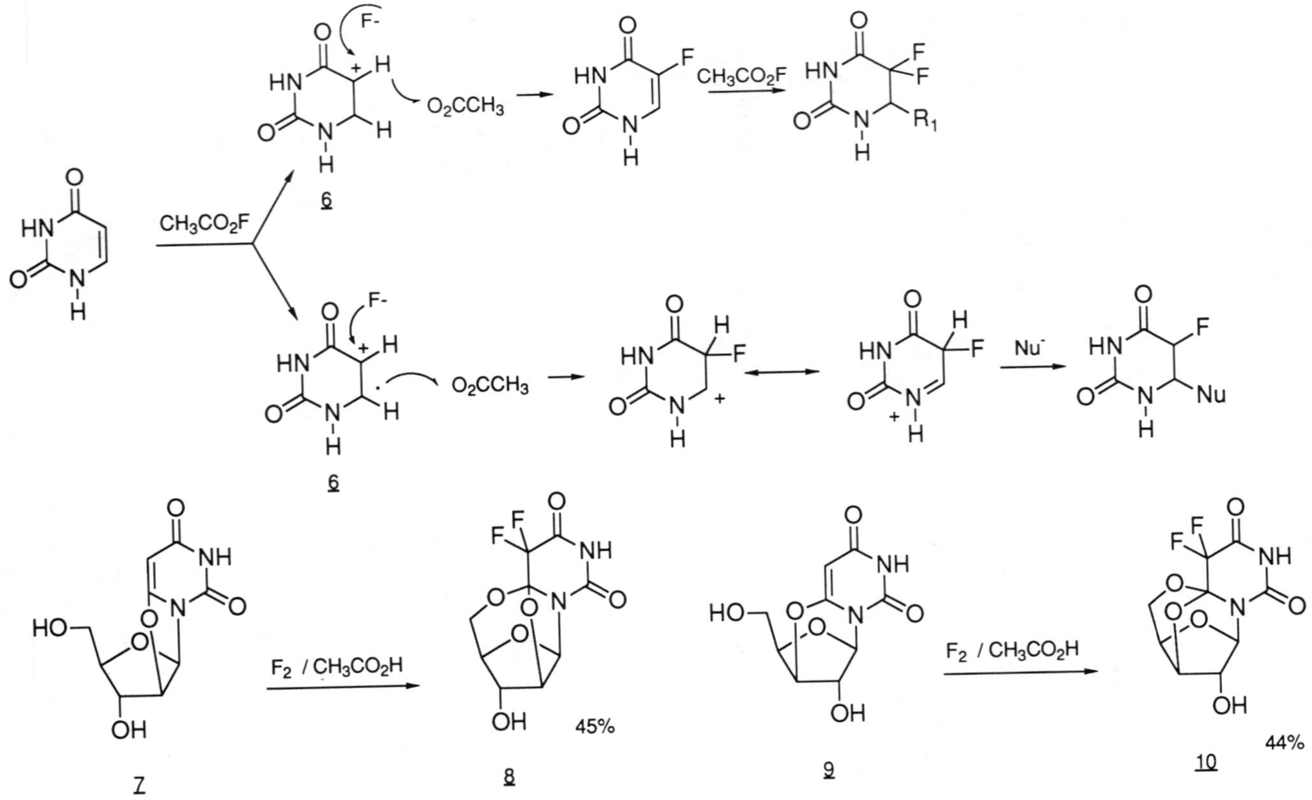
CH3CO2F
F-
O2CCH3
CH3CO2F
R1
6
F-
O2CCH3
Nu-
Nu
6
F2 / CH3CO2H
45%
F2 / CH3CO2H
44%
7
8
9
10

A direct comparison of the reactivity of fluorine and acetyl hypofluorite was made. The mechanism of the reaction was proposed to involve a single-electron transfer (SET) pathway to form a radical cation (**6**) which collapsed to form the fluorinated species. In the initial step there is a one-electron oxidation by the fluorinating agent to form a radical cation from uracil. Fluoride is subsequently incorporated either when a hydrogen atom is abstracted by the remaining radical portion of the fluorinating reagent or when a second oxidative single-electron transfer step occurs. In either case a stabilized cationic intermediate is formed which reacts with the nucleophilic solvent.[7]

11.2.2 6-*O*-Dicyclo-5,5-difluoro-5,6-dihydrouracils

When 6:2'-cyclouracil arabinoside (**7**) and 6:3'-cyclouracil arabinoside (**9**) were treated with fluorine in acetic acid at 40°C.[8] Careful control of the temperature was crucial to the success of the reaction. Fluorination of **7** formed 6:2',6:5'dianhydro-5,5-difluoro-6,6-dihydroxy-5,6-dihydrouracil arabinoside (**8**) in 45% yield and 6:3',6:5'-dianhydro-5,5-difluoro-6,6-dihydroxy-5,6-dihydrouracil xyloside (**10**) in 44% yield from **9**.

11.2.3 5-Fluoro-2',3'-dideoxy-3'-fluorouridine

5-Fluoro-2'-deoxyuridine is a potent cytotoxic agent in cell culture; however, in vivo it is substantially degraded to the free pyrimidine base, 5-fluorouracil, already described as cytotoxic itself but substantially less so than 5-fluoro-2'-deoxyuridine. Additionally, 5-fluoro-2'-deoxyuridine is very specific for DNA where as 5-fluorouracil is often incorporated into RNA. 5-Fluoro-2'-deoxyuridine was converted into 2,3'-anhydro-1-(2'-deoxy-β-D-*lyxo*-furanosyl)-5-fluorouracil (**11**) by known methods.[6] The anhydro compound was opened with anhydrous hydrogen fluoride in the presence of aluminum trifluoride to form 5-fluoro-2',3'-dideoxy-3'-fluorouridine (**12**) in 30% yield after deprotection. The overall activity of the compounds was diminished by 3-fluorination but the stability of the nucleosidic linkage was improved toward cleavage.

11.2.4 5'-Deoxy-4',5-difluorouridine

5'-Deoxy-4',5-difluorouridine was previously reported to have superior anti-tumor activity to 5-fluorouracil, presumably because it is the release of the base by uridine phosphorylase which generates the cytotoxic agent.[9] Tumor specificity is higher because of the higher levels of phosphorylase in tumor cells. Fluorination was effected with hydrogen fluoride–pyridine at −50°C by reaction of 1-(5′-deoxy-2′,3′-bis(benzyloxy)-β-D-*erythro*-pent-4′-enofuranosyl)-5-fluorouracil (**13**) to form 5′-deoxy-4′,5-difluoro-2′3′-bis(benzyloxy)-β-D-uridine (**14**) in 86% yield.[10] This compound was more easily hydrolyzed than 5′-deoxy-5-fluorouridine with the result that the activity improved tenfold. Other workers have shown that the carbohydrate is not necessary for the compounds to retain their activity.[2]

13   → HF - py   14   86%

11.2.5 6-Aza-5-fluorouracil

As described several times previously, selective fluorination with molecular fluorine is highly dependent on the reaction conditions. When 1-(2′,3′,5′-tri-*O*-acetyl-β-D-*ribo*-furanosyl)-6-azauracil (**15**) was treated with fluorine directly only a 0.3% yield of the desired 6-aza-5-fluoronucleoside could be isolated.[11] However, when 6-azauracil (**15**) was treated with fluorine in the presence of oxygen difluoride, the 6-aza-5-fluorouracil (**16**) was formed in 55% yield.[12] The 5-fluoronucleoside may be isolated in 20% yield when formed under similar conditions.

15   → F₂ / OF₂   16   55%

11.2.6 4-*O*-(Difluoromethyl)-5-fluorouracil

Although many nucleoside antiviral agents have activity against herpes simplex I (HSV-I), activity against herpes simplex II (HSV-II) is much more difficult to realize. The difluoromethyl group was introduced in 56% yield by treating the protected nucleoside (**17**) with difluorocarbene generated in situ by treatment of bis(trifluoromethyl)mercury $[(CF_3)_2Hg]$ with sodium iodide. Simultaneous difluoromethylation of the carbohydrate hydroxyls was not observed.[13]

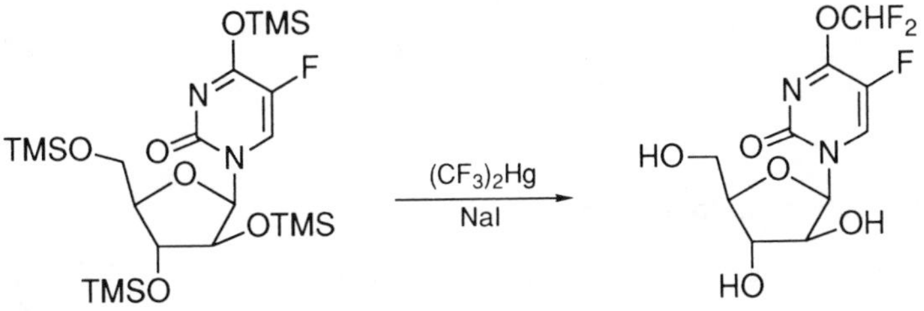

$$\underline{17} \xrightarrow[\text{NaI}]{(CF_3)_2Hg} \underline{18}$$

11.3 5-FLUOROALKYL-SUBSTITUTED PYRIMIDINES

11.3.1 5-Trifluoromethyluracil

5-Trifluoromethyluridine has been demonstrated to have antiviral and anti-cancer activity but better methods for the synthesis of 5-trifluoromethyluracil (**19**) are still under investigation. The direct trifluoromethylation of uracil has been reported.[14] Uracil **1** was allowed to react with bis(trifluoromethyl)mercury $[(CF_3)_2Hg]$ in an aqueous medium in the presence of azoisobutyronitrile. The desired trifluoromethylated material (**19**) was formed in 56% yield. This method also proceeded with bis(pentafluoroethyl)mercury $[(CF_3CF_2)_2Hg]$ albeit in lower yield, 30%. The synthesis of 5-trifluoromethyluracil was reported previously by the fluorination of uracil 5-carboxylic acid (**20**) with sulfur tetrafluoride.[15] Kolbe electrolysis of a trifluoroacetic acid solution of uracil (**1**) with a nickel anode and an iron cathode also formed 5-trifluoromethyluracil.[16]

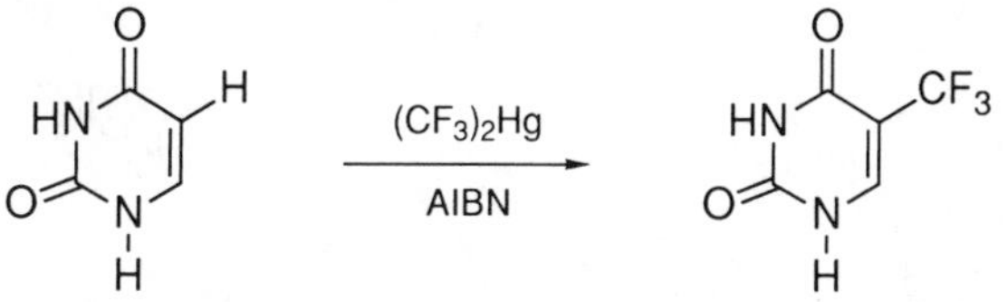

$$\underline{1} \xrightarrow[\text{AIBN}]{(CF_3)_2Hg} \underline{19} \quad 56\%$$

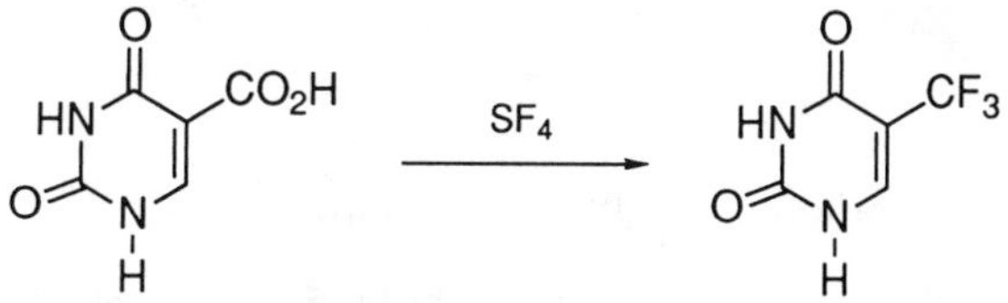

$$\underline{20} \xrightarrow{SF_4} \underline{19}$$

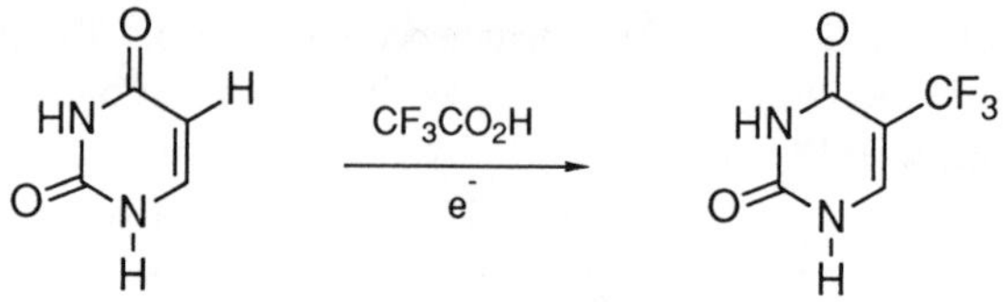

11.3.2   5-Trifluoromethyluridine

Treatment of uridine (21) with bis(trifluoromethyl)mercury [(CF$_3$)$_2$Hg] was also effective for introducing the trifluoromethyl group, albeit in lower yield, 11%.[14] The yields of 5-trifluoromethyluridine (22) were considerably better when 5-iodouridine (23) was treated with trifluoromethyliodide in the presence of copper powder in hexamethylposphoramide. The desired 5-trifluoromethyluridine could be isolated in 37–54% yield[17] and the procedure was useful for preparing both protected and unprotected uridines.[18] This technique was extended to the preparation of 5-pentafluoroethyluridine and also 1-β-D-arabino-furanosyl-5-(trifluoromethyl)uracil. The arabino compound displayed good antiviral activity but the pentafluoroethyl compound was inactive.

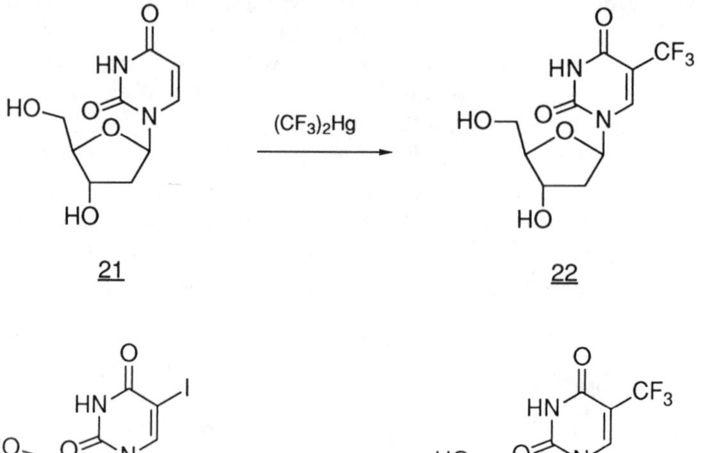

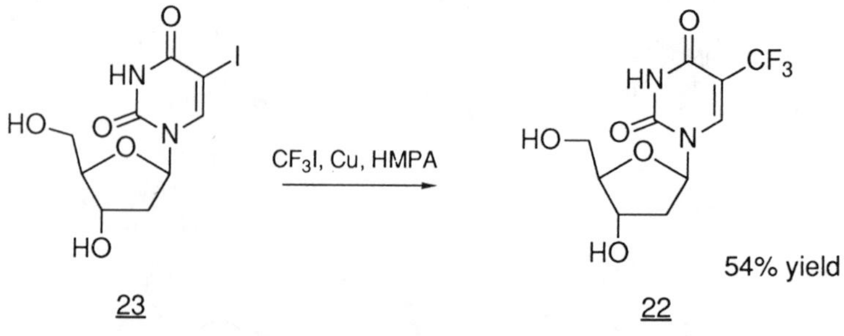

11.3.3   5-Monofluoromethyl- and 5-Difluoromethyluracil

The fluorination of the corresponding alcohol (25) and aldehyde (27) prepared from the methyluracils (24) with diethylaminosulfur trifluoride yielded 5-

monofluoromethyl- (**26**) and 5-difluoromethyluracil (**28**) in 76% and 70% yield respectively.[19]

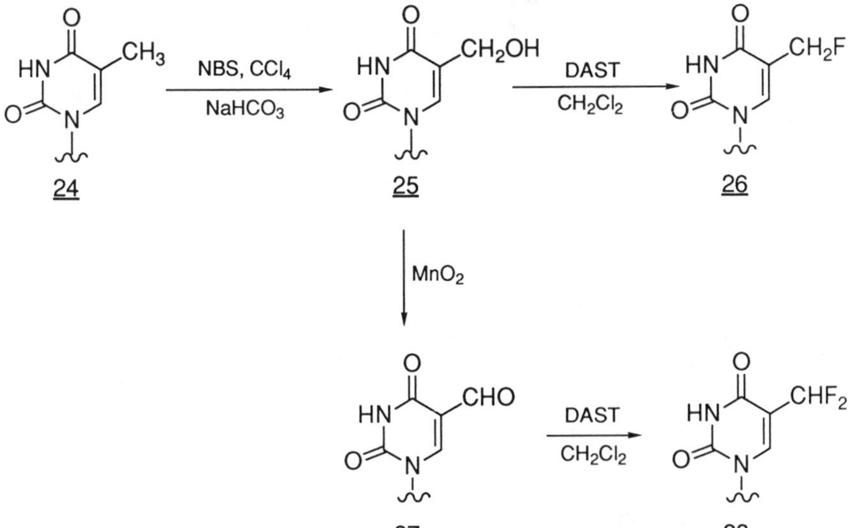

11.3.4 Total Synthesis

When 2-bromo-3,3,3-trifluoropropene, bis(triphenylphosphine)palladium(II) chloride, urea, triethylamine, and dimethylformamide are heated together at 100°C for 10 hours under 40 atmospheres of carbon monoxide, 5-trifluoromethyl-5,6-dihydrouracil (**29**) is formed in 26% yield.[20] 1,3-Dimethyl-5-trifluoromethyl-5,6-dihydrouracil (**30**) is formed by the reaction of, N',N-dimethylurea in 70% yield. The reaction proceeds nicely with 2,3-dibromo-1,1,1-trifluoropropane as well. 5-Trifluoromethyluracil (**19**) is prepared, in yields as high as 82%, from 5-trifluoromethyl-5,6-dihydrouracil (**29**) by treatment with bromine in acetic acid followed by heating in dimethylformamide solution.[21]

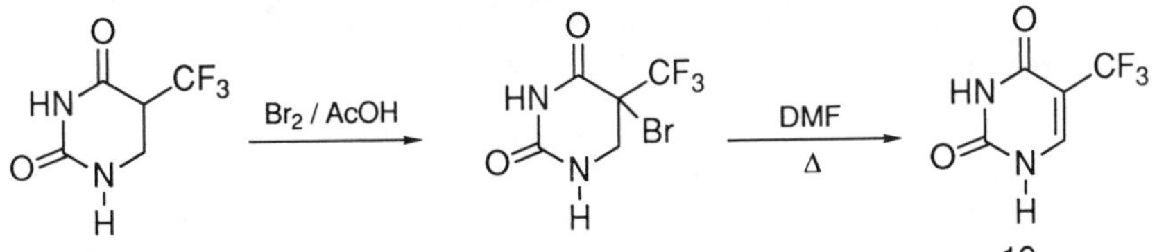

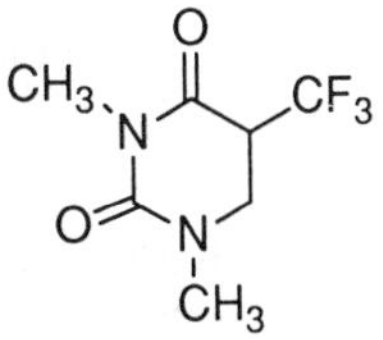

The yields have been improved when trifluoromethylacrylic acid (**31**) prepared by the reaction of carbon monoxide with 2-bromo-3,3,3-trifluoropropene catalyzed by bis(triphenylphosphine)palladium(II) chloride heated with urea in the presence of acetic anhydride.[22] Since the reaction tends to stop after the first addition, acetic anhydride was added to improve the yield of the ring closure step. The yields are as high as 67% by this method. N,N-dimethyl-5-trifluoromethyluracil (**32**) has good activity against ascetic mastocarcinoma MM2 of mice.

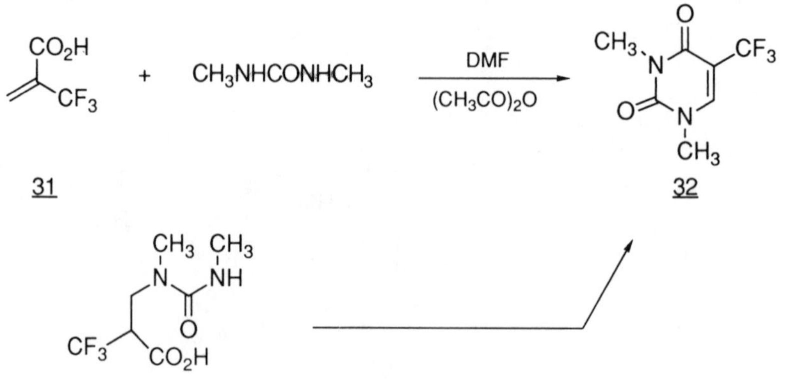

11.3.5 5-Fluoroalkyl and 5-Fluoroalkenyl Pyrimidines

Although the preparation of pentafluoroethyluracil (**34**) was briefly mentioned previously, the aim of these preparations was not clearly stated. Increased activity against the somewhat less susceptible herpes simplex virus II (HSV-II) was sought. The target compound (**34**) was prepared in 84% yield by fluorination of the corresponding hydroxyethyl pyrimidine (**33**) with diethylminosulfur trifluoride and triethylamine.23

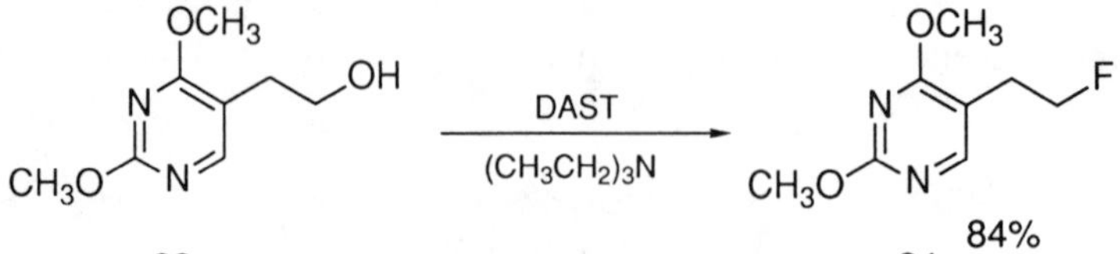

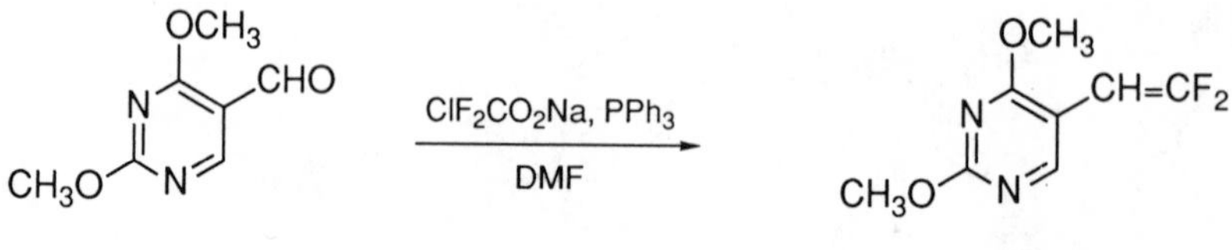

In a related study 5-(2,2-difluorovinyl)uracil was prepared. It exhibited activity comparable to BVDU against HSV-II and HSV-I as well as several tumor lines. The material was prepared via the addition of lithium chlorodifluoroacetate to a hot solution of triphenylphosphine in dry dimethylformamide containing 2,4-dimethoxy-5-formylpyrimidine (**35**). The yield of the protected fluorovinyl pyrimidine (**36**) was 56% after chromatography.[24]

11.3.6 (*E*)-5-(3,3,3-Trifluoro-1-propenyl)-2′-deoxyuridine

2-Chloromercuri-2′-deoxyuridine (**37**) was allowed to react with 3,3,3-trifluoropropene in the presence of lithium tetrachloropalladate.[25] The desired (*E*)-5-(3,3,3-trifluoro-1-propenyl)-2′-deoxyuridine (**38**) was formed in 26% yield along with 5-(3,3,3-trifluoro-2-methoxyprop-1-yl)-2′-deoxyuridine (**39**) in 59% yield. The methoxy compound could be converted to the desired trifluoropropene in 65% yield by treatment with trifluoroacetic anhydride. (*E*)-5-(3,3,3-Trifluoro-1-propenyl)-2′-deoxyuridine was very effective against HSV-I but was not at all active against HSV-II.

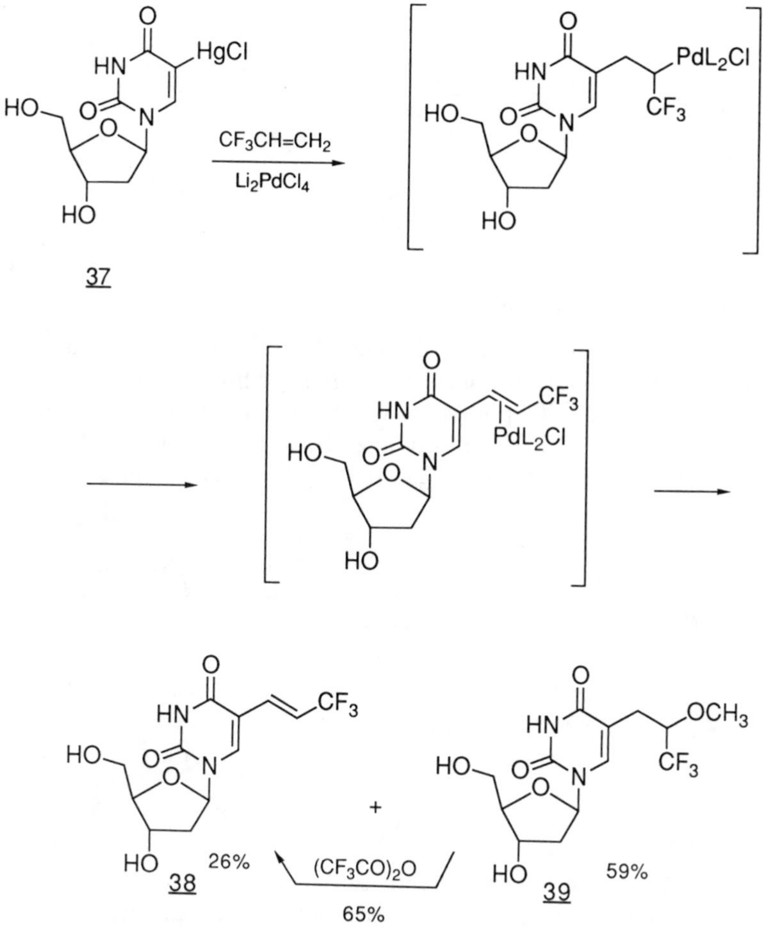

11.4 PURINES

11.4.1 2-Fluoroadenine

2-Fluoroadenine nucleosides are resistant to deamination by the catabolic enzyme deaminase. Thus it is possible to improve the lifetime of adenosine analogs in vivo.[26] The nonaqueous diazotization of 2,6-diamino-9-(2,3,5-tri-*O*-acetyl-β-D-*ribo*-furanosyl)purine (**40**) with *t*-butylnitrite in hydrogen fluoride–pyridine solution afforded a 48% yield of 2-fluoroadenosine triacetate (**41**).[27] The reaction was general and yielded 6-chloro-2-fluoro-9-(2,3,5-tri-*O*-acetyl-β-D-*ribo*-furanosyl)purine in 85% yield and 2,6-difluoro-9-(2,3,5-tri-*O*-acetyl-β-D-*ribo*-furanosyl)urine in 66%. The transformation was very sensitive to the concentration of hydrogen fluoride; for example, 70% hydrogen fluoride in pyridine was ineffective, whereas a concentration between 45 and 65% worked well.

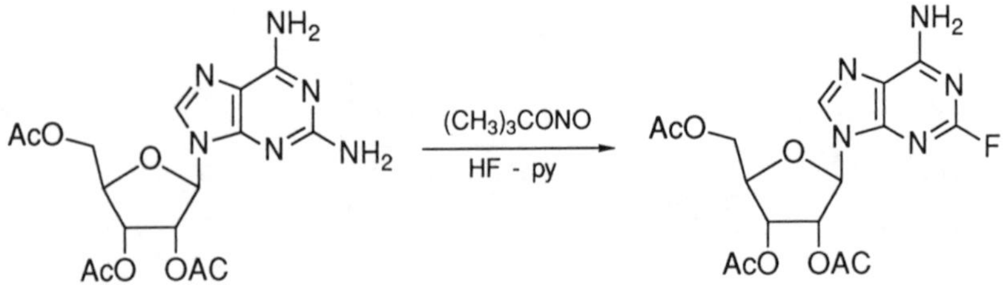

11.4.2 2-Fluoro-8-aza-adenosine

Fluorination of 8-aza-adenosine was undertaken to suppress deamination, as described earlier. 2-Amino-8-aza-adenosine (**42**) prepared from the 2-thiomethyl compound was diazotized with potassium nitrite in fluoroboric acid. In situ dediazoniation formed the desired 2-fluoro-8-aza-adenosine (**43**) in 21% yield.[28] The compounds showed good activity against P388 leukemia in mice.

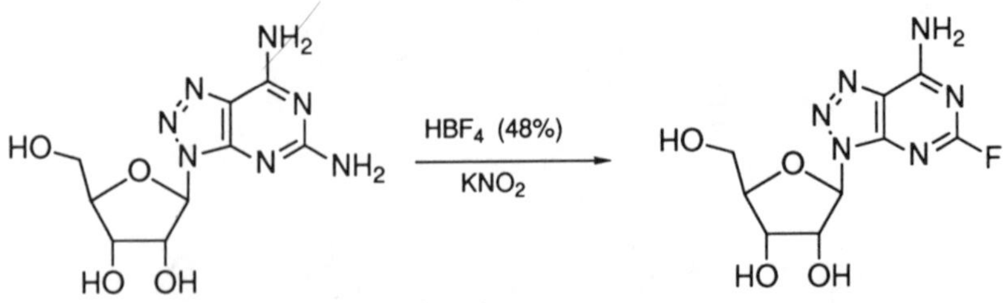

11.4.3 2-Fluoroformycin

Formycin has some in vivo anticancer activity but is subject to deamination by adenosine deaminase, whereupon the in vivo cytotoxicity is greatly reduced.

5,7-Diamino-3-(β-D-*ribo*-furanosyl)pyrazolo (4,3-*d*)pyrimidine (**44**) was dissolved in hydrogen fluoride–pyridine (60% hydrogen fluoride) and treated with *t*-butyl nitrite. The desired 2-fluoro compound (**45**) was isolated in 17% yield.[29] Fluorination diminished the activity tenfold, relative to the parent formycin, against L1210 cells. Apparently the loss of potency is related to the failure of di- and triphosphate of 2-fluoroformycin to form.

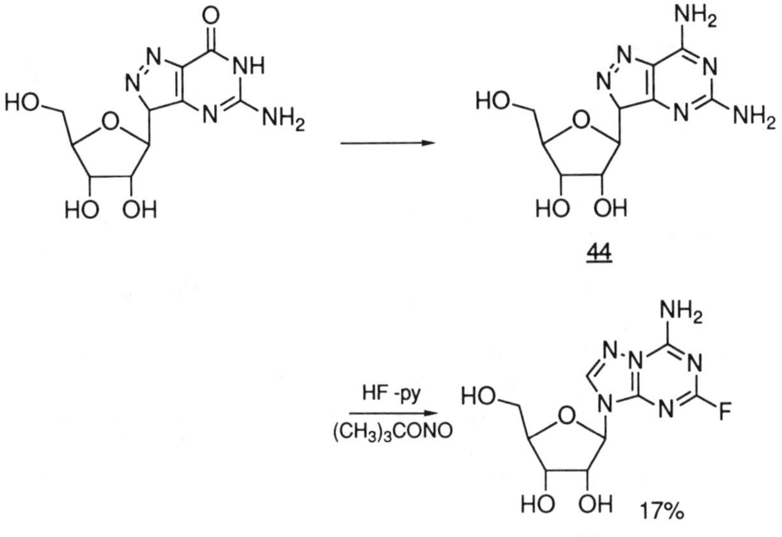

44

45

11.4.4   8-Trifluoromethyl Adenosine and Inosine

The 2′,3′,5′-tri-*O*-acetyl-8-iodoadenosine (**46**) was subjected to trifluoromethylation with trifluoromethyl iodide and copper as was described for the pyrimidines earlier. The principal product was reduction of the iodide. Removal of unreacted copper powder by filtration, demonstrating the solubility of the trifluoromethylating reagent, suppressed the reduction reaction[30] and yielded (**47**) in 64% yield. The procedure was general, yielding 8-trifluoromethyl inosine (**49**) from 2′,3′,5-tri-*O*-acetyl-8-bromoinosine (**48**) in 42% yield. The compounds had only modest anticancer activity.

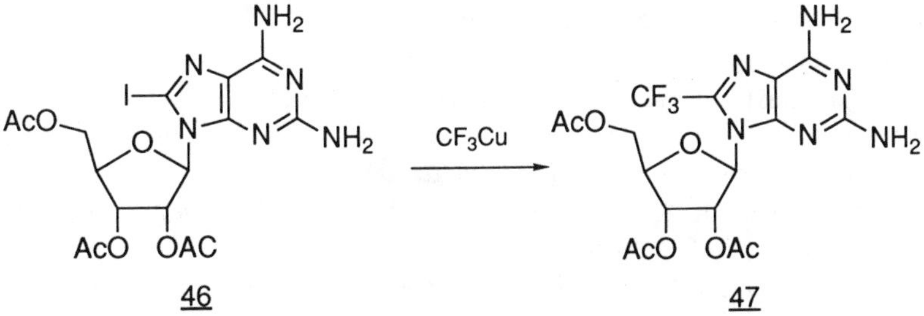

46

47

48 → CF₃Cu → 49

REFERENCES

1. D. E. Bergstrom and D. J. Swartling, in *Fluorine-Containing Molecules. Structure, Synthesis and Applications* (ed. J. F. Liebman, A. Greenberg, and W. R. Dolbier, Jr.), VCH, Deerfield Beach, Fla., 1988.

2. (a) N. M. Lucey and R. S. McElhinney, *J. Chem. Res. (S)* **1985**, 240; (b) A. Hoshi, in *Fluoropyrimidines in Cancer Therapy* (eds. K. Kimura, S. Fujii, M. Ogawa, G. P. Bodey, and P. Alberto), Elsevier Science, New York, 1984.

3. D. Cech and A. Holy, *Collect. Czech. Chem. Commun.* **41**, 3335, 1976.

4. D. Cech, L. Hein, R. Wuttke, M. Janata, A. Otto, and P. Langen, *Nucl. Acids Res.* **2**, 2177–2182, 1975.

5. C. Y. Shiue, A. P. Wolf, and M. Friedkin, *J. Labelled Compd. Radiopharm.* **21**, 865–873, 1984.

6. Y. Kobayashi, I. Kumadaki, and A. Makazato, *Tetrahedron Lett.* **24**, 1055, 1983.

7. G. M. V. Visser, S. Boele, V. W. Halteren, G. H. J. N. Knops, J. D. M. Herscheid, G. A. Brinkman, and A. Hoekstra, *J. Org. Chem.* **51**, 1466–1471, 1986.

8. I. Kumadaki, M. Nakazawa, Y. Kobayashi, T. Maruyama, and M. Honjo, *Tetrahedron Lett.* **24**, 1055, 1983.

9. (a) R. D. Armstrong and R. B. Diasio, *Cancer Res.* **40**, 3333, 1980; (b) W. Bollag and H. P. Hartmann, *Eur. J. Cancer* **16**, 427, 1980.

10. S. Ajmera, A. R. Bapat, E. Stephanina, and P. V. Danenberg, *J. Med. Chem.* **31**, 1094–1098, 1988.

11. J. Farkas, *Collect. Czech. Chem. Commun.* **48**, 2676, 1983.

12. J. Wrubel and R. Mayer, *Z. Chem.* **24**, 253, 1984.

13. J. Reefschlaeger, C. -D Pein, and D. Cech, *J. Med. Chem.* **31**, 393–397, 1988.

14. B. Schwarz, D. Cech, and J. Reefschlager, *J. Prakt. Chem.* **326**, 985–993, 1984.

15. M. P. Mertes and S. E. Saheb, *J. Pharm. Sci.* **52**, 508, 1963.

16. L. Hein and D. Cech, *Z. Chem.* **11**, 415–416, 1977.

17. Y. Kobayashi, I. Kumadaki, and K. Yamamoto, *J. Chem. Soc. Chem. Commun.* **1977**, 536.

18. T. -S. Lin and Y. -S. Gao, *J. Med. Chem.* **26**, 598, 1983.

19. J. Matulic-Adams, K. Takahashi, T. -C. Chou, H. Gadler, R. W. Price, A. R. V. Reddy, T. I. Kalman, and K. A. Watanabe, *J. Med. Chem.* **31**, 1642–1647, 1988.

20. T. Fuchikami and I. Ojima, *Tetrahedron Lett.* **23**, 4099, 1982.

21. T. Fuchikami and A. Yamanouchi, *Chem. Lett.* **1984**, 1595–1598.

22. T. Fuchikami, A. Yamanouchi, and I. Ojima, *Synthesis* **1984**, 766.

23. H. Griengl, E. Wanek, W. Schwarz, W. Streicher, B. Rosenwirth, and E. DeClercq, *J. Med. Chem.* **30**, 1199–1204, 1987.

24. M. Bobek, I. Kavai, and E. DeClercq, *J. Med. Chem.* **30**, 1494–1497, 1987.

25. D. E. Bergstrom, J. L. Ruth, P. A. Reddy, and E. DeClercq, *J. Med. Chem.* **27**, 279, 1984.

26. R. W. Brockman, Y. -C. Cheng, F. M. Schabel, Jr., and J. A. Montgomery, *J. Med. Chem.* **26**, 1483, 1984.

27. M. J. Robbins and B. Uznanski, *Can. J. Chem.* **59**, 2608, 1981.

28. J. A. Montgomery, A. T. Shortnacy, and J. A. Secrist, *J. Med. Chem.* **26**, 1483, 1983.

29. J. A. Secrist, A. T. Shortnacy, and J. S. Montogemery, *J. Med. Chem.* **28**, 1740, 1985.

30. Y. Kobayashi, K. Yamamoto, T. Asai, M. Nakano, and I. Kumakaki, *J. Chem. Soc. Perkin (1)* **1980**, 2755.

12

FLUORINATED AROMATICS

12.1 INTRODUCTION

Fluorinated aromatics have found wide use in an enormous variety of molecules which may have antibiotic or sedative activity or importance as agrochemicals or even as radiochemical imaging agents. Fluorinated aromatics are a particular synthetic challenge to prepare efficiently or in high yield. There are no simple methods for the introduction of fluorine into the aryl nucleus. The best known method is of course diazotization of an aromatic amine followed by dediazoniation in the presence of fluoride-containing counterion. Selective electrophilic fluorination procedures are difficult to find. As such most syntheses of bioactive materials start with unfunctionalized fluorinated aromatic materials upon which the functionality is later arrayed. We consider the introduction of the trifluoromethyl or difluoromethyl group as well, since the synthesis of the compounds has features related to the preparation of the aryl fluorides themselves.

12.2 GENERAL METHODS FOR THE FLUORINATION OF AROMATICS

12.2.1 Background

The use of general methods for the introduction of fluorine into organic compounds was recently reviewed.[1] The use of fluorine, trifluoromethyl hypofluorite, acetyl hypofluorite, cesium fluorosulfate, xenon difluoride, and N-fluoropyridinium triflate for the fluorination of aromatic substrates was reported. Additionally, some reference has been made to classic dediazoniative procedures.

Since they are the most generally studied and also the best established, the Balz–Schiemann reactions are discussed first.

12.2.2 Balz–Schiemann Reactions

As electrophilic fluorination reagents are derived from the reaction of molecular fluorine with some carrier molecule, scaleup of these methods is generally not practical. The classic Balz–Schiemann reaction has been applied to the synthesis of fluorophenylephrines.[2] 3-Methoxy-2-nitrobenzaldehyde (**1**), protected as the dimethylacetal, was reduced to the amine. The amine was diazotized and dediazonized in a photochemical variant of the Balz–Schiemann reaction. The overall yield of this process of 14% was not desirable.[3] Compounds such as **4** prepared by this method are useful for the study of the adrenergic properties of phenylephrine.[4]

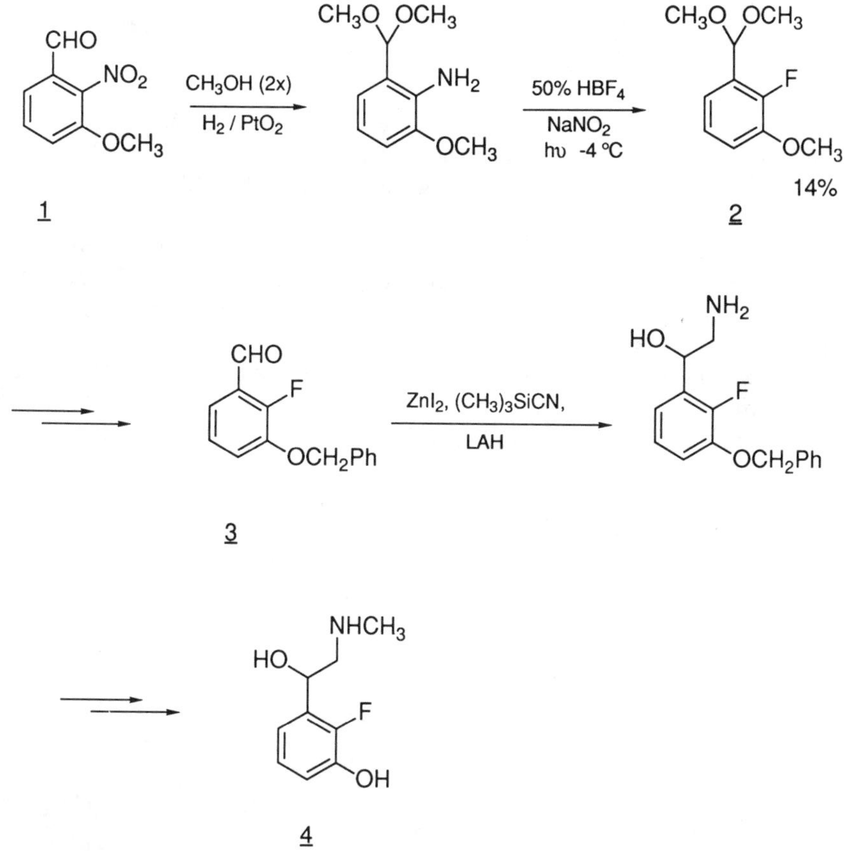

The Balz–Schiemann reaction has also been applied to heterocyclic synthesis.[5] The starting amine may be appended to the pyridyl nucleus, as in the case of 6-chloro-2-(4-acetyl-1-piperazinyl)-3-pyridinediazonium tetrafluoroborate (**6**), which was prepared from the 3-aminopyridine (**5**) by diazotization in 42% tetrafluoroboric acid in ethanol with aqueous sodium nitrite. Heating under reflux resulted in dediazoniative fluorination to form (**7**). It was determined that

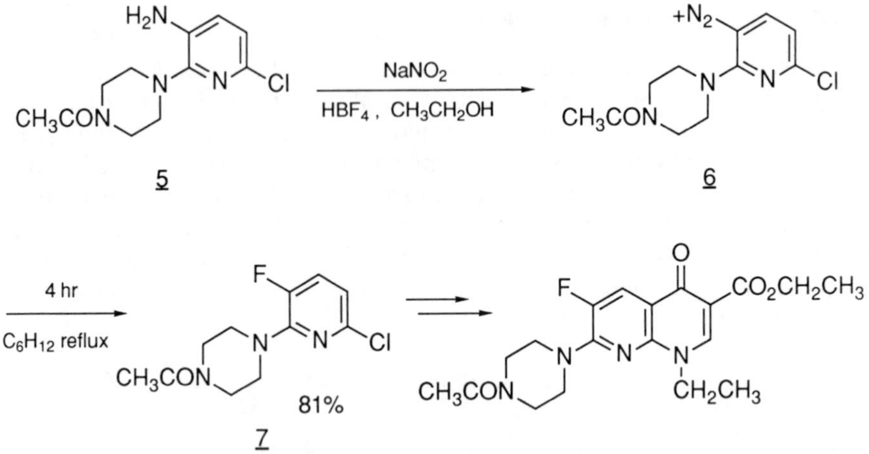

5 → (NaNO$_2$, HBF$_4$, CH$_3$CH$_2$OH) → **6**

4 hr, C$_6$H$_{12}$ reflux → **7** (81%) →

the yield of the fluorinated material isolated on thermal decomposition was dependent on the temperature of the reaction and on the time employed. As can be seen from Table 12.1, cyclohexane gave the highest yields with the best reaction times.

This approach has also been applied to the naphthyridine ring system itself. The direct dediazoniation of 3-(ethoxycarbonyl)-1-ethyl-1,4-dihydro-7-methoxy-4-oxo-1,8-naphthyridine-6-diazonium tetrafluoroborate (**8**) was unsuccessful.[6] However, replacing the 7-methoxy group by the 4-acetyl-piperazinyl unit as in **9** yielded a substance that could successfully be diazotized with 65% hexafluorophosphoric acid and sodium nitrite. Thermal decomposition in n-heptane at 80–100°C yielded the desired fluorinated material (**10**) in 36% yield.

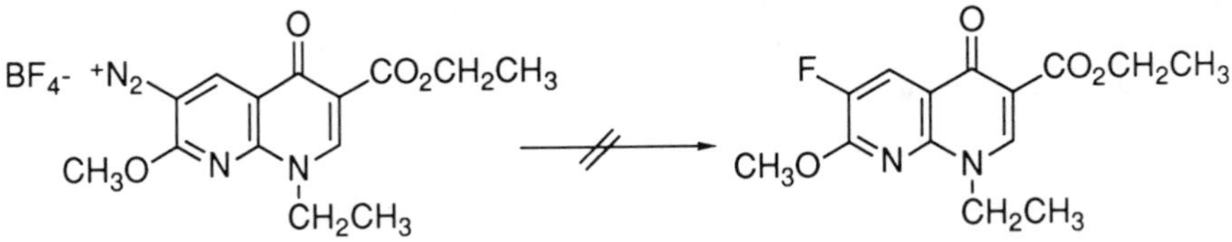

8

TABLE 12.1   Dediazoniative Fluorination of 3-Aminopyridine 5.

Solvent Used	Temperature (°C)	Time (hours)	Yield (%)
Hexanes	50–90	10.5	62
Carbon tetrachloride	77	18	36
Ethyl acetate	77	10.5	73
Cyclohexane	81	3.5	81
Isopropyl acetate	89	3	45
n-Heptane	98	0.5	64
Toluene	110	0.3	65

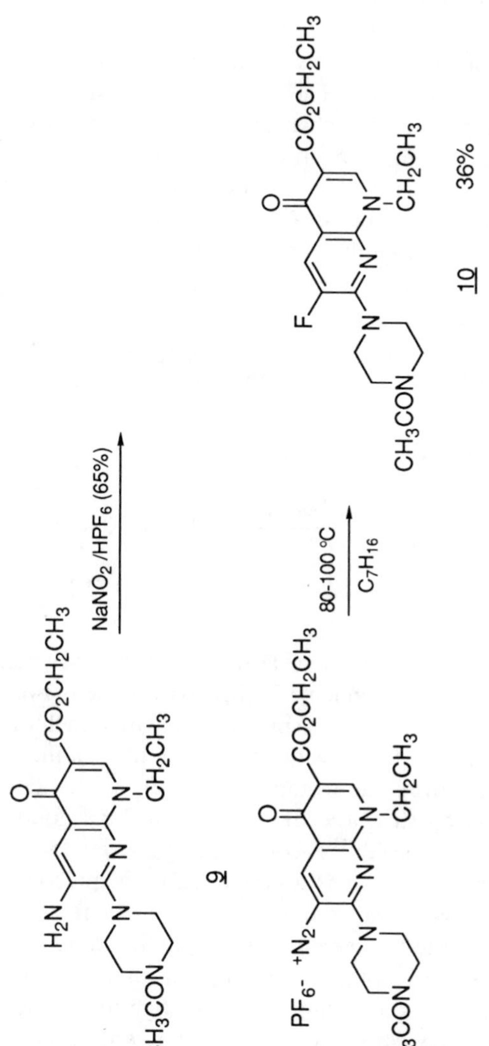

H2N
CH3CON
CO2CH2CH3
CH2CH3
9
NaNO2 /HPF6 (65%)
PF6- +N2
80-100 °C
C7H16
F
10
36%

Introduction of fluorine into molecules of this type resulted in pronounced enhancement of activity against both Gram-negative and Gram-positive bacteria.

An alternative approach to this simple dediazoniative strategy was required for the introduction of fluorine into the aromatic nucleus when ^{18}F-containing materials were employed. Processes employing ^{18}F must be very quick since the half-life is only 90 minutes. These methods must also be very efficient radiochemically; for example, when tetrafluoroborate is used as the fluoride source only a single atom is incorporated. When simple diazonium ions with nonfluorine counterions (**11**) were decomposed in the presence of a fluoride ion source, the yield of fluoroaromatics (**12**) isolated was less than 1%.[7] Additionally, the intermediate salts were difficult to purify.

$$C_4H_9\!-\!\langle\rangle\!-\!NH_2 \xrightarrow[\text{NaBPh}_4]{\text{HCl, NaNO}_2} C_4H_9\!-\!\langle\rangle\!-\!N_2^+\ BPh_4^-$$

$$90\%$$
$$\underline{11}$$

$$\xrightarrow{48\%\ HF}\ C_4H_9\!-\!\langle\rangle\!-\!F$$

$$<1\%$$
$$\underline{12}$$

Conversion of aminotamoxifen, (**13**) to 1-(4-piperidinoaza)phenyl-1-(4-(2-(N,N-dimethylamino))ethoxy)phenyl-2-phenyl-1-butane (**14**), an aryltriazene that may itself be decomposed in the presence of hydrogen fluoride to form an aryl fluoride, followed by treatment with aqueous hydrogen fluoride or pyridinium hydrogen fluoride, forms fluorotamoxifen (**15**) in 35–28% yield.[8] Fluoro-tamoxifen prepared by this procedure proved to be a good estrogen receptor binding agent. Radiolabeled ^{18}F-containing material for studying the tissue distribution of estrogen binding sites might easily be prepared by the preceding methods. This approach was employed for the ^{18}F-radiolabeled synthesis of spiroperidol (**16**) for the in vivo study of dopamine receptors.[9]

In closely related work, 2-fluorohexestrol was prepared for use as an estrogen receptor binding agent. A very thorough investigation of the conditions for the introduction of fluorine was made assuming that with radiolabeled fluoride ion, fluoride should be utilized efficiently.[9b] In model studies it was determined that ortho substituent markedly reduce the yield of fluoride when meta and para substituents are tolerated well. As can be seen in Table 12.2, the substituents strongly affect the outcome of the fluorination reaction.

Decomposition of the triazene **17** prepared from 2′-aminohexestrol formed the 2′-fluorohexestrol (**18**) in 25–43% yield but the reaction was accompanied by formation of a ring-closed product (**19**) in considerable amount (46%), from intramolecular alkylation.[7]

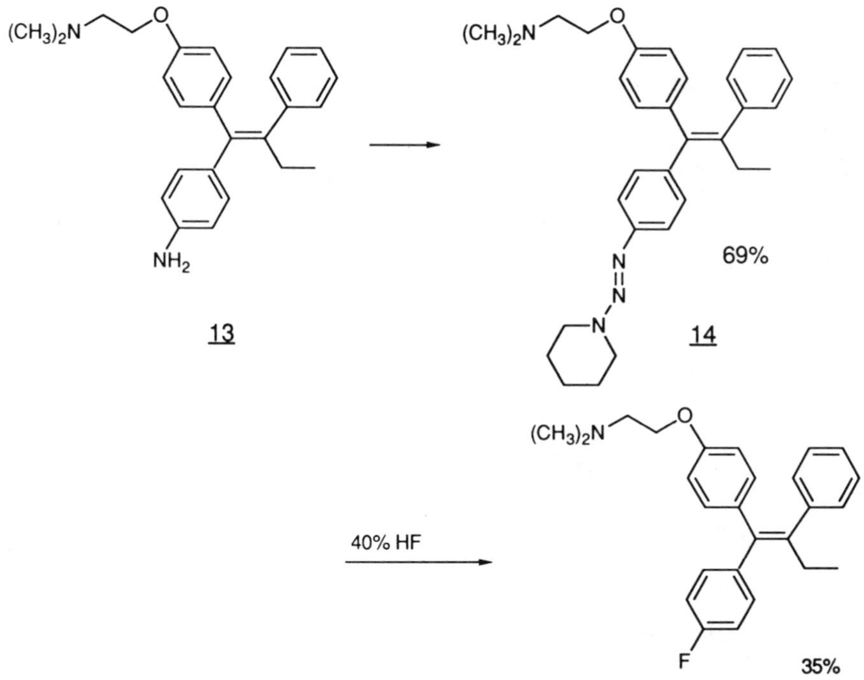

13 14 15

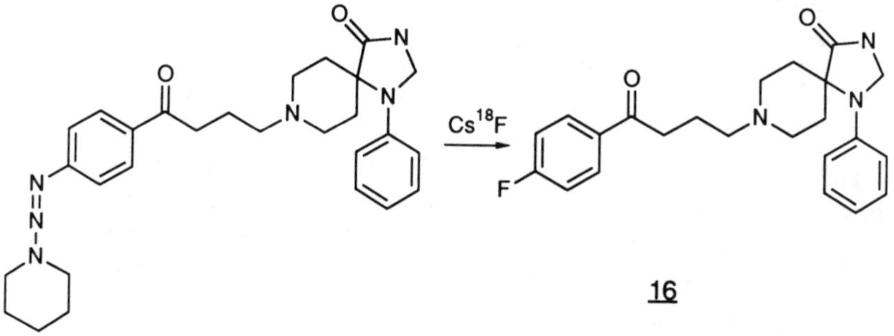

16

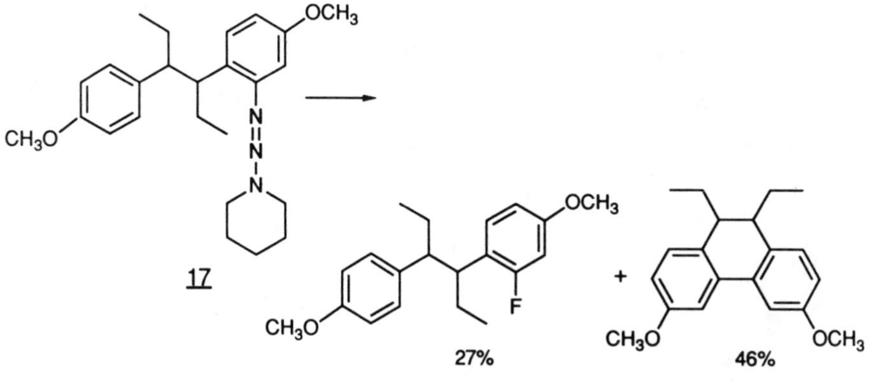

17 18 19

TABLE 12.2   The Synthesis of 2′-Fluorohexestrol, 18.

Triazene	Reaction Condition	Yield (%)
4-CH$_3$	Anhydrous HF–THF	15
	HF–py	15
	48% aqueous HF	27
4-CH$_3$(CH$_2$)$_3$	48% aqueous HF	16
2-(CH$_3$)2CH	HF–py–benzene	22
3-CH$_3$O	HF–py–benzene	19
	48% aqueous HF	17
	Anhydrous HF–benzene	18
	Anhydrous HF–CH$_2$Cl$_2$	trace
2-CH$_3$O-5-phenyl	48% aqueous HF	0
	HF–py–benzene	0
2-CH$_3$SO$_3$	48% aqueous HF	0
2-CH$_3$O	48% aqueous HF	0
2-CH$_3$O-5CH$_3$	48% aqueous HF	0

12.2.3   Direct Fluorination

Molecular Fluorine

As mentioned previously, the direct fluorination of aromatics is problematic. However, for imaging of dopamine receptors in normal and diseased tissues, 6-fluorodopamine was prepared by the direct fluorination of dopamine with fluorine-18 gas in hydrogen fluoride.[10] Although the 6-fluoro compound was formed in 21% yield along with the 2-fluoro material in 12%, and with a small amount, 17%, of the fluorodopamine, the reaction did not demonstrate strict electrophilic substitution character. Earlier studies had suggested that fluorine could behave as an electrophile but was generally unselective.[11]

Remarkably, when the solvent was changed to trifluoroacetic acid containing sodium acetate a much better regiocontrol was possible. Selectivity for the 2 position was greater than 88%, and the product 2-fluorodopa, was obtained. However, the overall yields were less (Fig. 12.1).

F$_2$ / HF	34%	5%	16%
AcOF/TFA	89%	10%	1%

FIGURE 12.1

Trifluoroacetyl Hypofluorite[12]

The same observation generally applied to the fluorination of phenylalanine and tyrosine as well.

Acetyl Hypofluorite

Acetyl hypofluorite was also an effective reagent for the fluorination of the false neurotransmitter metaraminol. 6-(Acetoxymercurio)-N-t-BOC-metaraminol (**20**) was treated with acetyl hypofluorite followed by removal of the BOC group to form 6-fluorometaraminol (**21**) in up to 42% radiochemical yield.[13] Preliminary biological data indicate this material possesses high selectivity for adrenergic nerves.

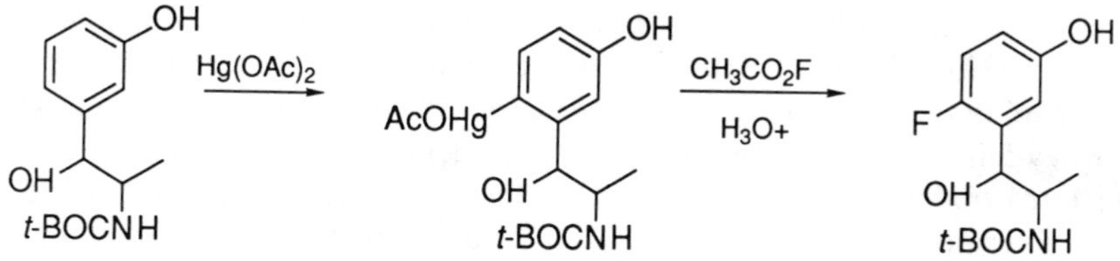

$$\underline{20} \qquad \qquad \underline{21}$$

N-Fluorosulfonimides

Finally, typical of the N-fluoro reagents is N-fluoro(bistrifluoromethyl)sulfonimide.[14] Although the reagent requires the use of liquid fluorine, it is stable for long periods at 22°C. Yields of fluorinated material with activated aromatic

TABLE 12.3 Aromatic Fluorinations with N-Fluoro(bistrifluoromethyl)sulfonimide.

Substrate	Products	Yield (%)
Nitrobenzene	—	0
Benzene	Fluorobenzene	50
Toluene	2-Fluorotoluene	74
	3-Fluorotoluene	4
	4-Fluorotoluene	22
Anisole	2-Fluoroanisole	69
	4-Fluoroanisole	24
m-toluol	2-Fluoro-5-methyl phenol	44
	4-Fluoro-3-methyl phenol	56
p-toluol	2-Fluoro-4-methyl phenol	80
Naphthalene	1-Fluoronapthalene	80
	2-Fluoronapthalene	7

rings ranged as high as 80% for the highly reactive *p*-toluol to no reaction with nitrobenzene (see table 12.3).

12.3 FLUORINATED AROMATIC BUILDING BLOCKS

12.3.1 Fluorinated Phenols

As mentioned earlier, very commonly the preparation of fluorinated aromatic compounds is based upon the elaboration of simple fluorinated substrates. One such simple starting material was 2-fluoro-6-methoxy-phenol (**22**). When treated with hexamethylene tetraamine and trifluoroacetic acid in a modified Duff reaction, 2-fluoro-3-hydroxy-4-methoxybenzaldehyde (2-fluoroisovanillin) (**23**) was formed in 75% yield.[15] However, treatment of the phenol with formaldehyde and N,N-dimethylamine led to formation of N,N-dimethyl-3-methoxy-4-hydroxy-5-fluorobenzylamine (**24**) in 95% yield. This material was readily converted under standard conditions to the aldehyde, 5-fluorovanillin (**25**), a useful intermediate for the synthesis of new drugs with selective action on the adrenergic nervous system.

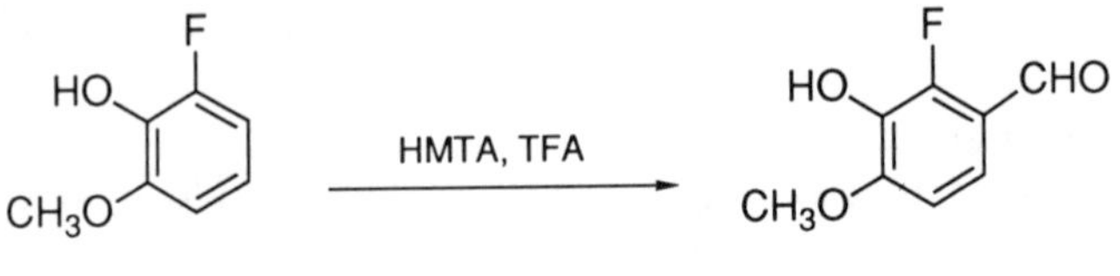

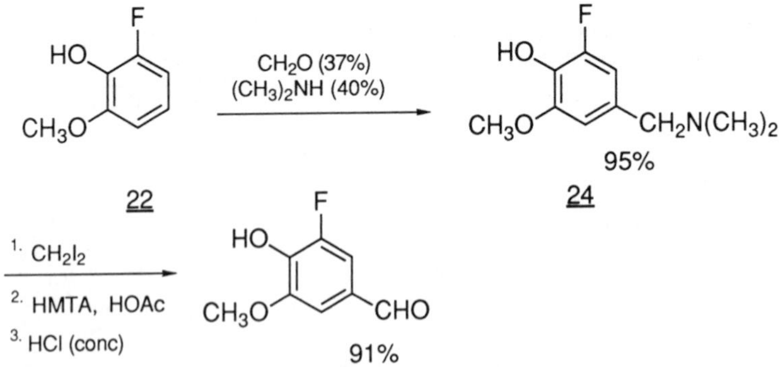

The related synthesis of fluorinated phenylephrine was affected by lithiation with *sec*-butyllithium of 2-fluorophenol, protected as the *t*-butyl-dimethylsilyl ether, **26**. The carbanion so formed was quenched with dimethyl-formamide to yield 2-fluoro-3-hydroxybenzaldehyde, **27**, in 86% yield. Remarkably, lithiation occurs ortho to fluorine.

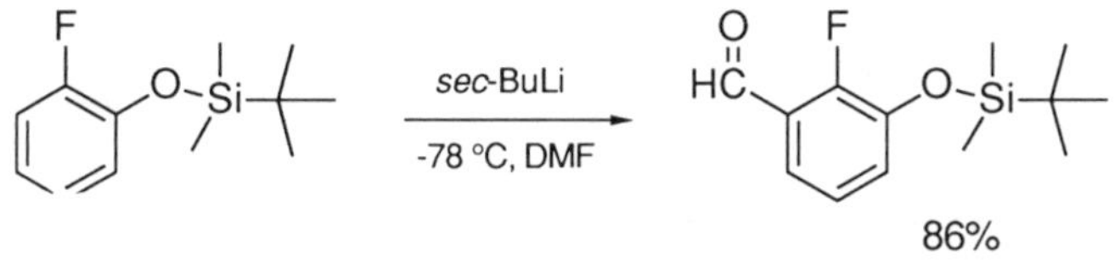

26 → 27

86%

12.3.2 Fluorinated Aldehydes

Directed metalation was also employed in the functionalization of the dimethyl-acetal of 3-fluorobenzaldehyde (**28**). Treatment of **28** with *sec*-butyllithium followed by addition of carbon dioxide led to formation of the hydroxyphthalide (**29**) in 86% yield.[16] This procedure was also effective in the reaction of the oxazoline (**30**) and *p*-difluorobenzene to form upon completion of the reaction sequence the hydroxyphthalides **31** and **34** in 98% and 93% yield respectively. The product hydroxyphthalides were further transformed to fluorinated analogs of daunomycinone.

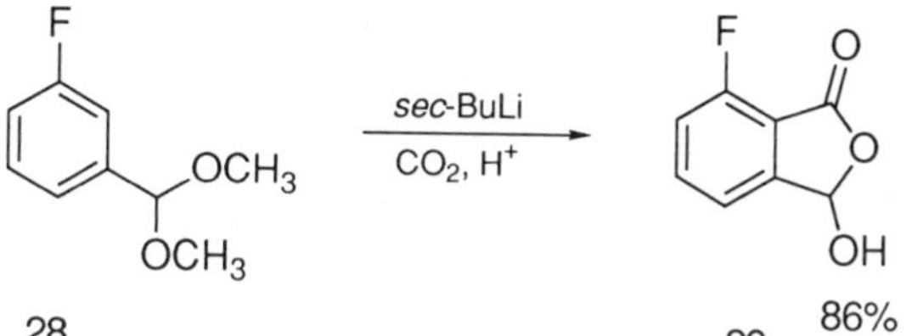

28 → 29   86%

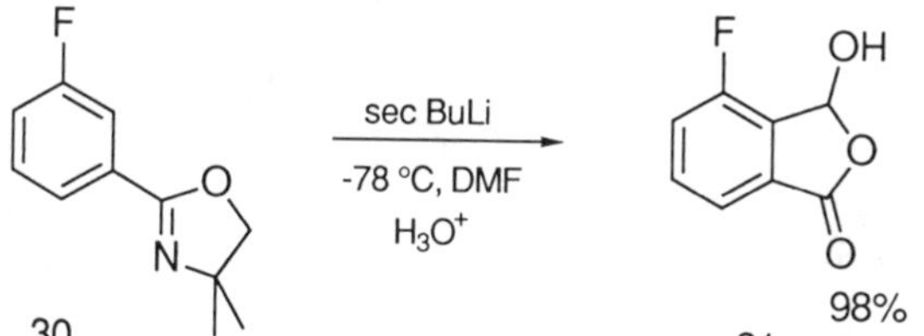

30 → 31   98%

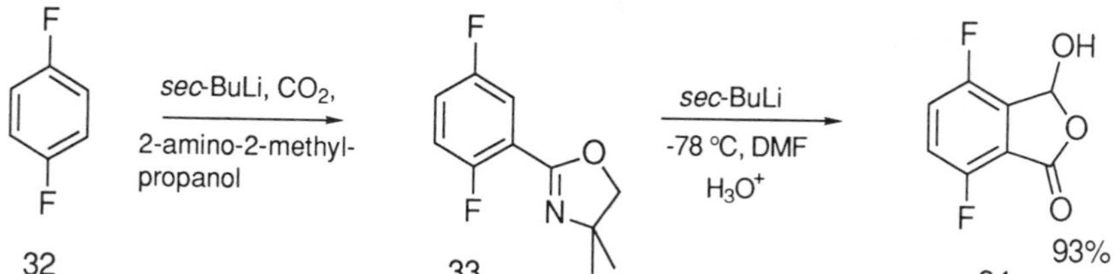

32 → 33 → 34   93%

12.3.3 Fluorinated Acids

The respective fluoronitrobenzoic acids were coupled to the *t*-butyl esters of L-glutamic acid and were then reduced to the *p*-aminofluorobenzamides. The

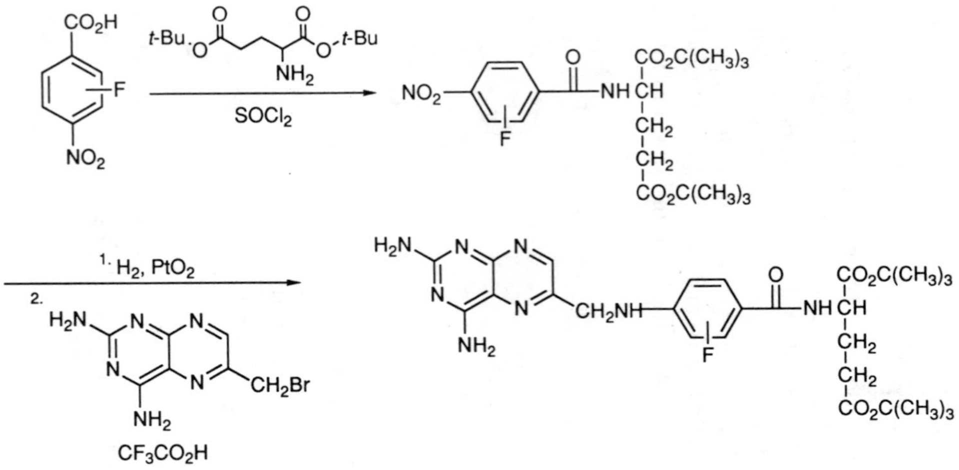

FIGURE 12.2

amino compounds were condensed with 6-(bromomethyl)-2,4-pteridine diamine to form on deprotection a 40% yield of the fluoroaminopterins (Fig. 12.2).[16] Both the 2'-fluoro and 3'-aminopterins were more active than methotrexate against L1210 cells and the human stomach cancer HuTu80 cells.

Fluorination of methotrexate has also been employed as a tool for studying the binding of the material to dihydrofolate reductase.[17] When bound to the enzyme, each of the fluorines in **35** is in a different chemical environment. It was possible to observe the effect of added NADH or NADPH to the bound methotrexate analog.

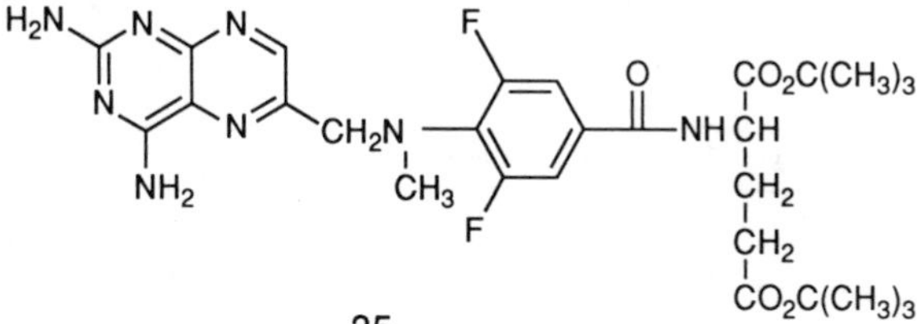

35

Anisole was acylated with 4-fluorobenzoyl chloride to form the unsymmetric benzophenone **36** for use in the preparation of fluorotamoxifen **38**, a good

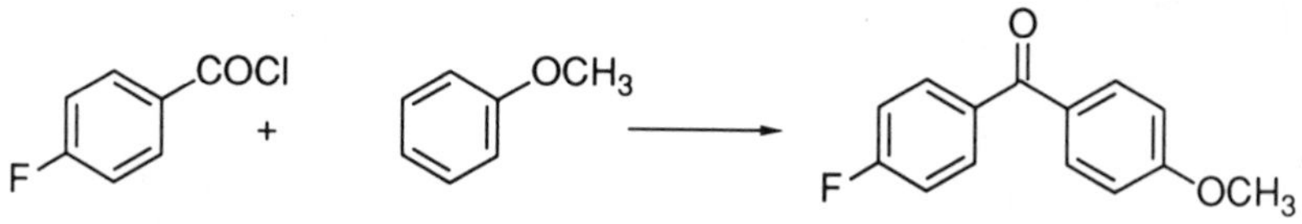

36

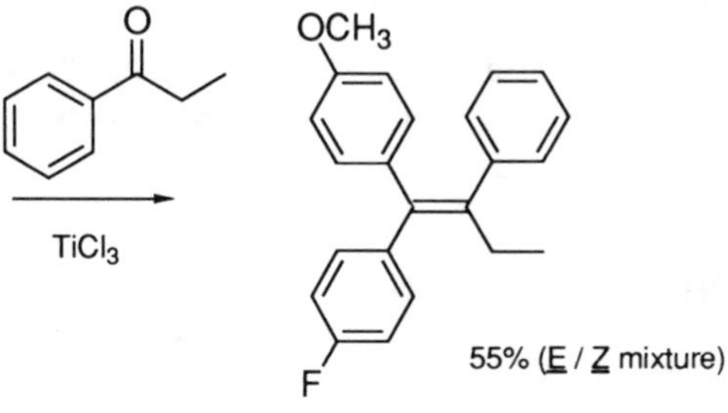

55% (E / Z mixture)

37

36%

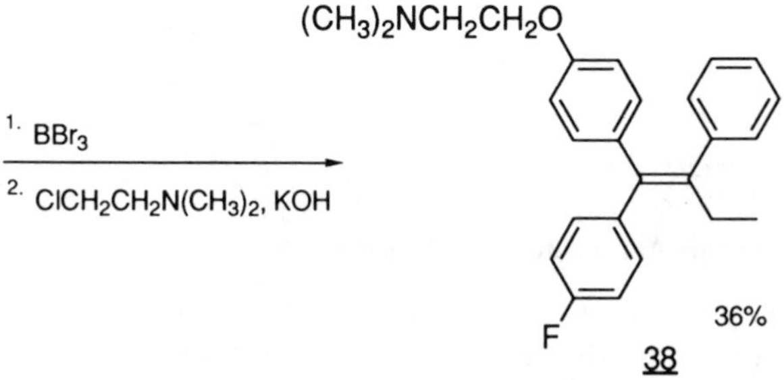

38

estrogen receptor binding agent, in 75% yield. McMurry coupling of propiophenone in 55% yield formed the desired triphenylethylene derivative (37). The 2-(dimethylamino)ethyl ether (38) was prepared in 36% yield. This synthesis is not easily adapted to preparing radiolabeled material.[8]

12.3.4 Fluorinated Anilines

Fluorinated aniline (39) was used as a building block for the preparation of fluorinated analogs of nalidixic acid (40). The incorporation of fluorine in the molecule showed a 16-fold increase in antibacterial potency.[18]

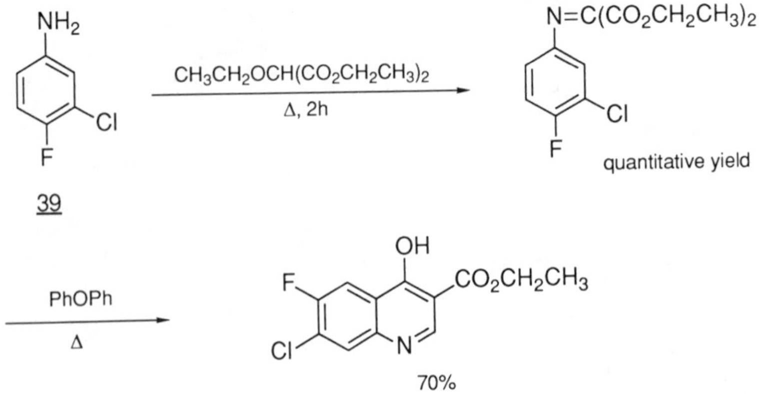

12.3.5 Fluorinated Chalcones

Fluorinated chalcones, such as 41 have been converted by simple cyclization reactions to a series of fluorinated flavinoids, such as 42 with antiviral, antifungal, and antibacterial activity.[19]

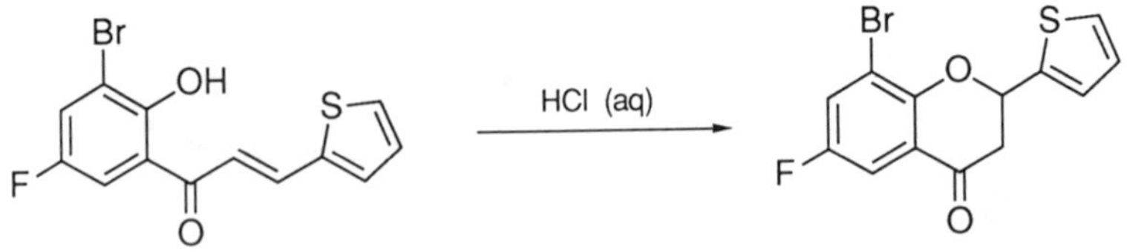

12.4 FLUOROALKYL AROMATICS

12.4.1 Trifluoromethylated Aromatic Building Blocks

Trifluoromethyl groups generally have an effect on biological activity of the materials into which they are substituted, particularly when these substitutents are introduced into CNS agents, such as 2'-trifluoromethyl-1-methyl-4-phenyl-1,2,5,6-tetrahydropyridines (**45**).[20] This material was prepared by transmetallation of 2-bromobenzotrifluoride (**43**) with butyllithium at $-70°C$ and addition to 1-methyl-4-pyridine (**44**).

Other trifluoromethylaromatics such as **46** have been employed as anti-convulsants and muscle relaxants. Some of these materials were highly effective sleep-inducing substances for *Cebus* monkeys.[21]

12.4.2   Introduction of the Trifluoromethyl Group

The introduction of the trifluoromethyl group into the aromatic nucleus can be extraordinarily difficult, just as direct introduction of fluorine onto the aromatic ring is difficult. Chlorotrifluoromethyl diazirine (**47**) has been used as a source of chlorotrifluoromethyl carbene. This carbene reacts with cyclic dienes to yield on ring expansion trifluoromethylated six-membered ring aromatic products.[22] When a solution of **47** and pyrrole was sealed under vacuum in a Pyrex tube and heated to 120°C for 2 hours a 35% yield of 3-trifluoromethylpyridine (**48**) was isolated.

Alternatively, it is possible to prepare substituted aromatics which contain the trifluoromethyl group, although the trifluoromethyl group is not directly substituted on the aromatic ring. Aromatic materials where the trifluoromethyl group is directly substituted are only very slowly decomposed in biological systems. Therefore, the use of the 3,3,3-trifluoropropyl group was explored since linearly alkylated materials may be more easily degraded than branched substances.[23]

Trifluoromethylpropene was bubbled into a suspension of benzene and aluminium trichloride, but only 1,1-difluoroindane (**49**) was formed in low yield.

The use of protic acid seemed more promising. Nafion-H gave only low yields of 3,3,3-trifluoromethylpropylated materials; however, tetrafluoroboric acid was quite efficient with a 59% yield of the alkylated material (**50**) isolated. Interestingly, boron trifluoride in a 1:1 molar stoichiometry with benzene was effective, proceeding in 57% yield. Boron trifluoride was a general catalyst, yielding alkylated products with chlorobenzene (43%), benzotrifluoride (13%), and phenol (20%).

REFERENCES

1. J. Mann, *J. Chem. Soc.* **16**, 381–436, 1987.

2. K. L. Kirk, O. Olubad, K. Buchhold, G. A. Lewandowski, F. Gusovsky, D. McCulloh, J. W. Daly, and C. R. Creveling, *J. Med. Chem.* **29**, 1982–1988, 1986.

3. K. L. Kirk, *J. Org. Chem.* **41**, 2373, 1976.

4. F. Gusovsky, E. T. McNeal, O. Olubajo, K. L. Kirk, C. R. Creveling, and J. W. Daly, *Eur. J. Pharm.* **136**, 317–324, 1987.

5. J. Matsumoto, T. Miyamaoto, A. Minamida, and Y. Nishimura, *J. Heterocycl. Chem.* **21**, 673–674, 1984.

6. J. Matsumoto, T. Miyamot, T. Minamida, Y. Nishimura, H. Egawa, and H. Nishimura, *J. Med. Chem.* **27**, 292–300, 1984.

7. K. S. Ng, J. A. Katzenellenbogen, and M. R. Kilbourn, *J. Org. Chem.* **46**, 2520–2528, 1981.

8. (a) J. Shani, A. Gazit, T. Livshitz, and S. Biran, *J. Med. Chem.* **28**, 1504, 1985; (b) H. Hoving, B. Crysell, and P. F. Leadlay, *Biochemistry* **24**, 6163, 1985.

9. (a) M. R. Kilbourn, M. J. Welch, C. S. Dence, T. J. Tewson, H. Saji, and M. Maeda, *Int. J. Appl. Radiat. Isot.* **35**, 591–598, 1984; (b) M. J. Welch, M. E. Raichle, M. R. Kilbourn, and M. A. Mintun, *Ann. Neurol.* **15** (suppl.), 577–578, 1984.

10. G. Firnau, R. Chirakal, and E. S. Garnet, *J. Nucl. Med.* **25**, 1228, 1984.

11. F. Cacace, P. Giacomello, and A. P. Wolf, *J. Am. Chem. Soc.* **102**, 3511, 1980.

12. H. H. Coenen, K. Frankern, P. Kling, and G. Stöcklin, *Appl. Radiat. Isot.* **39**, 1243–1250, 1988.

13. S. G. Mislankar, D. L. Gildersleeve, D. M. Wieland, C. C. Massin, G. K. Mulholland, and S. A. Toorongian, *J. Med. Chem.* **31**, 326–366, 1988.

14. S. Singh, D. D. Desmarteau, S. S. Zyber, M. Witz, and H. -N. Huang, *J. Am. Chem. Soc.* **109**, 7194–7196, 1987.

15. M. T. Clark and D. D. Millar, *J. Org. Chem.* **51**, 4072–4073, 1986.

16. G. W. Morrow, J. S. Swenton, J. A. Filppi, R. L. Wolgemuth, *J. Org. Chem.*, **52**, 713–719, 1987.

17. G. Fendrich, and R. H. Abeles, *Biochemistry* **21**, 6685, 1982.

18. H. Koga, A. Itoh, S. Murayami, S. Suzue, and T. Irikura, *J. Med. Chem.* **23**, 1358–1363, 1980.

19. (a) S. P. Sachchar, and A. K. Singh, *Proc. Natl. Acad. Sci. India* **56**, 28–34, 1986; (b) S. P. Sachchar, N. N. Tripathi, and A. K. Singh, *Indian J. Chem.* **26B**, 493–495, 1987.

20. N. J. Riachi, P. K. Arora, L. M. Sayre, and S. I. Harik, *J. Neurochem.* **1988**, 1319–1321.

21. W. J. Houlihan, J. H. Gogerty, E. A. Ryan, and G. Schmitt, *J. Med. Chem.* **28**, 28–31, 1985.

22. Y. Kobayashi, T. Nakano, H. Iwasaki, and T. Kumadaki, *J. Fluorine Chem.* **18**, 533 536, 1981.

23. Y. Kobayashi, T. Nagai, I. Kumadaki, M. Takashi, and T. Yamauchi, *Chem. Pharm. Bull.* **32**, 4382–4387, 1984.

INDEX

Acetamidomalonic ester, 26
Acetylcholine esterase, *see* Enzyme,
    acetylcholine esterase
N-Acetyl-3-fluoroglutamic acids, 9
N-Acetyl-4-fluoroglutamic acid, synthesis of, 30
N-Acetyl-3-hydroxyglutamic acids, 9
Acetyl hypofluorite, 121, 146, 223, 241
Acids, unsaturated, bromofluorination of in
    liquid hydrogen fluoride, 87
N-Acyl-α-amino acid, 42
2-Acylamino-3,3,3-trifluoropropionitriles, 36
Acylase, 9
α-Acyl histidine esters, 34
N-Acyl trifluoroacetaldimine, 36
Adenosine deaminase, *see* Enzyme, adenosine
    deaminase
F-Adenosyl-L-methionine decarboxylase, *see*
    Enzyme, *F*-adenosyl-L-methione
    decarboxylase
Adenoylsuccinate synthetase and lyase, *see*
    Enzyme, adenoylosuccinate synthetase and
    lyase
Adrenergic nervous system, 242
AgF, 27
Agrochemicals, 234
Alanine, 19
β-Alanine, antimetabolite of, 58
Alanine racemase, *see* Enzyme, alanine
    racemase
Aldol condensations, 152
Alkyl carbamates, 89

Alkyl pinanediol boronates, 15
Allergies, 7
Allothreonine, 9
Amberlyst[R] A-26 (F⁻ form), 140
Amidocarbonylation, 42
Amino acid decarboxylases, *see* Enzyme, amino
    acid decarboxylases
Amino acids, analogs of, 15
γ-Aminobutyric acid, 7
α-Amino-β,β-difluoroisobutyric acid, synthesis
    of using 1,1′-difluoroacetone, 42
3-Amino-4-fluorobutanoic acid,
    synthesis of, 35
4-Amino-2-fluorobutanoic acid, 40
    synthesis of, 40
4-Amino-2-fluorobutanoic acid (fluoro (GABA),
    GABA antimetabolite, 58
4-Amino-2-fluorocrotonic acid, synthesis of, 37
4-Amino-2-fluorohexanoic acid, 40
    synthesis of, 40
4-Amino-5-fluoropentanoic acid:
    GABA-transaminase inhibitor, 55
    synthesis of, 29
5-Amino-2-fluoropentenoic acid, synthesis of, 37
Amino group, protection by hydrogen fluoride,
    20
2-Amino-3-hydroxy-4,4,4-trifluorobutyric acid,
    50
Aminomalonates, 31
Aminotamoxifen, 238
Ammonolysis, 14

Angiotensin converting enzyme, *see* Enzyme, angiotensin converting enzyme
1,2-Anhydro-3,4:5,6-di-O-isopropylidene-1-C-nitro-D-mannitol, 142
Antibacterial, 8
Antibiotics:
  aminoglycoside, 135
  anthracycline, 144
Anti-convulsants, 247
Antihypertensives, 128
Antimony trifluoride, 27, 30
Anti-viral activity, 137
Arachidonic acid, 163
Aromatic amino acid decarboxylase (AADC), *see* Enzyme, aromatic amino acid decarboxylase (AADC)
L-Aromatic-α-amino acid decarboxylase, *see* Enzyme, L-aromatic-α-amino acid decarboxylase
Aromatic amino acids, ring-fluorinated, 25
N-Aroyl ethylene imine, 40
Aryltriazene, 238
Aspartyl proteases, *see* Enzyme, aspartyl proteases
6-Aza-5-fluorouracil, 224
Aziridine carboxylate, 17
Aziridines, 16, 142
  ring-opening reactions of, 16, 67
Azirines, 16
  ring opening to form ketones, 120
1-Azirines, 18
Azlactones, 45
Aztreonam, 58

Balz–Schiemann reactions, 235
  in steroid synthesis, 197
Benzotrifluoride, 248
Benzyl 2,3-anhydro-4-deoxy-4-fluoro-α-D-*lyxo*-pyranoside, 137
Benzyl 2,3-anhydro-4-deoxy-4-fluoro-β-L-*lyxo*-pyranoside, 137
Benzyl 4,6-O -benzylidene-2,3-benzoylepimino-2,3-dideoxy-α-D -*allo*-pyranoside with tetra-*n*-butylammonium fluoride, 145
1,1′-Bis(diphenylphosphino) ferrocene, 96
Bis(pentafluoroethyl)mercury, 225
Bis(trifluoromethyl)mercury, 225
BOC-L-leucinal, 39
Boron trifluoride, 248
Bromine fluoride, 204
N-Bromoacetamide, 14
Bromobenzenesulfonate, 136, 140
Bromodifluoromethylphosphonate, 117

Bromofluorination, 14
2-Bromo-3-fluorocarboxylic acids, 14, 87
2-Bromo-3,3,3-trifluoropropene, 227
Brosylate, 136
(*E*)-Butenyl α-fluoro-α-trimethylsilylacetate, 154
t-Butyl monofluoroacetate, 38
Butynyl-Co A dehydrogenase, *see* Enzyme, Butynyl-Co A dehydrogenase

Calcium, complexation of, 117
Cancerostatic, 8
Carbohydrates, branched fluorinated, 147
Carbon–fluorine bond length, 2
Carbon–fluorine bond strength, 2
Carboxypeptidase A, *see* Enzyme, carboxypeptidase A
Catechol amines, 7
Cell development, 7
Central nervous system, 8
Cesium fluoride, in hexamethylphosphoramide or dimethylformamide, 168
Cesium fluoride-18, 133
Cesium fluoroxysulfate, 205
$CF_2{=}CFCF_3$, 45
CF_3Br, 31
$CF_3CH{=}CH_2$, 42
CF_3OF, 19
CH_2FCl, 31
CH_2FCN, 35
$CH_3CHFCOCH_3$, 46
CHF_2Cl, 31
$CHFCl_2$, 31
Chloramphenicol, 17
Chlorodifluoroacetic acid, in prostaglandin synthesis, 181
Chlorodifluoromethane, 33, 77, 92, 179
  in prostaglandin synthesis, 179
Chlorofluoromethanephosphonate ester, 116
Chloromethylboronates, 15
N-Chlorosuccinimide, 20
2-Chloro-1,1,2-trifluoroethyl-diethylamine, (Yarovenko reagent), 14, 193
Chlorotrifluoromethylcarbene, 247
Chlorotrifluoromethyldiazirine, 247
Cis- and *trans*-4-fluoro-L-proline, synthesis of, 29
Claisen condensation, of ethyl fluoroacetate, 152
Claisen rearrangement, fluoroacetamide acetal, 154
Cobalt carbonyls, as catalysts, 42
Curtius rearrangement, 33
2,3-Cyclic sulfates, 139
2,3-Cyclic sulfites, 139

Cycloserine, 58
β-Cystathionase, *see* Enzyme, β-cystathionase
γ-Cystathionase, *see* Enzyme, γ-cystathionase
Cystine, 22
Cytoplasmic ATP citrate lyase, *see* Enzyme,
    cytoplasmic ATP citrate lyase
Cytotoxic, 8

DAST, *see* Diethylaminosulfur trifluoride
Daunosamine, 137
Deaminative fluorination, preparation of α-
    fluoro-carboxylic acids, 88
(*E*)-Dehydro difluoromethyl putrescines,
    ornithinedecarboxylase inhibitor, 84
(*E*)-Dehydro α-fluoromethyl putrescines,
    ornithinedecarboxylase inhibitor, 84
5'-Deoxy-4',5-difluorouridine, 223
1-(2-Deoxy-2-fluoro-β-D-*arabino* -furanosyl)-5-
    alkenylcytosine, 137
1-(2-Deoxy-2-fluoro-β-D-*arabino* -furanosyl)-5-
    alkyluracil, 137
1-(2-Deoxy-2-fluoro-β-D-*arabino* -furanosyl)-5-
    iodocytosine, 137
1-(2-Deoxy-2-fluoro-β-D-*arabino* -furanosyl)-5-
    methyluracil, 137
2-Deoxy-2-fluoro-arabinose, 137
6-Deoxy-6-fluoro carbohydrates, 135
1-Deoxy-1-fluoro-fructose, 155
1-Deoxy-1-fluoro-D-fructose, 135
4-Deoxy-4-fluoro-fructose, 155
6-Deoxy-6-fluoro-D-fructose, 155, 157
2-Deoxy-2-fluoro-furanoses, 137
2-Deoxy-2-fluoro-glucose (^{18}F), 132
2-Deoxy-2-fluoro-glucose, 132
4-Deoxy-4-fluoro-glucose, 155
6-Deoxy-6-fluoro-glucose, 148, 155
4-Deoxy-4-fluoro-*lyxo*-pyranoses, 137
2-Deoxy-2-fluoromannose, 134
    AH109A tumor imaging, 134
2-Deoxy-2-fluoro-pyranoses, 132
2-Deoxy-2-fluoro-D,L-ribitol, 152
2-Deoxy-2-fluoro-*ribo*- and *arabino*-D-, 152
6-Deoxy-6-fluoro-D-sorbose, 157
1-Deoxy-1-fluoro-sucrose, 155
1-Deoxy-6-fluoro-sucrose, 155
Diaminosuccinic acid, 15
Diazoketones, 73, 119
Dibromodifluoromethane, 80, 113
2,3-Dibromo-1,1,1-trifluoropropane, 227
Dichlorofluoromethane, 91
Dichlorofluoromethanephosphonate ester, 116
Dichloromethyllithium, 15
6-*O*-Dicyclo-5,5-difluoro-5,6-dihydrouracils, 223

2,3-Dideoxy-2-fluoro-3-*C*-methyl-5-*O*-
    trifluoroacetyl- *ribo*-pentofuranose, 154
Diethylaminodimethylamino-sulfur difluoride,
    193
Diethylaminosulfur trifluoride, 12, 67, 148, 209,
    226
    in fluoride synthesis, 190
    with 1-hydroxy-carbohydrates, 158
    hydroxyethyl pyrimidine, 228
    in the preparation of fluorinated acids, 87
    in prostaglandin, 166
    in steroid synthesis, 201
Diethyl(2-chloro-1,1,2-trifluoroethyl)amine, 165
Diethyl *N*-ethyl phosphinodifluoroacetamide, 115
Diethyl fluoromalonate, 38, 40
Diethyl lithiomethanephosphonate, with
    perchloryl fluoride, 115
Diethylphosphonoacetate, 46
Diethylpiperidyl methyl formamidomalonate, 52
P,*P*-Diethyl-*P'*,*P'*-diisopropyl
    fluoromethylenebisphosphonate, 115
Diethyl sodiofluorooxaloacetate, 93
1,1'-Difluoroacetone, 41
3,3-Difluoroalanine, 22
β,β-Difluoroalanine, 31, 55
    alanine racemase inhibition, 55
Difluoroamines, 69
β,β-Difluoroasparagine, 24
    asparagine dependent leukemias, 59
    potential carcinostatic, 59
β,β-Difluoroaspartic acid, 24
    glutamate-oxaloaceteate transaminase
        (aspartate aminotransferase) inhibitor, 59
    potential carcinostatic, 59
    potential inhibitor of aspartate utilizing
        coenzymes, 57
(3,3-Difluorobutyryl) pantetheine, 103
Difluorocarbene, 32
    in steroid synthesis, 210
Difluorochloromethane, 32
19,19-Difluorocholesterol acetate, 203
2,3-Difluorofumarate, by enzyme fumaraze, 105
24,24-Difluoro-25-hydroxy-vitamin D_3, 210
1,1-Difluoroindane, 248
Difluoroketone, 174
5,5-Difluorolysine, 10
    collagen biosynthesis inhibitor, 59
Difluoromethanephosphonate, acylation of
    anions, 115
α-Difluoromethyl-α-amino acids, 33
α-Difluoromethyl dehydroputrescine, 80
α-Difluoromethyldiphenylacetic acid, 92
α-Difluoromethyldopa, 55

α-Difluoromethyl dopamine, 77, 84
Difluoromethylene group, 32, 179
4-*O*-(Difluoromethyl)-5-fluoro-uracil, 224
α-Difluoromethylornithine, 55
α-Difluoromethylputrescine, 78, 84
5-Difluoromethyluracil, 226
Difluorooxetane, 95
16,16-Difluoro-PGF$_{2α}$ methyl ester, 173
Difluorophosphonoacetic acid, 113
19,19-Difluoroprevitamin D$_3$, 203
4,4-Difluoroproline, 11
Difluorostatine, 60
Difluorostatone, 126
2,2-Difluorosuccinic acid, 216
Difluorosulfonium salt, 20
19,19-Difluorotachysterol, 203
3,5-Difluoro-L-tryptamine, 26
3,5-Difluorotyramine, 75
4,4′-Difluorovaline, 28
5-(2,2-Difluorovinyl)uracil, 228
Dihydrofolate reductase, 245. *See also* Enzyme,
     dihydrofolate reductase
1α,25-Dihydroxy-26,26,26,27,27,27-hexafluoro-
     vitamin D$_3$, 214
(*E*)-2-(3,4-Dihydroxyphenyl)-3-fluoroallylamine,
     78
1α-25-Dihydroxy-vitamin D$_3$, 214
Diisopropylfluoromethane phosphonate, 114
2,3: 4,5-Di-*O*-isopropylidene-D-fructopyranose
     with DAST, 135
Dimethyl 2-fluoroheptyl phosphonate, 117
1,3-Dimethyl-5-trifluoromethyl-5,6-
     dihydrouracil, 227
N,*N*-Dimethyl-5-trifluoromethyluracil, 228
2,3-Diphenylaziridine, 68
Dopamine receptors, 238

Electrochemical fluorination, 1
Electronegativity, 2
Electrophilic additions, of fluorine to unsaturated
     carbohydrate centers, 145
Electrophilic fluorination, of enols and enolates
     in prostaglandins, 174
α-Elimination, 33
Enol ethers, additions to, 205
Enzymatic methods, dissacharides, 155
Enzymatic reductions, of fluorinated ketones and
     ketoesters, 110
Enzymatic synthesis, 83
     (*E*)-β-fluoromethylene GABA, 83
     (*E*)-β-fluoromethylene-*meta*-tyramine, 83
     2-fluorourocanic acid, 84
     4-fluorourocanic acid, 84

Enzymatic transformations, fluorinated
     carbohydrates, 155
Enzyme:
     acetylcholine esterase, 126
     activated irreversible inhibitors of
          corresponding carboxylases, 84
     adenosine deaminase, 230
     adenolsuccinate synthetase and lyase, 59
     F-adenosyl-L-methionine decarboxylase, 84
     alanine racemase, 55
     amino acid decarboxylases, 55
     angiotensin converting enzyme, 126
     aromatic amino acid decarboxylase (AADC),
          59
     L-aromatic-α-amino acid decarboxylase, 84
     aspartyl proteases, 126
     butynyl-Co A dehydrogenase, 103
     carboxypeptidase A, 126
     β-cystathionase, 55
     γ-cystathionase, 55
     cytoplasmic ATP citrate lyase, 104
     dihydrofolate reductase, 57
     folylpoly(α-glutamate)synthetase, 57
     fructose-1,6-diphosphate aldolase, 157
     fumarase, 105
     GABA transaminase, 55, 85
     glucose isomerase, 155
     glutamate decarboxylase, 57
     glutamate-oxaloaceteate transaminase
          (aspartate aminotransferase), 59
     glutamate racemase, 57
     glutamic acid decarboxylase, 55, 85
     glycerol-3-phosphate dehydrogenase, 111
     glyoxalase I, 103
     glyoxalases I and II, 104
     hexokinase, 155
     histidine ammonia lyase, 58
     horse liver alcohol dehydrogenase, 111
     irreversible inactivation of, 54
     irreversible inhibitors of pyridoxal phosphate-
          dependent, 7
     MAO type B, 85
     monoamine oxidase, 85
     ornithine decarboxylase, 55, 84
     phosphogluocoisomerase, 155
     phospholipase A$_2$, 129
     pyruvate–glutamate transaminase, 55
     rat liver ornithinedecarboxylase, 84
     tryptophanase, 55
     tryptophan synthetase, 55
     tyrosine aminotransferase, 58
     uridine phosphorylase, 224
     urocanase, 85

zinc metalloproteases, 126
Enzyme inhibitors, 7
Ephedrine, 9, 66, 73
Epoxide opening:
  with hydrogen fluoride/pyridine, 71
  in prostaglandin synthesis, 170
  reactions, 18
  in steroid synthesis, 198
Epoxides, 16
  anchimeric assistance, 137
  stereospecific trans opening of, 51
(−/+)-Erythro-2-fluorocitrates, 104
Estrogen binding sites, 238
Estrogen receptor binding, 239
Ethyl 2-acetamidoacrylate, 45
Ethyl α-acetamidoacrylate, 38. *See also* Ethyl 2-
  acetamidoacrylate
Ethyl bromodifluoroacetate, 39, 94
Ethyl bromofluoroacetate, 93, 153, 178, 182
Ethyl 2-chloro-3-keto-4,4,4-trifluorobutyroate,
  50
Ethyl 2,2-difluorohexanoate, 172
Ethyl 2,4-difluoro-3-oxo-butanedioate, 152
Ethyl fluoroacetate, 37, 38, 92, 152
  aldol condensation of, 152
Ethyl α-fluoroacrylate, 45
Ethyl α-fluorocrotonate, 46
Ethyl 2-fluorohexanoate, 184
Ethyl trifluoroacetate, 40
Ethyl trifluoroacetate in steroid synthesis, 213
Ethyl trifluoroacetoacetate, 50

F_2, *see* Molecular fluorine
^{18}F, 238
$FClCHCF_2NEt_2$, 14
$FClO_3$, *see* Perchoryl fluoride
^{18}F-labeled fluoroketones, 122
^{19}F NMR studies, 5-fluorotryptophan in, 58
Flavinoids, 246
Fluorinated aldehydes, 243
Fluorination, with replacement of hydrogen, 21
Fluorinative dehydroxylation:
  in carbohydrates, 147
  in prostaglandins, 165
  in steroid synthesis, 190
Fluorine, 1, 21, 74, 145, 189, 205
  fluoro desulfurization by, 22
Fluorine-18, 3, 240
Fluorine/hydrogen fluoride, fluorodesulfurization
  with, 74
Fluoro aziridine, 18
Fluoroacetonitrile, 33, 35, 82
5-Fluoro-6-acetoxy-5,6-dihydrouracil, 221

2-Fluoroadenine, 230
3-Fluoro-alanine, 22
3-Fluoro-D-alanine, antibacterial agents from, 58
3-Fluoro-D,L-alanine, 38
3-Fluoro-D-alanine-2D, alanine racemase
  inhibition, 55
α-Fluoro-β-alanine:
  from 5-fluorouracil, 58
  synthesis of, 37
5-Fluoro-6-alkoxy-5,6-dihydrouracil, 221
Fluoroalkylamine reagents, 165, 193. *See also*
  2-Chloro-1,1,2-trifluoroethyl-diethylamine,
  (Yarovenko reagent)
5-Fluoroalkyl and 5-fluoroalkenyl pyrimidines,
  228
3-Fluoro-2-amino esters, 17
Fluoroaminopterins, 245
Fluoroandrostanones, 194
Fluoroasparagine, 38
  as glycosylation inhibitor, 60
D,L-*threo*-β-Fluoroasparagine, antileukemic
  activity, 59
β-Fluoroaspartates, mammalian cell cytotoxicity,
  59
β-Fluoroaspartic acid, 15, 38
  adenoylosuccinate lyase inhibition by, 59
  adenoylosuccinate synthetase inhibition by, 59
2-Fluoro-8-aza-adenosine, 230
Fluoroboric acid, 25, 26, 74
N-Fluoro(bistrifluoromethyl)sulfonimide, 241
1-Fluoro-3-bromo-butane, 97
3-Fluoro-2-butanone, 46
6-Fluoro-caproic acid, 97
α-fluorocarbonyl derivative, using acetyl
  hypofluorite, 121
16-Fluoro–9α-carboprostacyclin, 172
β-Fluoro-carboxylic acids, 88
ω-Fluoro-carboxylic acids, 101
Fluorochloromethane, 33
3-Fluorocinnamic acid, 94
2-Fluorocitrate, 93
25-Fluoroclionosterol, 192
Fluoro-deoxy-glycerol phosphates, 111
2-Fluoro-L-daunosamine, 144
10-Fluorodecanoic acid, 101
1-Fluorodehydroxy-chloramphenicol, 21
Fluorodehydroxylation, 8, 9
3-Fluoro-3-deoxycitrate, 104
3′-Fluoro-3′-deoxythymidine, 149
5-Fluoro-2′-deoxyuridine, 220
Fluorodesulfurization, 20
3-Fluoro-2-deuterio-D-alanine, cycloserine
  analogs, 58

5-Fluoro-2′,3′-dideoxy-3′-fluorouridine, 223
24(*R*)-Fluoro-1α,25-dihydroxycholesterol, 216
24(*R*)-Fluoro-1α,25-dihydroxy-vitamin D₃, 215
4-Fluoro-2,6-dimethyl resorcinol, 94
2-Fluorodopamine, 76
6-Fluorodopamine, 75, 240
5-Fluoro-D₂-PGI₂, 166
4-Fluoro-17β-estradiol, 207
16-Fluoro-estrone, 205
4-Fluoro- and 2-fluoro-estrone, 197
Fluoroformates, 89
2-Fluoroformycin, 230
4-Fluoro-glutamic acid, 14, 23, 45
    E. coli, growth inhibitor, 58
    synthesis of, 30, 45
γ-Fluoro-glutamic acid:
    glutamic acid antimetabolite, 56
    synthesis of, 38
3-Fluoro-L-glutamic acids, transaminases
        inhibitors, 55
Fluorohalomethylation, 31
2-Fluorohexestrol, 239
2-Fluorohistamine, 75
4-Fluorohistamine, 74
2-Fluoro-histidine, antimetabolic and antiviral
        properties, 58
4-Fluoro-D,L-histidine, 26
4-Fluoro-histidine, soporific and anaesthetic
        effects, 58
Fluorohydrin, 171
α-Fluoro-hydrocortisone acetate, 2
α-Fluoro-2-hydroxycinnamic acid, 94
3-Fluoro-1-hydroxypropane-2-one, 109
1α-Fluoro-25-hydroxy-vitamin D₃, 200
2β-Fluoro-1α-hydroxy-vitamin D₃, 199
4-Fluoroindole, 52
5-Fluoroindole, 52
6-Fluoroindole, 52
γ-Fluoroisoleucine, synthesis of, 46
2-Fluoroisovanillin, 242
α-Fluoroketones, 119
    hydrolytic enzymes inhibitors, 126
5-Fluoro-6-keto-PGE₁ methyl ester, 183
5-Fluorolysine and 4-fluoroarginine, *E. coli*
        inhibition, 58
Fluoromalonates, 45
6-Fluoro-metaraminol, 241
Fluoromethane phosphonates, 113
Fluoromethionine, 31
γ-Fluoromethotrexate, 57
2-Fluoro-4-methoxy-benzyl chloride, 26
2-Fluoro-6-methoxy-phenol, 242
α-Fluoromethyl-α-amino acids, 31

α-Fluoromethyl-dehydroornithine, synthesis of,
        36
α-Fluoromethyl dehydroputrescine, 82
16-Fluoromethyl-15-deoxy-16-hydroxy PGE₂,
        185
Fluoromethylated glycines, 31
    synthesis of, 30
Fluoromethylene biphosphonates, 115
β-Fluoromethylene dopamine, MAO type B
        inhibition by, 85
(*E*)-β-Fluoromethylene glutamic acid, 28, 85
(*E*)-β-Fluoromethylene-*meta*-tyramine,
        monoamine oxidase and, 85
(*E*)-β-Fluoromethylene-*m*-tyrosine, activation by
        aromatic amino acid decarboxylase
        (AADC), 61
16-Fluoromethylene PGE₁, 179
Fluoromethyl glyoxal, 95, 103
    by glyoxalases I and II glutathione (GSH),
        104
α-Fluoromethyl putrescine, 73
Fluoromethyl putrescines, 84
5-Fluoro-3-methylvaleric acid, 97
3-Fluorophenylalanine, antibacterial activity, 58
Fluoromethylphenyl sulfoxime, 180
3-Fluoro-*p*-nitro-phenylalanine, 21
4-Fluoro-nortestosterone, 207
Fluorooxaloacetic acid, 92
5-Fluoro PGF₂α, 183
10α-Fluoro PGF₂α, 168
10β-Fluoro PGF₂α, 170
13-Fluoro PGF₂α, 176
13(*E*)- and 13(*Z*)-14-Fluoro-PGF₂α, 177
(7*R*)-7-Fluoro PGI₂, 168
(7*S*)-7-Fluoro-PGI₂, 166
3-Fluorophenylalanine, 21
m-Fluorophenylalanine, peptides containing, as
        inhibitors of *Candida albicans,* 58
2-Fluoro-2-phenyl-cyclohexylamine, 69
Fluorophenylephrines, 235
3-Fluoro-phenylpyruvic acid, 21
(*Z*)-Fluorophospho-enolpyruvate, 105
3-Fluoro-1,2-propanediol, 109
3-Fluoro-propionyl-Co A, 103
5-Fluoroprostacyclins, 176
10β-Fluoroprostaglandin F₂α methyl ester, 170
Fluoropyruvate, 105
3-Fluoropyruvates, 21, 104
Fluoropyruvic acid, 92, 95
    toxicity of, 102
Fluoropyruvic esters, 88
26-Fluorositosterol, 192
9(10)-Fluorosteric acid, 101

2-Fluoro-2-substituted malonic acid diesters, microbial hydrolysis of, 102
N-Fluorosulfonimides, 241
Fluorosulfoximines, in prostaglandins, 180
α-Fluoro sulphenyl fluoride, 23
Fluorotamoxifen, 238
L-4-Fluorothreonine, 13
Fluorotryptophan, 58
4-Fluorotryptophan, 52
5-Fluorotryptophan, 52
    plasma albumin binding in ^{19}F NMR studies, 58
6-Fluorotryptophan, 52
4-, 5-, and 6-Fluorotryptophans, with tyrosine aminotransferase, 56
3-Fluorotyramine, 75
2-Fluorotyrosine, 26
3-Fluoro-L-tyrosine, 26
2- and 3-Fluorotyrosines, microorganism growth inhibition by, 58
5-Fluorouracil, 220
2-Fluorourocanic acid, urocanase inhibitor, 85
Fluorovaline, 20
β-Fluorovaline, 12
5-Fluorovanillin, 242
1α-Fluoro-vitamin D$_3$, 200
3β-Fluoro-vitamin D$_3$, 192
6-Fluoro-vitamin D$_3$, 210
Fluoroxytrifluoromethane, 73
Folylpoly(α-glutamate)synthetase, see Enzyme, folypolyl(α-glutamate)synthetase
Freon, 31
Fried, J., 2, 187
Fructose-1,6-diphosphate aldolase, see Enzyme, fructose-1,6-diphosphate aldolase
Fumarase, see Enzyme, fumarase

GABA, 7
GABA-transaminase, see Enzyme, GABA transaminase
Gastric ulcers, 7
Gem-difluorocompounds, 10
Geminal difluorination:
    2,2-difluoro-carbohydrates, 151
    in steroid synthesis, 200
Glucose isomerase, see Enzyme glucose isomerase
Glutamate decarboxylase, see Enzyme, glutamate decarboxylate
Glutamate-oxaloaceteate transaminase (aspartate aminotransferase), see Enzyme, glutamate-oxaloaceteate transaminase (aspartate aminotransferase)

Glutamate racemase, see Enzyme, glutamate racemase
Glutamic acid decarboxylase, see Enzyme, glutamic acid decarboxylase
Glycerolipid metabolism, 111
Glycerol-3-phosphate dehydrogenase, see Enzyme, glycerol-3-phosphate dehydrogenase
Glycidic ester, fluorinated, 51
Glycosyl fluorides, 157
Glyoxalase I, see Enzyme, glyoxalase I
Glyoxalases I and II, see Enzyme, glyoxalases I and II

Halcinonid, 198
Halocarbohydrates, 132
Halogen replacement, metallic or non-metallic fluorides, 27
α-Haloketones, with yellow mercuric oxide in hydrogen fluoride/pyridine, 119
Hemotransmethyl GABA, 55
Henne, A. L., 1
Herbicide, 117
Herpes simplex virus I, 224
Herpes simplex virus II, 224, 228
Hexafluoroacetone, 41
    in steroid synthesis, 214
Hexafluorophosphoric acid, 236
Hexafluoropropene, 45
Hexafluorovaline, angiotensin II analogs, 60
    synthesis of using hexafluoroacetone, 41
Hexokinase, see Enzyme, hexokinase
Histamine, 7
Histidine ammonia lyase, see Enzyme, histidine ammonia lyase
Horner-Emmons reagent, fluorinated, 177
    in prostaglandin synthesis, 177
Horse liver alcohol dehydrogenase, see Enzyme, horse liver alcohol dehydrogenase
Human stomach cancer HuTu80, 245
Hydroformylation, 42
Hydrogenation, catalyic asymmetric of fluorinated olefins, 111
Hydrogen bonds, 8
Hydrogen cyanide, 36
    in diglyme, 199
Hydrogen fluoride, 1, 14
    anhydrous, 223
    aqueous with N-bromosuccinimide, 205
    bromofluorinated of unsaturated acids, 87
    in diglyme, 199
    in fluorinated acid preparation, 87
    liquid, 8

Hydrogen fluoride (*Continued*)
  to open epoxides, 142
  in steroid synthesis, 198
Hydrogen fluoride–fluoroboric acid, 22
Hydrogen fluoride–pyridine, 16, 67, 119, 158, 171, 230
  alkyl carbamates in, 89
  azirine carboxylates with, 18
  to 1-azirines, 69
  concentration and product ratio, 89
  diazo esters with, 89
  fluorodeamination of α-aminoesters, 89
  fluorodediazoniation with, 73
  fluorodehydroxylation using, 73
  ring opening of glycidic esters, 88
α-Hydroxy-β-amino amides, 51
Hydroxyaspartate, 9
Hydroxyaziridines, 69
2-Hydroxy-1,4-dibromobutanone, 67
Hydroxyglutamates, 9
Hydroxyglutamic, 9
25-Hydroxy-26,26,26,27,27,27-hexafluoro-vitamin D_3, 214
β-Hydroxyhistamine, 66
(*E*)-2-(3-Hydroxyphenyl)-3-fluoroallylamine, 78
1-Hydroxy-2-keto-4,4,4-trifluorobutane phosphoric acid, 122
25-Hydroxy-26,26,26-trifluoro-27-nor-vitamin D_3, 215
25-Hydroxy-26,26,26-trifluoro-vitamin D_3, 215
5-Hydroxytryptamine, 7
Hypersensitivity, 7

Imidazole, 34
Imidazole diazonium fluoroborate, 26
Imidazolesulfonates, 139
Inhibitors:
  dihydrofolate reductase, 57
  glutamic acid decarboxylase, 55
Iodine monofluoride, 203
Iodobenzenedifluoride, 122
p-Iodotoluenedifluoride, 208
Iπ repulsive, 2
Ireland ester enolate Claisen rearrangement, 154
Iσ, 2
L-Isoleucine, 19
(*R*)-2,3-*O*-isopropylidene-glyceraldehyde, 152

Kanamycin A, 135
Ketone, 11
  using sulfur tetrafluoride, 10
KF, *see* Potassium fluoride
KHF$_2$, *see* Potassium bifluoride
Kolbe electrolysis, 225

Lithium chlorodifluoroacetate, 229
Lithium enolates, fluorinated, 183
Lithium hexamethyldisilazide, 15

Malonate diesters, 33
Malonate esters, alkylation of, 33
MAO type B, *see* Enzyme, MAO type B
Mercuric fluoride, 27, 30
Mesylates, 141
Metabolic hydroxylation, 187
Metal fluorides, as fluorinating agents, 27
Metaraminol, 241
Methanesulfonates, 141
Methionine, 31
Methotrexate, 57, 245
4-Methoxy-α,α,α-trifluoroacetophenone, 212
Methyl 4,6-*O*-benzylidene-2,3-dideoxy-3-diallylamino-2-fluoro-altropyranoside, 143
Methyl 2,2-difluorobutyrate, 122
Methyl 18-fluorostearate, 101
  toxicity, 102
4-Methyliodobenzenedifluoride, 122
Methyl iododifluoroacetate, 95
Methyl 2,3-*O*-isopropylidene-D,L-glycerate, aldol condensation of, 152
3-Methyl-4,4,4-trifluorobutyrate, with MoO$_5$-pyridine-hexamethylphosphoramide complex, 99
5-Methyl-6,6,6-trifluoroleucine, synthesis of using 1,1,1-trifluoroacetone, 42
Michael acceptor, 45
Michael addition, 38, 55
  to diethyl fluoromalonate, 40
  to α-fluoroacrylate ester, 45
Midgley, T., 1
Moissan, H., 1
Molecular fluorine, 120, 204
  and aromatics, 240
  and enols, 21
  in steroid synthesis, 189
Molecular fluorine ^{18}F, in water, 146
Molybdenum hexafluoride, in prostaglandin synthesis, 172
Molybdenum hexafluoride–boron trifluoride, 172
Monoamine oxidase, *see* Enzyme, monoamine oxidase
α-Monofluoromethyl-α-amino acids, 33
α-Monofluoromethyldopa, 53
Monofluoromethylglutamic acids, 53
α-Monofluoromethylhistidine, gastric acid secretion inhibition, 53
α-Monofluoromethyl putrescines, 84
5-Monofluoromethyluracil, 226

Monofluorooxaloacetate, 38
Monofluoroputrescine, 67
Morpholinosulfur trifluoride, 168
Muscle relaxants, 247

Nalidixic acid, 246
2-Nitro-1-butene, 40
Nucleic acids, 220
Nucleocidin, 131

Olefins, Additions to, 203
Ornithine decarboxylase, *see* Enzyme, ornithine
    decarboxylase
17-Oxo-5β-fluoro-androsten-3α-ol acetate, 190

Penicillamine, 20
2-Pentafluorophenylpropionaldehyde, 52
3-Pentafluorophenylpropionic acid, 101
Pentafluorostyrene, 52
    hydrocarboxylation of, 101
Pentapeptide, 39
Perchloryl fluoride, 23, 117, 176, 177, 197,
    204, 207
    with diethyl lithiomethanephosphonate, 115
    mechanism of reaction, 24
    with prostaglandins, 174
Perfluoroalkyl Iodides, 34
Perfluorocarboxylic acid esters, 124
Perfluoro-2-methylpropene, 97
Perfluorosuccinimide, 83
PGI₂ with perchloryl fluoride, 175
Phenols, fluorinated, 242
Phenylalanine, 12
Phenyl azirines, 120
2-Phenyl-3,3-difluoroallylamine, 80
(Z)-2-Phenyl-3-fluoroallylamine, 79
3-Phenyl-3-fluoro-2-methylamino-propane, 74
Phenylhydrazone, 50
2-Phenyl-3-methyl aziridine, 68
1-Phenylsulfonyl-3,3,3-trifluoropropene, 45, 109
Phenyltetrafluorophosphine, 193
Phenylthioglycosides, to glycosyl fluoride, 158
4-Phenyl-1,2,4-triazoline-3,5, 191
Phosphogluocoisomerase, *see* Enzyme,
    phosphogluocoisomerase
Phospholipase A₂, *see* Enzyme, phospholipase
    A₂
Photochemical fluorination, using fluoroboric
    acid, 26
Photochemical trifluoromethylation, 34
Photofluorination, 19, 25
Piperidinosulfur trifluoride, 168, 210
Polyhalogenomethane derivatives, 33
Polyhydrogen fluoride-pyridine, 16

Positron emission tomography (PET), 3
    ¹⁸F Putrescines in, 84
Potassium bifluoride, 30, 137, 139, 142, 170,
    171, 199
    in steroid synthesis, 199
Potassium fluoride, 28, 91, 199
    3-fluoro-alanine, 27
    in steroid synthesis, 199
    N-substituted α-halo amides with, 91
    in triflate displacements, 168
Potassium fluoride-18-crown-6, 170
    in dimethylformamide, 195
Prostacyclin, 175
Pseudoephedrine, 67, 73
Putrescine, 7
Putrescines, ¹⁸F may be used in tumor
    localization in conjunction with positron
    emission tomography, 84
Pyranosyl fluorides, 146
Pyrazine, 71
Pyridine, 16
Pyridinium polyhydrogen fluoride, 205, 238. *See
    also* Hydrogen fluoride–pyridine
Pyridoxal enzymes, 55
Pyrimidines, 5-substituted, 220
Pyruvate-glutamate transaminase, *see* Enzyme,
    pyruvate–glutamate transaminase

Rabbit muscle aldolase, 128
Radiochemical imaging agents, 234
Radiopharmaceutical, 132
Rat liver ornithinedecarboxylase, *see* Enzyme,
    rat liver ornithinedecarboxylase
Reduction:
    catalytic of fluorinated olefine, 111
    of—CHClF group with *tri-n*-butyltin hydride,
        33
Reformatsky reaction, 39, 93
    to form 2,2-difluoro-3-hydroxy esters, 94
Reformatsky-type reactions in prostaglandins,
    182
Refrigerants, 1
Renin, competitive inhibitors of, 60
Rhodium carbonyls, as catalysts, 42
R_f radicals, electrophilic nature of, 35

Saccharomyces cervisiae, 99
Scheele, 1
Schiff bases, 31
Serotonin, 7
SF₄, 8. *See also* Sulfur tetrafluoride
Sigma withdrawing, 2
Silver fluoride, 29
Simons, J.H., 1

Single electron transfer (SET) pathway, 223
Sodium enolates, alkylation of, 33
Spiroperidol, 238
Sporaracin B, 150
Statine, 60
Steroids, bromofluorination of, 204
Steroid synthesis, toluenesulfonates in, 194
Sucrose synthetase, 155
Suicide inhibitors, 52
Sulfur tetrafluoride, 8, 66
  fluorodehydroxylation in hydrogen fluoride, 66
  in prostaglandin synthesis, 173
  toxicity of, 12
  uracil 5-carboxylic acid with, 225
Sulfur tetrafluoride–hydrogen fluoride system, 9
Swarts, F., 1

1,2,3-4-Tetra-*O*-acetyl-6-*O* -
  trifluoromethanesulfonyl-*gluco*-pyranose, 138
Tetra-*n*-butylammonium fluoride, 134, 136, 171, 194, 199
  with benzyl 4,6-*O*-benzylidene-2,3-
    benzoylepimino-2,3-dideoxy-α- D-*allo*-
    pyranoside, 145
  in THF, 169
Tetraethylammonium fluoride, 134, 143
Tetraethyl difluoromethylene biphosphonate, 116
Tetrafluoroboric acid, 235
4,5,6,7-Tetrafluoroindoles, 3-substituted, 52
Tetrafluoroputrescine, 83
4,5,6,7-Tetrafluorotryptophan, tryptophanyl
  hydroxamate and amino acyl *t*-RNA
  formation inhibition, 58
Tetraisopropylmethylenebisphosphonate, 117
Tetramethylammonium fluoride, 133
Thiolamino acids, 20
Thiols, 22
Threo and *allo*-2-amino-4,4,4-trifluoro-3-
  hydroxybutanoic acids, 50
Threo-1-fluorodehydroxy chloramphenicol, 17
Threo-3-fluoro-phenylalanine, 17
Threo-phenyl-2-methylamino-propanethiol, 73
Threonine, 9
Thromboxane A2, 95, 106
Torgov reaction, 211
Trans-4-amino-2-fluorocrotonic acid, 53
Trans-4-fluoroproline, trans-4-hydroxyproline by
  enzymatic hydroxylation, 57
Trans 4-fluorourocanic acid, from L-4-
  fluorohistidine, 56
Trans-3-methyl-L-proline, 19

Triethylamine tris(hydrogen fluoride), 143
Trifluoroacetic acid anhydride, 48
Trifluoroacetic anhydride, 82
1,1,1-Trifluoroacetone, 41
Trifluoroacetylhypofluorite, 241
3,3,3-Trifluoroalanine, 36, 53
  synthesis of, 36, 49
  using hexafluoroacetone, synthesis of, 41
4,4,4-Trifluorobutanoic acid, 47
γ,γ,γ-Trifluorocrotonate, 98
2,2,2-Trifluoroethanol, 122
  active site probe for horse liver alcohol
    dehydrogenase, 111
  yeast-promoted additions of, 122
4,4,4-Trifluoro-3-hydroxy-2-methoxy-imino-
  butanoate, 50
5,5,5-Trifluoroleucine, 45
  in proteins, 58
  synthesis of using 1,1,1-trifluoroacetone, 41
Trifluoromethanesulfonate esters:
  for carbohydrate synthesis, 132, 138
  in steroid synthesis, 195
Trifluoromethyl, 7, 10
Trifluoromethylacrylic acid, 96, 228
8-Trifluoromethyl adenosine, 231
18,18,18-Trifluoromethyl analog of estrone,
  estradiol, ethynyl estradiol, 211
2-Trifluoromethyl-1,4-benzoquinone, 211
3-Trifluoromethyl-γ-butyrolactone, 51
  resolution of, 51
2-Trifluoromethyl-1,3-cyclopentanedione, 211
5-Trifluoromethyl-2′deoxyuridine, 220
5-Trifluoromethyl-5,6-dihydrouracil, 227
Trifluoromethylglutamic acids, 55
2-Trifluoromethyl-histamine, 81, 82
4-Trifluoromethyl histamine, 81
2-Trifluoromethyl-L-histidine, 48
Trifluoromethyl hypofluorite, 19, 145, 147, 205, 231
8-Trifluoromethyl inosine, 231
Trifluoromethyl iodide, 34, 81
  trifluoromethyl radical from, 34
Trifluoromethyl malonic ester, 97
3-Trifluoromethylphenol, 211
3-Trifluoromethylpyridine, 247
5-Trifluoromethyl uracil, 225
5-Trifluoromethyluridine, 225, 226
Trifluoromethylurocanic acids, 81
6,6,6-Trifluoronorleucine, 45
  synthesis of, 48
5,5,5-Trifluoronorvaline, 42
  E. coli, growth inhibitor, 58
  synthesis of, 47
4,4,4-Trifluoro-3-oxobutanoate, 99

using baker's yeast, 99
3,3,3-Trifluoropropene, 42, 96, 248
bromination of, 96
hydroesterification and hydrocarboxylation of, 96
(*E*)-5-(3,3,3-Trifluoro-1-propenyl)-2'-deoxyuridine, 229
3,3,3-Trifluoropropionic acid, 122
4,4,4-Trifluorothreonines, synthesis of, 40
4,4,4-Trifluorovaline, 42
synthesis of using 1-phenylsulfonyl-3,3,3-trifluoropropene, 45
1α,24(*R*)25-Trihydroxy-vitamin D_3, 215
16,16-Trimethylene prostaglandins, PGE_2, 173
Tris(dimethylamino)sulfonium difluorotrimethylsilicate, 134
Tryptophanase, *see* Enzyme, tryptophanase
Tryptophane synthetase, *see* Enzyme, tryptophane synthetase
Tyramine, 7

Tyrosine aminotransferase, *see* Enzyme, tyrosine aminotransferase

Uridine phosphorylase, *see* Enzyme, uridine phosphorylase
Urocanase, *see* Enzyme, urocanase

van der Waal's radius, 2
Vicinal difluorides, in steroid synthesis, 204
Vicinal halofluorination, 203
Vitamin D_3, 187

Xenon difluoride, 31, 121, 146, 197, 205, 208
with silyl enol ethers, 121

Yarovenko reagent, 14

Zinc metalloproteases, *see* Enzyme, zinc metalloproteases